AF813244

Methods in Molecular Biology

Series Editor
John M. Walker
School of Life and Medical Sciences
University of Hertfordshire
Hatfield, Hertfordshire, UK

For further volumes:
http://www.springer.com/series/7651

For over 35 years, biological scientists have come to rely on the research protocols and methodologies in the critically acclaimed *Methods in Molecular Biology* series. The series was the first to introduce the step-by-step protocols approach that has become the standard in all biomedical protocol publishing. Each protocol is provided in readily-reproducible step-by-step fashion, opening with an introductory overview, a list of the materials and reagents needed to complete the experiment, and followed by a detailed procedure that is supported with a helpful notes section offering tips and tricks of the trade as well as troubleshooting advice. These hallmark features were introduced by series editor Dr. John Walker and constitute the key ingredient in each and every volume of the *Methods in Molecular Biology* series. Tested and trusted, comprehensive and reliable, all protocols from the series are indexed in PubMed.

Germline Development in the Zebrafish

Methods and Protocols

Edited by

Roland Dosch

Institute for Developmental Biochemistry, University of Göttingen Medical Center
Georg-August-Universität, Göttingen, Germany

Humana Press

Editor
Roland Dosch
Institute for Developmental
Biochemistry
University of Göttingen Medical Center
Georg-August-Universität
Göttingen, Germany

ISSN 1064-3745 ISSN 1940-6029 (electronic)
Methods in Molecular Biology
ISBN 978-1-0716-0969-9 ISBN 978-1-0716-0970-5 (eBook)
https://doi.org/10.1007/978-1-0716-0970-5

© Springer Science+Business Media, LLC, part of Springer Nature 2021
This work is subject to copyright. All rights are reserved by the Publisher, whether the whole or part of the material is concerned, specifically the rights of translation, reprinting, reuse of illustrations, recitation, broadcasting, reproduction on microfilms or in any other physical way, and transmission or information storage and retrieval, electronic adaptation, computer software, or by similar or dissimilar methodology now known or hereafter developed.
The use of general descriptive names, registered names, trademarks, service marks, etc. in this publication does not imply, even in the absence of a specific statement, that such names are exempt from the relevant protective laws and regulations and therefore free for general use.
The publisher, the authors, and the editors are safe to assume that the advice and information in this book are believed to be true and accurate at the date of publication. Neither the publisher nor the authors or the editors give a warranty, expressed or implied, with respect to the material contained herein or for any errors or omissions that may have been made. The publisher remains neutral with regard to jurisdictional claims in published maps and institutional affiliations.

This Humana imprint is published by the registered company Springer Science+Business Media, LLC, part of Springer Nature.
The registered company address is: 1 New York Plaza, New York, NY 10004, U.S.A.

Preface

Reproduction is a fundamental principle of biological systems. Multicellular organisms use gametes for their reproduction, which originate from primordial germ cells during embryogenesis. The events of this cell lineage leading to the formation of functional gametes, i.e., egg and sperm, are summarized under the theme of germline development, which is the topic of this book.

The zebrafish (*Danio rerio*) emerged as an eminent model to study the molecular mechanism of vertebrate biology. Zebrafish research already made critical contributions to our knowledge regarding the cardiovascular or the nervous system and their failure during disease. Research on the teleost germline has a long history originating in aquaculture. In recent years however, studies in the zebrafish took off with numerous innovative technologies, which permitted a closer look at the basic molecular processes of the mysterious germline.

The aim of this edition in *Methods in Molecular Biology* was therefore to combine many of these approaches and thereby facilitate a quick start for newcomers to the zebrafish, but also a compendium for experienced researchers. The 29 chapters of this book represent only a snapshot of the excellent methods available for research on the fish germline, because readers will find additional exciting research in the current literature. The chapters are roughly subdivided into four categories of (I) cultivating and manipulating germ cells, (II) imaging of germline processes and the molecular analysis of their (III) protein or (IV) RNA. Some of the chapters are difficult to categorize as they use an interdisciplinary approach or use other organisms like *Xenopus* to study the function of proteins usually expressed in the fish oocytes (Chapter 2) or investigate an important aspect of the ovary in the Medaka fish (Chapter 17). The methods also encompass a wide spectrum from the biophysical quantification of tissue tension in follicles (Chapter 10) to generating primordial germ cells in the petri dish (Chapter 7), which is a standard in mammalian germline research. We also included a historical review on the process of zygotic genome activation as an introduction for the various RNA-based technologies (Chapter 25).

I could continue forever to highlight the fascinating aspects of each contribution, but then this preface would reach the size of a separate chapter. I want to take this opportunity to thank the Editor-in-Chief Prof. John Walker for his guidance and patience to finish this book. I also would like to thank all the authors, whose interesting contributions provided the basis for this issue on the fish germline.

Göttingen, Germany — *Roland Dosch*

Contents

Contributors

NANA AI • *Faculty of Health Sciences, Centre of Reproduction, Development and Aging (CRDA), University of Macau, Macau, China*

JULIEN BOBE • *Laboratory of Fish Physiology and Genomics, INRA UR1037, Rennes Cedex, France*

ASFA SABRIN BORBORA • *Department of Biological Sciences, Tata Institute of Fundamental Research, Mumbai, India*

JOAN CERDÀ • *IRTA-Institute of Biotechnology and Biomedicine (IBB), Universitat Autònoma de Barcelona, Bellaterra, Cerdanyola del Vallès, Spain*

FRANÇOIS CHAUVIGNÉ • *IRTA-Institute of Biotechnology and Biomedicine (IBB), Universitat Autònoma de Barcelona, Bellaterra, Cerdanyola del Vallès, Spain*

EMMANUELLE COM • *Protim, Inserm U1085, Irset, Rennes Cedex, France*

ZSOLT CSENKI-BAKOS • *Department of Aquaculture, Szent István University, Gödöllő, Páter Károly u. 1., Hungary*

FABIO M. D'ORAZIO • *Institute of Cancer and Genomics Sciences, College of Medical and Dental Sciences, University of Birmingham, Birmingham, UK; MRC London Institute of Medical Sciences and Faculty of Medicine, Imperial College, London, UK; Institute of Clinical Sciences, Faculty of Medicine, Imperial College, London, UK*

VLADIMIR DESPIC • *Department of Pharmacology, Weill Cornell Medicine, New York, NY, USA*

CHAITANYA DINGARE • *Department of Developmental Biology of Vertebrates, Institute of Cell Biology and Neuroscience, Goethe University Frankfurt am Main, Frankfurt am Main, Germany; Laboratory of Comparative Developmental Dynamics, Department of Genetics, University of Cambridge, Cambridge, UK*

ROLAND DOSCH • *Institute for Developmental Biochemistry, University of Göttingen Medical Center, Georg-August-Universität, Göttingen, Germany*

CHRISTIAN DULLIN • *Institute for Diagnostic and Interventional Radiology, University Medical Center Göttingen, Göttingen, Germany; Italian Synchrotron "Elettra", Trieste, Italy; Max-Planck-Institute for Experimental Medicine, Göttingen, Germany*

YANIV M. ELKOUBY • *Department of Developmental Biology and Cancer Research, The Hebrew University of Jerusalem, Faculty of Medicine, Institute for Medical Research – Israel-Canada (IMRIC), Jerusalem, Israel*

ALBA FERRÉ • *IRTA-Institute of Biotechnology and Biomedicine (IBB), Universitat Autònoma de Barcelona, Bellaterra, Cerdanyola del Vallès, Spain*

EUGENIO F. FORNASIERO • *Department of Neuro- and Sensory Physiology & Center for Biostructural Imaging of Neurodegeneration (BIN), Institut für Neuro- und Sinnesphysiologie, University Medical Center Göttingen, Göttingen, Germany*

WEI GE • *Faculty of Health Sciences, Centre of Reproduction, Development and Aging (CRDA), University of Macau, Macau, China*

LOUISA HABICH • *Institute of Human Genetics, University Medical Center Göttingen, Göttingen, Germany*

YAVOR HADZHIEV • *Institute of Cancer and Genomics Sciences, College of Medical and Dental Sciences, University of Birmingham, Birmingham, UK*

Christina L. Hansen • *Laboratory of Genetics, University of Wisconsin – Madison, Madison, WI, USA*
Carl-Philipp Heisenberg • *Institute of Science and Technology Austria, Klosterneuburg, Austria; Life Sciences Institute, Zhejiang University, Hangzhou, China*
Ákos Horváth • *Department of Aquaculture, Szent István University, Gödöllő, Páter Károly u. 1., Hungary*
Honghui Huang • *Key Laboratory of Freshwater Fish Reproduction and Development, Ministry of Education, State Key Laboratory Breeding Base of Eco-Environments and Bio-Resources of the Three Gorges Reservoir Region, School of Life Sciences, Southwest University, Chongqing, China*
Kunio Inoue • *Department of Biology, Graduate School of Science, Kobe University, Kobe, Hyogo, Japan*
Hisyam Ismail • *Skin Research Institute of Singapore, Agency for Science Technology and Research, Singapore, Singapore*
Yoshiko Iwasaki • *Department of Marine Biosciences, Tokyo University of Marine Science and Technology, Tokyo, Japan*
Aleksandra Jasiulewicz • *Institute of Cancer and Genomics Sciences, College of Medical and Dental Sciences, University of Birmingham, Birmingham, UK*
Youngnam N. Jin • *College of Pharmacy, University of Utah, Salt Lake City, UT, USA; Medical Research Institute, Wuhan University, Wuhan, China*
Petra A. Klemmt • *Department of Developmental Biology of Vertebrates, Institute of Cell Biology and Neuroscience, Goethe University Frankfurt am Main, Frankfurt am Main, Germany*
Tímea Kollár • *Department of Aquaculture, Szent István University, Gödöllő, Páter Károly u. 1., Hungary*
Vladimir Korzh • *International Institute of Molecular and Cell Biology in Warsaw, Warsaw, Poland*
Gudrun Kracht • *Department of Developmental Biochemistry, Zentrum Biochemie, University Medical Center Göttingen, Göttingen, Germany*
Maciej Łapiński • *International Institute of Molecular and Cell Biology in Warsaw, Warsaw, Poland*
Esther Shuk-Wa Lau • *School of Life Sciences, The Chinese University of Hong Kong, Shatin, Hong Kong, China*
Regis Lavigne • *Protim, Inserm U1085, Irset, Rennes Cedex, France*
Virginie Lecaudey • *Department of Developmental Biology of Vertebrates, Institute of Cell Biology and Neuroscience, Goethe University Frankfurt am Main, Frankfurt am Main, Germany*
Qian Li • *National Engineering Center for Biochip at Shanghai, Shanghai Biochip Limited Corporation, Shanghai, People's Republic of China*
Woei Chang Liew • *Reproductive Genomics Group, Temasek Life Sciences Laboratory, Singapore, Singapore*
Lin Liu • *School of Life Sciences, The Chinese University of Hong Kong, Shatin, Hong Kong, China; School of Life Sciences, South China Normal University, Guangzhou, China*
Jelena Lujić • *Department of Biomedical Sciences, Center for Reproductive Genomics, Cornell University, Ithaca, NY, USA*
Sunit Mandad • *Department of Neuro- and Sensory Physiology & Center for Biostructural Imaging of Neurodegeneration (BIN), Institut für Neuro- und Sinnesphysiologie,*

University Medical Center Göttingen, Göttingen, Germany; Bioanalytical Mass Spectrometry Group, Max Planck Institute of Biophysical Chemistry, Göttingen, Germany

ZORAN MARINOVIĆ • *Department of Aquaculture, Szent István University, Gödöllő, Páter Károly u. 1., Hungary*

SINNAKARUPPAN MATHAVAN • *Vision Research Foundation, Sankara Nethralaya, Chennai, India*

JENNIFER L. MATTHEWS • *Zebrafish International Resource Center, University of Oregon, Eugene, OR, USA*

JERÓNIMO MIRANDA • *Helmholtz-Zentrum München, Neuherberg, Germany*

YUICHIRO MISHIMA • *Department of Frontier Life Sciences, Faculty of Lifesciences, Kyoto Sangyo University, Kyoto, Japan*

FERENC MÜLLER • *Institute of Cancer and Genomics Sciences, College of Medical and Dental Sciences, University of Birmingham, Birmingham, UK*

AVISHAG MYTLIS • *Department of Developmental Biology and Cancer Research, The Hebrew University of Jerusalem, Faculty of Medicine, Institute for Medical Research – Israel-Canada (IMRIC), Jerusalem, Israel*

SREELAJA NAIR • *Department of Biological Sciences, Tata Institute of Fundamental Research, Mumbai, India*

ALINA NIEDZWETZKI • *Department of Developmental Biology of Vertebrates, Institute of Cell Biology and Neuroscience, Goethe University Frankfurt am Main, Frankfurt am Main, Germany*

TOSHIYA NISHIMURA • *Division of Marine Life Science, Graduate School of Fisheries Sciences, Hokkaido University, Hakodate, Japan; Division of Biological Sciences, Graduate School of Science, Nagoya University, Nagoya, Japan*

LÁSZLÓ ORBÁN • *Reproductive Genomics Group, Temasek Life Sciences Laboratory, Singapore, Singapore; Frontline Fish Genomics Research Group, Department of Animal Sciences, Georgikon Campus, Szent István University, Keszthely, Hungary; Centre for Comparative Genomics, Murdoch University, Murdoch, WA, Australia*

FRANCISCO PELEGRI • *Laboratory of Genetics, University of Wisconsin – Madison, Madison, WI, USA*

ROSHAN PRIYARANGANA PERERA • *Institute of Developmental Biochemistry, University Medical Center, Göttingen, Germany*

RANDALL T. PETERSON • *College of Pharmacy, University of Utah, Salt Lake City, UT, USA*

CHARLES PINEAU • *Protim, Inserm U1085, Irset, Rennes Cedex, France*

MICHAL RABANI • *Silberman Institute of Life Sciences, Hebrew University, Jerusalem, Israel*

MARTA F. RIESCO • *Spanish Institute of Oceanography (IEO), Planta de Cultivos el Bocal, Monte, Santander, Spain*

VANESA ROBLES • *Spanish Institute of Oceanography (IEO), Planta de Cultivos el Bocal, Monte, Santander, Spain, Cell Biology Area, Department of Molecular Biology, University of León, León, Spain*

NADIA ROSTAM • *Institute of Human Genetics, University of Göttingen Medical Center, Georg-August-Universität, Göttingen, Germany; Department of Developmental Biology, Johann-Friedrich-Blumenbach Institute of Zoology and Anthropology, Göttingen Center of Molecular Biosciences, Göttingen, Germany; Department of Biology, College of Science, University of Sulaimani (UoS), Sulaimaniyah, Iraq*

HUA RUAN • *Key Laboratory of Freshwater Fish Reproduction and Development, Ministry of Education, State Key Laboratory Breeding Base of Eco-Environments and Bio-Resources of*

the Three Gorges Reservoir Region, School of Life Sciences, Southwest University, Chongqing, China
JOLLY M. SAJU • *Reproductive Genomics Group, Temasek Life Sciences Laboratory, Singapore, Singapore*
PRABHA SAMPATH • *Skin Research Institute of Singapore, Agency for Science Technology and Research, Singapore, Singapore; Department of Biochemistry, Yong Loo Lin School of Medicine, National University of Singapore, Singapore, Singapore*
DENHÍ SCHNABEL • *Instituto de Biotecnología UNAM, Cuernavaca, Morelos, Mexico*
KELLEE R. SIEGFRIED • *Biology Department, University of Massachusetts Boston, Boston, MA, USA*
JOCELYN S. STEINFELD • *Biology Department, University of Massachusetts Boston, Boston, MA, USA*
YONGHUA SUN • *State Key Laboratory of Freshwater Ecology and Biotechnology, Institute of Hydrobiology, Innovation Academy for Seed Design, Chinese Academy of Sciences, Wuhan, China; College of Advanced Agricultural Sciences, University of Chinese Academy of Sciences, Beijing, China*
MINORU TANAKA • *Division of Biological Sciences, Graduate School of Science, Nagoya University, Nagoya, Japan*
ANNA CHUNG-KWAN TSE • *School of Life Sciences, The Chinese University of Hong Kong, Shatin, Hong Kong, China*
KEH-WEEI TZUNG • *Reproductive Genomics Group, Temasek Life Sciences Laboratory, Singapore, Singapore; Institute of Molecular and Cell Biology (IMCB), A*STAR, Singapore, Singapore*
BÉLA URBÁNYI • *Department of Aquaculture, Szent István University, Gödöllő, Páter Károly u. 1., Hungary*
DAVID G. VALCARCE • *Spanish Institute of Oceanography (IEO), Planta de Cultivos el Bocal, Monte, Santander, Spain*
ZOLTAN M. VARGA • *Zebrafish International Resource Center, University of Oregon, Eugene, OR, USA*
CECILIA LANNY WINATA • *International Institute of Molecular and Cell Biology in Warsaw, Warsaw, Poland; Max Planck Institute for Heart and Lung Research, Bad-Nauheim, Germany*
PENG XIA • *Institute of Science and Technology Austria, Klosterneuburg, Austria*
DING YE • *State Key Laboratory of Freshwater Ecology and Biotechnology, Institute of Hydrobiology, Innovation Academy for Seed Design, Chinese Academy of Sciences, Wuhan, China; College of Advanced Agricultural Sciences, University of Chinese Academy of Sciences, Beijing, China*
XIAOGUI YI • *Key Laboratory of Freshwater Fish Reproduction and Development, Ministry of Education, State Key Laboratory Breeding Base of Eco-Environments and Bio-Resources of the Three Gorges Reservoir Region, School of Life Sciences, Southwest University, Chongqing, China*
OZLEM YILMAZ • *Laboratory of Fish Physiology and Genomics, INRA UR1037, Rennes Cedex, France; Institute of Marine Research, Austevoll Research Station, Storebø, Norway*
GORO YOSHIZAKI • *Department of Marine Biosciences, Tokyo University of Marine Science and Technology, Tokyo, Japan*

Chapter 1

Separation of Oocyte and Follicle Layer for Gene Expression Analysis in Zebrafish

Nana Ai, Lin Liu, Esther Shuk-Wa Lau, Anna Chung-Kwan Tse, and Wei Ge

Abstract

Zebrafish ovarian follicles are mainly composed of the oocyte and a thin layer of follicle cells. Recent studies have demonstrated extensive cell-cell interactions between the oocyte and surrounding follicle layer and that the two compartments communicate mostly through paracrine factors. To understand the paracrine communication within the follicle, it is essential to know the spatial expression patterns of genes in the two compartments. However, since the follicle layer is extremely thin and the oocytes are enormous in size in fish, it is often difficult to detect gene expression by traditional methods such as in situ hybridization. Separation of the oocyte and surrounding follicle layer followed by RT-PCR detection provides a sensitive way to reveal the expression of individual genes in the two compartments of the follicle. This chapter introduces a method for mechanic separation of the oocyte and follicle layer at full-grown stage for expression analysis. Since fish have similar follicle structure, this method may also be used in other species as well.

Key words Zebrafish ovary, Oocyte, Follicle layer, Gene expression

1 Introduction

In the ovary of vertebrates, each follicle contains three major cellular compartments: oocytes, internal granulosa cells, and external thecal cells. Unlike mammalian follicles that contain multiple layers of follicle cells, the follicles of fish like zebrafish only have a single layer of granulosa and thecal cells [1].

Follicle growth and maturation, or folliculogenesis, is a popular issue for research in reproductive physiology. As one of the most dynamic physiological processes, folliculogenesis is tightly controlled by a variety of regulatory factors, including endocrine hormones, paracrine growth factors, and steroids. Although pituitary gonadotropins, namely, follicle-stimulating hormone (FSH) and luteinizing hormone (LH), are master hormones responsible for ovarian development and function, increasing evidence suggests

Roland Dosch (ed.), *Germline Development in the Zebrafish: Methods and Protocols*, Methods in Molecular Biology, vol. 2218, https://doi.org/10.1007/978-1-0716-0970-5_1, © Springer Science+Business Media, LLC, part of Springer Nature 2021

that various local factors also play critical roles in the ovary, often mediating or modulating gonadotropin activities. For instance, growth differentiation factor 9 (GDF9), a member of the transforming growth factor β (TGF-β) superfamily [2], is specifically expressed in the oocyte, and it acts on the surrounding follicle cells to control folliculogenesis in mammals [3, 4].

To understand the modes of paracrine regulation within the follicle, it is essential to know the spatial patterns of expression of key genes in the two compartments. In a previous study, Chattoraj et al. separated the oocyte from the follicle layer in Indian major carp (*Labeo rohita*) to study the influence of melatonin on oocyte maturation [5]. We modified the protocol and applied the method in the zebrafish, which allows us to study gene expression within the follicle and the interaction between the oocytes and follicle layers. As shown by our previous studies, the expression of FSH and LH receptors (*fshr* and *lhcgr*) was demonstrated in both the intact follicles and isolated follicle layers, but not in the denuded oocytes [6]. In contrast to *fshr* and *lhcgr*, growth differentiation factor 9 (*gdf9*) was expressed exclusively in the oocytes [7].

Using this method, we have analyzed spatial expression patterns of a variety of ovarian factors and receptors in the two compartments of zebrafish follicles, including activin-inhibin-follistatin system [8], epidermal growth factor (EGF) family and the receptor (EGFR) [9], Kit ligands and receptors [10], growth differentiation factor 9 (GDF9) and bone morphogenetic protein (BMP) family [7, 11], and pituitary adenylate cyclase-activating polypeptide-growth hormone-insulin-like growth factor (PACAP-GH-IGF) system [12, 13].

Herein, we describe the method of separating oocyte and follicle layer at full-grown (FG) stage in zebrafish followed by analysis of gene expression in the two compartments. The method may also be used in other teleost species.

2 Materials

2.1 Zebrafish

Wild-type (WT) zebrafish of AB strain was used in the study. Zebrafish were maintained at 28 °C in the ZebTEC multilinking zebrafish system (28 °C, pH 7.3, conductivity of 400 mS/cm, and 14-h light/10-h dark photoperiod) with a Tritone automatic feeder system. The animals were handled according to the Animal Protection Act enacted by the Legislative Council of Macao Special Administrative Region under Article 71(1) of the Basic Law, and the experimental protocols were approved by the Research Ethics Panel of the University of Macau. All solutions were prepared with ultrapure water (UPW) (18 MΩ-cm at 25 °C).

2.2 Isolation and Separation of Follicles

1. *Cortland medium*: The chemicals used to prepare Cortland medium with or without Ca^{2+}/Mg^{2+} are listed in Table 1 [14]. Weight 7.25 g NaCl, 1.00 g $NaHCO_3$, 0.38 g KCl, 0.23 g $MgSO_4{\cdot}7H_2O$ (with Ca^{2+}/Mg^{2+}), 0.41 g $Na_2HPO_4{\cdot}H_2O$ or 0.36 g Na_2HPO_4, 0.23 g $CaCl_2{\cdot}H_2O$ or 0.26 g $CaCl_2{\cdot}2H_2O$ (with Ca^{2+}/Mg^{2+}), 0.11 g $MgCl_2{\cdot}6H_2O$ or 0.23 g $MgCl_2$ (with Ca^{2+}/Mg^{2+}), 1.00 g BSA, 5.20 g HEPES, and 1.00 g glucose, and place them in 1-L autoclaved glass bottle. Add 10 mL penicillin-streptomycin (10,000 units/mL penicillin and 10,000 μg/mL streptomycin) and UPW up to 1 L. Mix to dissolve the chemicals and adjust pH to 7.7. Store at 4 °C.
2. *MS-222 reagent*: Place 12.5 g MS-222 (tricaine methanesulfonate) powder into a 500-mL glass bottle, and add UPW up to 500 mL. Mix well to dissolve. The stock solution at concentration of 25 mg/mL is stored at 4 °C.
3. *Dissecting tools*: Ophthalmic scissors and blunt-end forceps were used for isolating follicles from zebrafish, and pointed forceps was used to separate oocyte and follicle layer (Fig. 1).

2.3 RNA Extraction and RT-PCR

1. *RNA extraction*: Total RNA is isolated from denuded oocytes, follicle layers, and intact follicles, respectively. The chemicals used are TRIzol with polyacryl carrier (5 μL in 1 mL of TRIzol) for extraction, chloroform, isopropanol, 75% ethanol in DEPC-water, and DEPC-water to dissolve RNA.
2. *Reverse transcription*: Oligo-dT in DEPC-water (0.5 μg/μL), 2 mM dNTP, M-MLV reverse transcriptase kit (5× first strand buffer, 0.1 M DTT and 200 U/μL M-MLV), 72 °C heating block, and 37 °C water bath.
3. *PCR*: The primers used in PCR are listed in Table 2. PCR is performed on the Bio-Rad CFX Real-Time PCR Detection System in a total volume of 10 μL, containing 2 μL diluted RT reaction mix, 1× PCR buffer, 0.2 mM of each dNTP, 2.5 mM $MgCl_2$, 0.75 U Taq polymerase, 0.2 μM of each primer, and autoclaved UPW. The numbers of cycles used are 25 for *ef1a* and 35 for *lhcgr* and *fshr*. The reaction profile consists of 95 °C for 30 s, 60 °C for 30 s, 72 °C for 1 min, and 80 °C for 7 s for signal detection. The signals are quantified by the Image Lab software from Bio-Rad.

3 Methods

3.1 Isolation of FG Follicles from Zebrafish

1. Anesthetize adult female zebrafish (*see* **Note 1**) with MS-222 (0.25 mg/mL) for 5 min, and decapitate them before dissection.

Table 1
Normal and Ca^{2+}/Mg^{2+}-free Cortland medium

Chemicals		
Normal	**Ca^{2+}/Mg^{2+}-free**	**Weight**
NaCl	NaCl	7.25 g
$NaHCO_3$	$NaHCO_3$	1.0 g
KCl	KCl	0.38 g
$MgSO_4 \cdot 7H_2O$	–	0.23 g
$Na_2HPO_4 \cdot H_2O$ MW 160/Na_2HPO_4 MW 142	$Na_2HPO_4 \cdot H_2O$ MW 160/Na_2HPO_4 MW 142	0.41 g/ 0.36 g
$CaCl_2 \cdot H_2O$ MW 129/$CaCl_2 \cdot 2H_2O$ MW 147	–	0.23 g/ 0.26 g
$MgCl_2 \cdot 6H_2O$ MW 203/$MgCl_2$ MW 95	–	0.107 g/ 0.23 g
BSA	BSA	1 g
HEPES	HEPES	5.2 g
+/– Glucose	+/– Glucose	1 g
Penicillin-streptomycin (10,000 units/mL penicillin and 10,000 μg/mL streptomycin)	Penicillin-streptomycin (10,000 units/mL penicillin and 10,000 μg/mL streptomycin)	10 mL
UPW	UPW	To 1000 mL
Adjustment the pH to 7.7 with NaOH	Adjustment the pH to 7.7 with NaOH	

2. Open the fish at the abdomen from the anal pore with ophthalmic scissors, and remove the ovaries carefully (Fig. 2).
3. Place the dissected ovaries in a 15-cm Petri dish containing Cortland medium.
4. Remove fat tissues and other unwanted tissues by forceps.
5. Separate the follicles carefully under a stereomicroscope using forceps, and select the FG follicles ($\geq$0.65 mm in diameter).
6. Transfer the FG follicles to another dish with Ca^{2+}/Mg^{2+}-free Cortland medium.

3.2 Pretreatment of FG Follicles

Unlike ovarian follicles of mammals, fish follicles contain only one thin layer of follicle cells (granulosa and theca cells). The follicle cells are tightly attached to the oocyte, making it extremely difficult to see the follicle layer and separate it from the oocyte (Fig. 3). To loosen the attachment of the follicle cells to the oocyte so as to

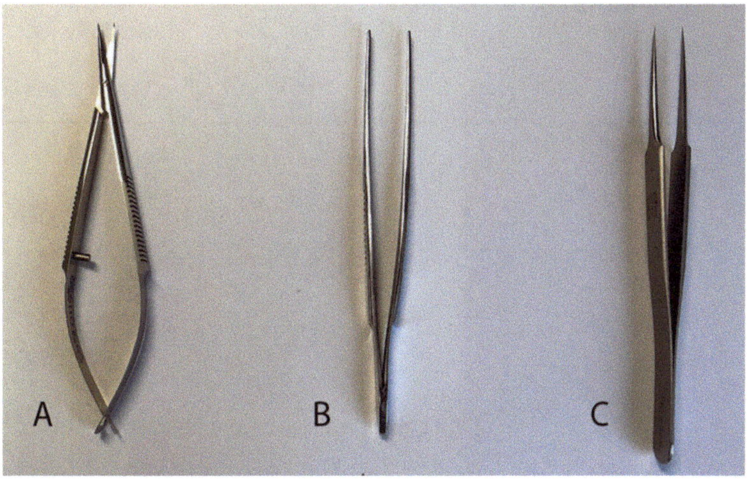

Fig. 1 Dissecting tools used for separating the oocyte and follicle layer. (**A**) Ophthalmic scissors. (**B**) Blunt-end ophthalmic forceps. (**C**) Pointed ophthalmic forceps

Table 2
Primers used in the protocol

Name	Sequence
gdf9 F	GAGTCTGTTGAACCCGACG
gdf9 R	GCAGGTGGATGTCCTTCTTA
ef1a F	GGCTGACTGTGCTGTGCTGATTG
ef1a R	CTTGTCGGTGGGACGGCTAGG
lhcgr F	TCGAGTGTCCGGAGATCTG
lhcgr R	GGAGACGTTCGGCAGGTTAT
fshr F	GCTGTGCTTTATTCTTGGCTGCT
fshr R	ACCTTGTTGCCCAGACAGAC

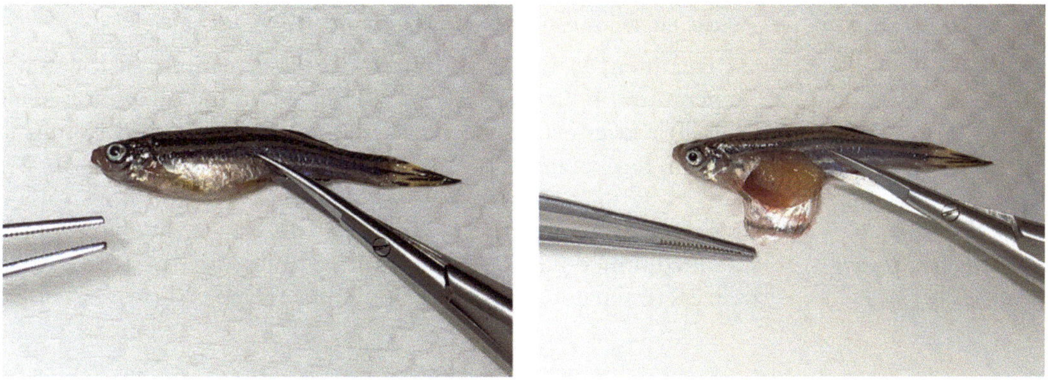

Fig. 2 Dissection of the ovary. The ophthalmic scissors were used to open the abdomen and the blunt-end forceps were used to remove the ovaries

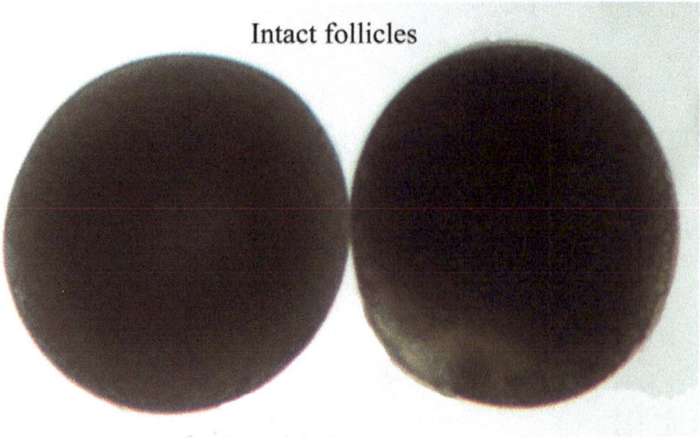

Fig. 3 Intact follicles before pretreatment in Ca^{2+}/Mg^{2+}-free Cortland medium

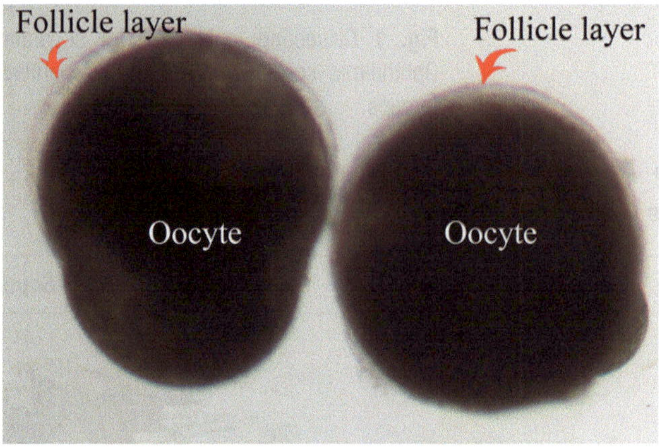

Fig. 4 Follicles after pretreatment at −20 °C for 15 min. The follicle layer is visible on the oocyte surface (red arrows)

allow for mechanic separation of the two compartments, the follicles need to be pretreated.

1. Incubate the follicles in Ca^{2+}/Mg^{2+}-free Cortland medium. This step of pretreatment results in loosening and separation of follicle layer and oocyte, making the follicle layer visible on the surface of oocyte (Fig. 4).
2. Treatment in Ca^{2+}/Mg^{2+}-free Cortland medium at low temperature (4 °C and −20 °C) can facilitate the separation. The follicle layer usually becomes visible after 60, 30, and 15 min at room temperature, 4 °C and −20 °C, respectively (*see* **Note 2**).

3.3 Mechanic Separation of Oocyte and Follicle Layer

After pretreatment of the FG follicles with Ca^{2+}/Mg^{2+}-free Cortland medium, e.g., at −20 °C for 15 min, use two forceps or tweezers to separate the oocyte and follicle layer as shown in Fig. 5.

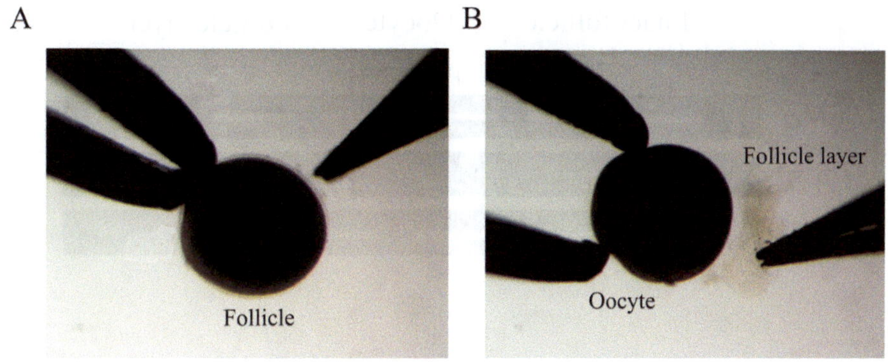

Fig. 5 The procedure of separating follicle layer from the oocyte. Two kinds of forceps were used. The blunt-end forceps were used to hold the follicle (**A**), while the pointed one was used to rip off the follicle layer (**B**)

1. Use the blunt-end forceps to grasp and hold the follicle gently.
2. Use the pointed one to peel off the follicle layer as a whole without breaking the oocyte (Fig. 5) (*see* **Notes 3** and **4**).

3.4 Purity Check by Molecular Markers gdf9 *and* lhcgr

Total RNA was isolated from individual denuded oocyte and isolated follicle layer with TRIzol for RT-PCR detection (*see* **Note 5**).

1. Add 200 μL TRIzol to each isolated oocyte or follicle layer in a microtube.
2. Shake the tubes for ~15 min on Eppendorf ThermoMixer to ensure complete tissue lysis.
3. Add 50 μL chloroform to each tube, vortex, and shake the tubes for 2 min.
4. Leave the tubes at room temperature for 8 min followed by centrifugation (20,000 × *g*) for 15 min at 4 °C.
5. Transfer 80 μL supernatant from each sample to a new tube and add 100 μL isopropanol, vortex, and incubate at −20 °C for 30 min, followed by centrifugation (20,000 × *g*) for 10 min at 4 °C.
6. Discard all supernatant without disturbing the pellet, wash the pellet with 300 μL 75% ethanol, and centrifuge at room temperature at 7500 × *g* for 5 min. Discard the supernatant and air-dry the pellet until it becomes transparent.
7. Convert the RNA into cDNA by reverse transcription.
8. To ensure clean separation, we used *gdf9*, an oocyte marker, and *lhcgr*, a follicle cell marker, to check all separated oocytes and follicle layers individually by RT-PCR.
9. Run agarose electrophoresis on RT-PCR products. The denuded oocytes should express *gdf9* but not *lhcgr*, whereas the follicle layer should express *lhcgr* but not *gdf9*.

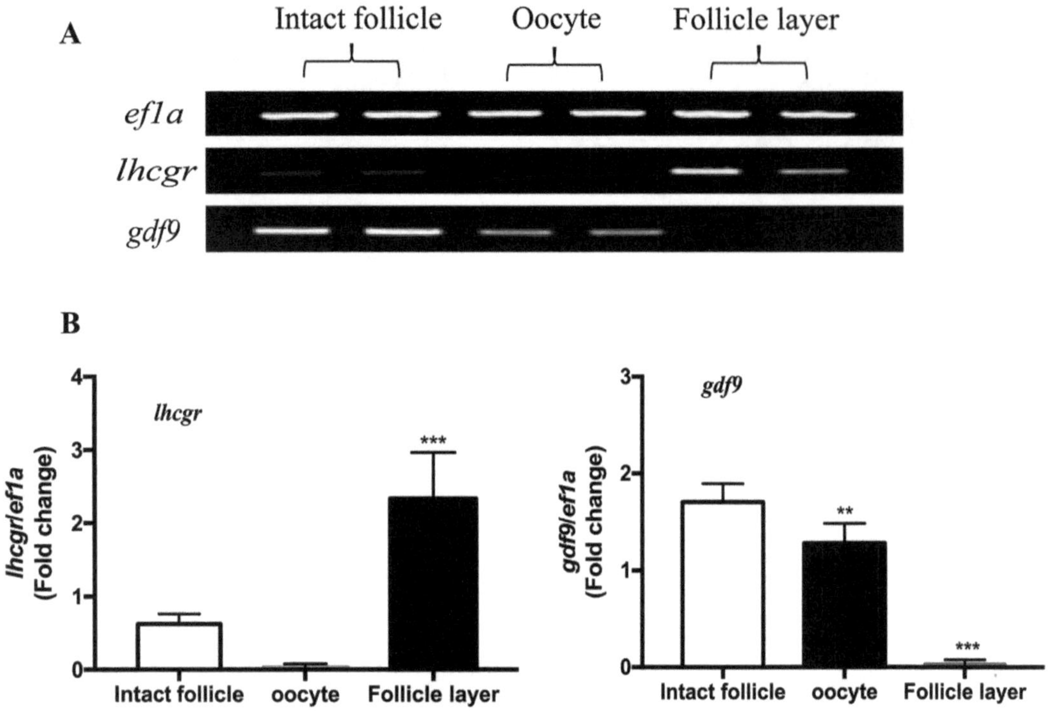

Fig. 6 Detection of expression in intact follicles, denuded oocytes, and isolated follicle layers. (**A**) The housekeeping gene (*ef1a*) was expressed in both compartments, whereas the follicle cell-specific gene (*lhcgr*) and oocyte-specific gene (*gdf9*) were only detected in the somatic follicle layer and oocyte, respectively. (**B**) Semiquantitative analysis of gene expression levels. The values are presented as means ± SD ($n = 3$). ***$P < 0.001$, **$P < 0.05$ vs. intact follicle group

10. Pool five clean samples (oocytes or follicle layers) together to examine expression of target genes of interest. As control, five intact follicles were pooled as one sample (Fig. 6).

4 Notes

1. We used 6-month-old zebrafish in the present study; however, spawning females of other ages should also be fine.
2. The step of pretreatment can be performed at different temperatures with cold temperatures facilitating the separation. However, what temperature to use depends on the purpose of experiments.
3. If the number of target genes for analysis is small, the assay can be performed on individual basis; otherwise, multiple follicle layers or denuded oocytes can be pooled to increase the yield of RNA (we normally pool five together).
4. Using the protocol described, it is relatively easy to obtain clean follicle layer without contaminating by oocyte contents.

However, it is more challenging to obtain clean oocytes without any follicle cells.

5. All media and tools used must be RNase-free, and the procedure should be performed in an RNase-free environment.

Acknowledgments

This study was supported by grants from the University of Macau (MYRG2015-00227-FHS, MYRG2016-00072-FHS, MYRG2017-00157-FHS, and CPG2014-00014-FHS) and The Macau Fund for Development of Science and Technology (FDCT089/2014/A2 and FDCT173/2017/A3) to W.G.

References

1. Selman K et al (1993) Stages of oocyte development in the zebrafish. J Morphol 218(2):203–224
2. McPherron AC, Lee SJ (1993) GDF-3 and GDF-9: two new members of the transforming growth factor-beta superfamily containing a novel pattern of cysteines. J Biol Chem 268(5):3444–3449
3. Hayashi M et al (1999) Recombinant growth differentiation factor-9 (GDF-9) enhances growth and differentiation of cultured early ovarian follicles. Endocrinology 140(3):1236–1244
4. Dong J et al (1996) Growth differentiation factor-9 is required during early ovarian folliculogenesis. Nature 383(6600):531–535
5. Chattoraj A et al (2005) Melatonin accelerates maturation inducing hormone (MIH): induced oocyte maturation in carps. Gen Comp Endocrinol 140(3):145–155
6. Eppig JJ et al (1997) Murine oocytes suppress expression of luteinizing hormone receptor messenger ribonucleic acid by granulosa cells. Biol Reprod 56(4):976–984
7. Liu L, Ge W (2007) Growth differentiation factor 9 and its spatiotemporal expression and regulation in the zebrafish ovary. Biol Reprod 76(2):294–302
8. Wang Y, Ge W (2003) Spatial expression patterns of activin and its signaling system in the zebrafish ovarian follicle: evidence for paracrine action of activin on the oocytes. Biol Reprod 69(6):1998–2006
9. Wang Y, Ge W (2004) Cloning of epidermal growth factor (EGF) and EGF receptor from the zebrafish ovary: evidence for EGF as a potential paracrine factor from the oocyte to regulate activin/follistatin system in the follicle cells. Biol Reprod 71(3):749–760
10. Yao K, Ge W (2013) Spatial distribution and receptor specificity of zebrafish Kit system—evidence for a Kit-mediated bi-directional communication system in the preovulatory ovarian follicle. PLoS One 8(2):e56192
11. Li CW, Ge W (2011) Spatiotemporal expression of bone morphogenetic protein family ligands and receptors in the zebrafish ovary: a potential paracrine-signaling mechanism for oocyte-follicle cell communication. Biol Reprod 85(5):977–986
12. Zhou R, Yu SM, Ge W (2016) Expression and functional characterization of intrafollicular GH-IGF system in the zebrafish ovary. Gen Comp Endocrinol 232:32–42
13. Zhou R et al (2011) Pituitary adenylate cyclase-activating polypeptide (PACAP) and its receptors in the zebrafish ovary: evidence for potentially dual roles of PACAP in controlling final oocyte maturation. Biol Reprod 85(3):615–625
14. Ulloa-Rodriguez P et al (2018) Patagonian blenny (Eleginops maclovinus) spermatozoa quality after storage at 4 masculineC in Cortland medium. Anim Reprod Sci 197:117–125

Chapter 2

The *Xenopus* Oocyte as an Expression System for Functional Analyses of Fish Aquaporins

François Chauvigné, Alba Ferré, and Joan Cerdà

Abstract

Aquaporins are membrane proteins present in all organisms that selectively transport water and small, uncharged solutes across biological membranes along an osmotic gradient. Recent gene editing technologies in zebrafish (*Danio rerio*) have started to uncover the physiological functions of the aquaporins in teleosts, but these approaches require methods to establish the effects of specific mutations on channel function. The oocytes of the South African frog *Xenopus laevis* are widely used for the expression of bacterial, plant, and animal aquaporins, and this heterologous system has contributed to numerous discoveries in aquaporin biology. This chapter focuses on techniques used for oocyte preparation and aquaporin expression and gives an overview of specific methods to determine water and solute permeability of the channels and their intracellular trafficking in oocytes.

Key words Water and solute permeability, Solute uptake, Aquaporin trafficking, Splice forms, Dominant-negative mutants

1 Introduction

The water-transporting proteins first discovered at the end of the twentieth century [1] are a group of membrane channels that belong to the aquaporin superfamily present in all domains of life. The aquaporins are small hydrophobic integral membrane proteins that form a pore allowing the bidirectional passage of water and small, uncharged solutes across biological membranes following an osmotic gradient [2]. In vertebrates, there are 17 different subfamilies of aquaporins, which can be classified into 4 major groups: the classical water-selective aquaporins (AQP0, AQP1, AQP2, AQP4, AQP5, AQP6, AQP14, and AQP15); the glycerol- and urea-transporting aquaporins, termed aquaglyceroporins (AQP3, AQP7, AQP9, AQP10, and AQP13); the AQP8 and AQP16 types; and the unorthodox intracellular aquaporins (AQP11 and AQP12) [3–5]. Some aquaporins can also transport a variety of substrates, such as carbon dioxide, nitric oxide, ammonia, hydrogen peroxide,

Roland Dosch (ed.), *Germline Development in the Zebrafish: Methods and Protocols*, Methods in Molecular Biology, vol. 2218, https://doi.org/10.1007/978-1-0716-0970-5_2, © Springer Science+Business Media, LLC, part of Springer Nature 2021

arsenite, silicic acid, boron, and antimonite [6, 7], reinforcing the notion of their important role in cell homeostasis.

While in mammals the importance of aquaporins in many physiological mechanisms is well established [8, 9], in teleosts, including the zebrafish (*Danio rerio*), this information is still scarce, although their role in different water and solute transport processes in somatic as well as germ cells has been proposed [10–14]. The zebrafish genome encodes 19 aquaporin paralogs, which show similar substrate permeabilities to their mammalian orthologs [15], but, as other teleosts, lacks the orthologs to mammalian AQP2, AQP5, and AQP6 and has *aqp15* as a putative pseudogene [3, 12]. Many of the paralogs are thus duplicated, and redundant expression of duplicated paralogs is observed in many tissues [15]. Recent genetic approaches to generate loss-of-function deletions of zebrafish aquaporins using CRISP/Cas9 have elucidated the subfunctionalization of one of the duplicates, such as Aqp0a in the lens [14], while other studies have shown that dominant missense mutations in Aqp3a, which influence pore permeability, can affect pigment cell behavior in the skin [13]. Gene editing technologies, and particularly those directed to introduce mutations that can affect aquaporin function, require methods to functionally characterize the modified channels and thus understand the mechanism of action of aquaporin failure.

The oocytes of the South African clawed frog *Xenopus laevis* have typically been employed as an expression system for the characterization of ion channels due to their ability to efficiently translate exogenous mRNA into proteins and their amenability for electrophysiological recordings [16]. The *Xenopus* oocyte expression system was also employed to demonstrate that the 28 kDa integral membrane protein isolated from human erythrocytes and renal tubules by Peter Agre and colleagues was indeed a membrane channel with water transport activity, leading to the discovery of AQP1, the first aquaporin [1]. Since then, *Xenopus* oocytes have become a very common system to functionally characterize aquaporins isolated from bacteria, plants, and animals. This system, although laborious, is less time-consuming than the reconstitution of recombinant functional aquaporins in proteoliposomes [17, 18] and may be more accurate than the expression and further measurements of aquaporin permeability in cultured cell lines [19]. However, *Xenopus* oocytes also present important limitations, such as the presence of a cellular environment, particularly signaling pathways, that may be very different from that in the cell where the aquaporin of interest is normally expressed. Consequently, considerable caution is needed when studying channel intracellular trafficking by secondary messengers in oocytes. In addition, there may be differences between invertebrates and vertebrates in the types of glycosylation of aquaporins, as noted for ion channels [20], which may affect the function and trafficking of the aquaporin [21].

This chapter describes the standard protocols for the synthesis of in vitro transcribed complementary aquaporin RNA (cRNA) and its injection into the oocyte cytoplasm, the determination of water and solute permeability of the expressed channels, procedures for the biochemical fractionation of oocytes and subsequent protein detection, and fluorescent methods for the subcellular localization of the channel.

2 Materials

2.1 DNA and cRNA Preparation

1. cDNA clones coding for wild-type and/or modified (splice forms, mutants, chimeric constructs) aquaporins.
2. Commercial kits for DNA miniprep of desired constructs (e.g., GenElute™ Plasmid Miniprep Kit, Sigma-Aldrich) and linear plasmid DNA purification by gel electrophoresis (e.g., GenElute™ Gel Extraction Kit, Sigma-Aldrich).
3. Oocyte expression vector pT7Ts [22] (*see* **Note 1**).
4. Reagents for plasmid linearization and capped cRNA synthesis: restriction enzymes, RNaseOUT™ Recombinant Ribonuclease Inhibitor, NTP mix, 10× T7 transcription buffer, T7 RNA polymerase, and Cap Analog ($m^7G(5')ppp(5')G$).
5. Commercial kit for RNA purification (e.g., GenElute™ Mammalian Total RNA Miniprep Kit with on-column DNase I digestion, Sigma-Aldrich).
6. Equipment to estimate RNA quality and concentration (e.g., NanoDrop® ND-1000, Thermo Fisher Scientific™).

2.2 Xenopus laevis Oocyte Extraction and Microinjection

1. *Xenopus laevis* ovarian lobes (*see* **Note 2**).
2. Osmometer.
3. Modified Barth's solution (1× MBS): 88 mM NaCl, 1 mM KCl, 2.4 mM $NaHCO_3$, 0.82 mM $MgSO_4$, 0.33 mM $Ca(NO_3)_2$, 0.41 mM $CaCl_2$, 10 mM HEPES and 25 μg/ml gentamycin, pH 7.5. Osmolality should be set at 200 mOsm.
4. Collagenase solution: 1× MBS + 1 mg/ml collagenase (Type I, Sigma-Aldrich) (*see* **Note 3**).
5. Rotator shaker.
6. A *Xenopus* oocyte microinjection workstation (Fig. 1) comprising:
   (a) Stereo microscope with external cold light source and equipped with a calibrated eyepiece.
   (b) Micromanipulator and micropipette holder.
   (c) Micropipette puller.
   (d) Microinjector.

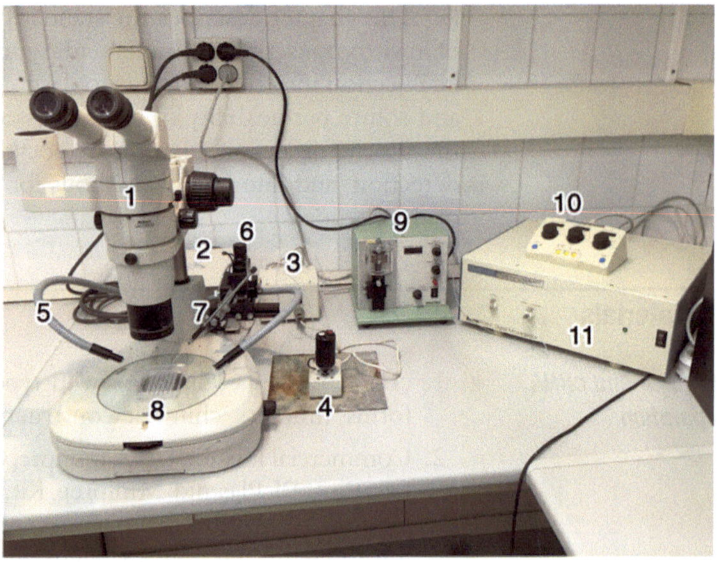

Fig. 1 *Xenopus* oocyte microinjection workstation. In our laboratory the workstation is composed of a stereo microscope Nikon SMZ1000 equipped with a calibrated eyepiece and transmitted lighting (1), an external cold light Volpi Intralux 4000–1 light source (2) with flexible light guides (5), the power transformer of the motor-driven coarse manipulator Narishige MM-188NE (3) and control unit (joystick controller) (4), the drive unit and adaptor of the micromanipulator (6), micropipette holder (7), the micropipette puller Narishige PC-10 (9), and the user interface module of the Sutter Instruments XenoWorks digital microinjector (10) and compressor module (11). A Terasaki plate used for oocyte microinjection is also shown on the stereo microscope (8)

7. Small temperature-controlled incubator set at 18 °C.
8. Microinjection borosilicate glass capillaries of 0.6-mm inner diameter and 90 mm in length.
9. Microloader pipette tips.
10. Parafilm™ M laboratory wrapping film.
11. 72-well crystallization Terasaki plates (Greiner Bio-One) (Fig. 1).
12. 12- and 6-well cell culture plates with flat bottom.
13. Sterile graduated transfer 3-ml plastic.
14. Jeweler's forceps, Dumont No. 5.

2.3 Determination of Oocyte Water and Solute Permeability

1. Stereo microscope with transmitted lighting equipped with a high-resolution digital camera.
2. Computer running calculation software (e.g., Microsoft Excel).
3. Image analysis software (e.g., NIS-Elements AR 4.30.02 software, Nikon).

4. Sterile 35 × 10 mm vented cell culture dishes.
5. Glycerol and urea.
6. [1,2,3-^{3}H]Glycerol (50 Ci/mmol) and [^{14}C]urea (58 mCi/mmol) (American Radiolabeled Chemicals, Inc.).
7. Scintillation vials.
8. Liquid scintillation cocktail (e.g., Ultima Gold XR, PerkinElmer).
9. Liquid scintillation counter for ^{3}H and ^{14}C measurements.

2.4 Extraction of Plasma and Total Membrane Proteins from Oocytes

1. Microcentrifuge.
2. HbA buffer: 20 mM Tris-HCl, pH 7.4, 5 mM $MgCl_2$, 5 mM NaH_2PO_4, 1 mM EDTA, 80 mM sucrose, and protease inhibitors (Mini EDTA-free; Roche) [23].
3. MES buffered saline for silica (MBSS): 20 mM MES, 80 mM NaCl, pH 6.0 [23].
4. LUDOX® CL-X colloidal silica (Sigma-Aldrich) at 1% in MBSS.
5. Polyacrylic acid (Sigma-Aldrich) at 0.1% in MBSS.
6. 1× Laemmli sample buffer [24]: 30 mM Tris-HCl, pH 6.8, 1% SDS, 5% glycerol and 0.0025% bromophenol blue supplemented with protease inhibitors, 2 mM Na_3VO_4, 2 mM NaF, and 0.1 M DL-dithiothreitol (DTT) (*see* **Note 4**).

2.5 Co-immunoprecipitation (Co-IP)

1. RIPA buffer: 50 mM Tris-HCl, pH 8, 150 mM NaCl, 1 mM NaF, 1 mM Na_3VO_4, 1% Triton X-100, 0.5% sodium deoxycholate, 0.1% SDS, and protease inhibitors.
2. Dynabeads™ M-280 Tosylactivated (Invitrogen, Thermo Fisher Scientific).
3. Buffer B: 0.1 M sodium phosphate buffer (100 mM Na_2HPO_4, 100 mM NaH_2PO_4, pH 7.4).
4. Buffer C: 3 M ammonium sulfate in buffer B.
5. Buffer D: PBS containing 0.5% BSA.
6. Buffer E: PBS containing 0.1% BSA.
7. PBST: PBS with 0.05% Tween-20.

2.6 Immunofluorescence Microscopy

1. Phosphate buffer saline (1× PBS): 137 mM NaCl, 2.7 mM KCl, 10 mM Na_2HPO_4, 1.8 mM KH_2PO_4, pH 7.5.
2. 4% paraformaldehyde (PFA) in PBS.
3. Ethanol.
4. Xylene.
5. Paraplast Plus® for oocyte embedding (Sigma-Aldrich).
6. Paraffin oven (e.g., Memmert UN30pa).

7. Histology tissue cassettes.
8. Embedding molds.
9. Microtome (e.g., LEICA RM 2255).
10. UltraStick™/UltraFrost™ adhesion microscope slides (Electron Microscopy Sciences).
11. Blocking solution: PBS containing 5% normal goat serum and 0.1% bovine serum albumin (BSA).
12. Primary commercial or custom-made antibodies.
13. Fluorescent dye-labeled goat or donkey anti-rabbit, anti-mouse, or anti-goat secondary antibodies (depending on primary antibodies used).
14. Cover glass.
15. Fluoromount™ aqueous mounting medium (Sigma-Aldrich).

3 Methods

The methods that are described here for the heterologous expression of aquaporin mRNAs in *Xenopus* oocytes are essentially the same as those employed for the expression of ion channels and receptors. A summarized chronological protocol is depicted in Fig. 2.

3.1 cRNA Synthesis

1. Subclone the aquaporin cDNA(s) into the pT7Ts vector (*see* **Note 5**).
2. Prepare DNA miniprep of desired constructs using standard kits.
3. Linearize plasmid: Digest 2.5 μg of DNA during 2 h at 37 °C at a restriction site after the poly-C (either *Xba*I, *Sal*I, or *Eco*RI, depending on the restriction sites present within the construct sequence). Purify linear plasmid DNA (~1.3 μg) by gel electrophoresis. Elute DNA with 32 μl DNase-/RNase-free water.
4. Synthesize capped cRNA: Mix 28 μl linear plasmid (~2.2 μg), 5 μl 100 mM DTT, 5 μl 5 mM Cap Analog [$m^7G(5')ppp(5')G$], 40 units RNaseOUT™ Recombinant Ribonuclease Inhibitor, 5 μl 10 mM NTP mix, 5 μl 10× T7 transcription buffer, and 20 units of T7 RNA polymerase. Incubate for 1 h at 37 °C.
5. Purify the cRNA using the GenElute™ Mammalian Total RNAs Miniprep Kit with on-column DNase I digestion. Elute the cRNA in 40 μl of DNase-/RNase-free, and estimate quality and concentration using the NanoDrop® ND-1000. Usually the total yield of cRNA is of 8–20 μg. Aliquots of 5–10 μl are stored at −80 °C until use.

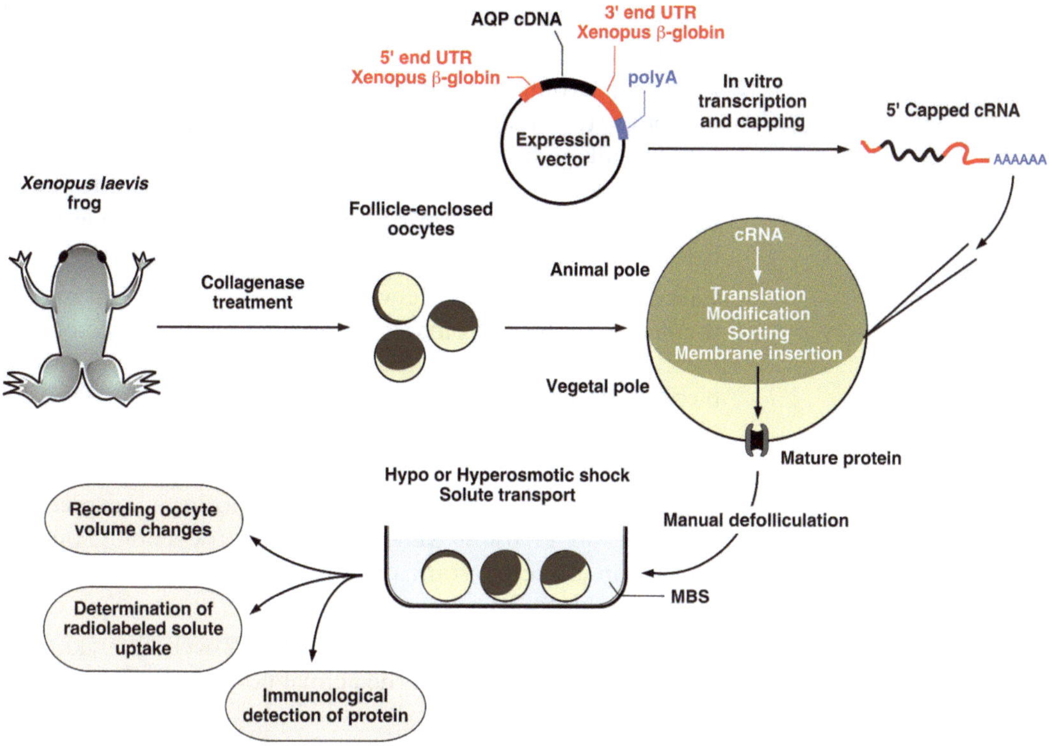

Fig. 2 Schematic representation of the workflow of the *Xenopus* oocyte expression system to evaluate the substrate permeability and intracellular trafficking of heterologously expressed aquaporins

3.2 Ovarian Follicle Isolation and Microinjection

1. Collect ovarian lobes by surgical laparotomy (*see* **Note 2**).
2. Rinse ovarian lobes three times with 30 ml 1× MBS. Open the lobes with closed forceps to avoid any damage of the ovarian follicles.
3. Digest extracellular tissues and separate follicle-enclosed oocytes from the ovary by collagenase treatment. Transfer the ovarian pieces (the equivalent of 3 ml) to a 15-ml Falcon tube containing 7 ml of collagenase solution. Wrap the tube with aluminum foil to protect from light, and shake it on a rotator shaker with gentle rotation during 2 h at room temperature until individual follicles are isolated. Adjust the time if necessary to avoid over-digestion and low oocyte survival. Wash the follicles 5–6 times with 1× MBS and incubate at 18 °C in a Petri dish for a few hours.
4. Follicle selection: Under a stereo microscope, select healthy (showing contrasted white and dark hemispheres, vegetal and animal poles, respectively) stage IV–V follicle-enclosed oocytes (1.0–1.1-mm diameter), and transfer them to a Petri dish with fresh MBS (*see* **Note 6**).

5. Selected follicle-enclosed oocytes can be injected after 1 h of equilibration in MBS at 18 °C or the following day.
6. Prepare microinjection pipettes: Fix a glass capillary in the micropipette puller, and adjust parameters to produce an elongated shank with a gradual uniform taper and a submicron tip (use step 1 in the Narishige PC-10 puller). Under the stereo microscope, break the tip of the needle at a 45° angle using forceps to increase the tip inner diameter to ~20 μm.
7. Calibration: Fill the needle with RNase-free water and mount the needle in the micromanipulator. Set the microinjector to a fixed pressure (300–800 hPa) and deposit a drop on a clean piece of Parafilm. Using the stereo microscope at 8× magnification, measure the volume, and adjust the pressure to 50 nl (i.e., the drop should measure 40 mm in diameter).
8. Dilute the cRNA coding for the aquaporins to express to 1–25 ng/50 nl using RNase-free water (depending on the translation efficiency, some cRNAs can be injected at 0.2 ng/ 50 nl) (*see* **Note 7**).
9. Inject oocytes: Dispense oocytes in the 72-well Terasaki plate (one oocyte per well) with the animal pole facing the left side. Fill the microinjection needle with 5 μl of the cRNA solution, and inject 50 nl in the equatorial zone of each oocyte, the needle facing the direction to the animal pole. Injection is carried out sequentially by moving the plate with one hand while moving the pipette in and out of oocytes at a 45° angle with the other hand. Control oocytes are injected with RNase-free water or are non-injected (*see* **Note 8**). Once all oocytes are injected, dispense them in 12- and 6-well cell culture plates, in which control and aquaporin-injected oocytes are in different wells, and incubate in fresh MBS at 18 °C.
10. At 24 h after injection, peel off the follicular cells from the oocytes using forceps under the stereo microscope. The follicular layer gets stuck to the plastic of the Petri dish and breaks to liberate the oocyte, which helps the defolliculation procedure. Denuded oocytes are stored overnight in MBS at 18 °C.

3.3 Measurement of Oocyte Water and Solute Permeability

Here, we will describe a protocol for the determination of water and solute permeability of aquaporins expressed in *Xenopus* oocytes using volumetric (oocyte swelling) methods. Aquaporin-mediated solute permeability in oocytes can also be determined more accurately by measuring the uptake of radiolabeled compounds. These procedures will also be described.

3.3.1 Volumetric Assays and Calculation of Oocyte Permeability Coefficients

1. Healthy oocytes are selected on the same day as the assays are done.

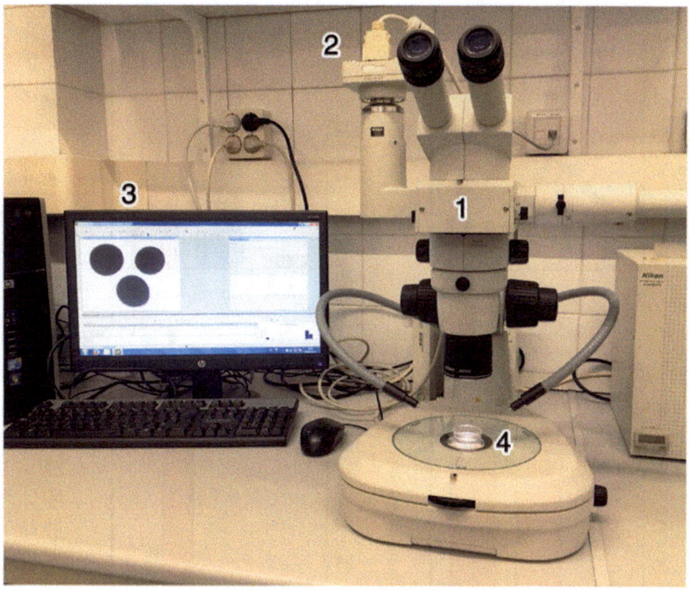

Fig. 3 Workstation for *Xenopus* oocyte volume recordings. The workstation is composed of a stereo microscope Nikon SMZ800 (1) equipped with a high-resolution Nikon DS-Fi2 camera (2) and the computer HP Compaq 6000 Pro running Microsoft Windows 7 Prod with Microsoft Excel 2010 and the Nikon NIS-Elements AR 4.30.02 software (3). The picture also shows a 35 × 10 mm vented Petri dish on the stereo microscope (4), where oocytes are immersed in diluted MBS or isotonic MBS containing solutes and illuminated with transmitted light

2. Transfer one to four oocytes to a 35 × 10 mm vented cell culture dish containing tenfold diluted MBS (~20 mOsm) for determination of water permeability (*see* **Note 9**), or with isotonic MBS in which NaCl is replaced by 160 mM of the solutes (typically glycerol or urea), to measure solute permeability (*see* **Note 10**). The osmolarity of the MBS + solute is measured for each experiment with an osmometer and adjusted to 200 mOsm with NaCl if necessary.
3. Place the dish under a stereo microscope with transmitted light (oocytes appear completely dark) and equipped with a high-definition camera connected to the image analysis software (Fig. 3). To measure water permeability, set the system to capture serial images of the oocytes every 2 s during the first 20 s and to calculate the area of the oocytes. To determine solute permeability, serial images are taken at 5-s intervals during 1 min.
4. Transfer the recorded data into an excel sheet for the calculation of the osmotic water permeability (P_f) or glycerol and urea permeability (P_{gly} and P_{urea}, respectively) coefficients. The P_f is calculated taking into account the time-course changes in the

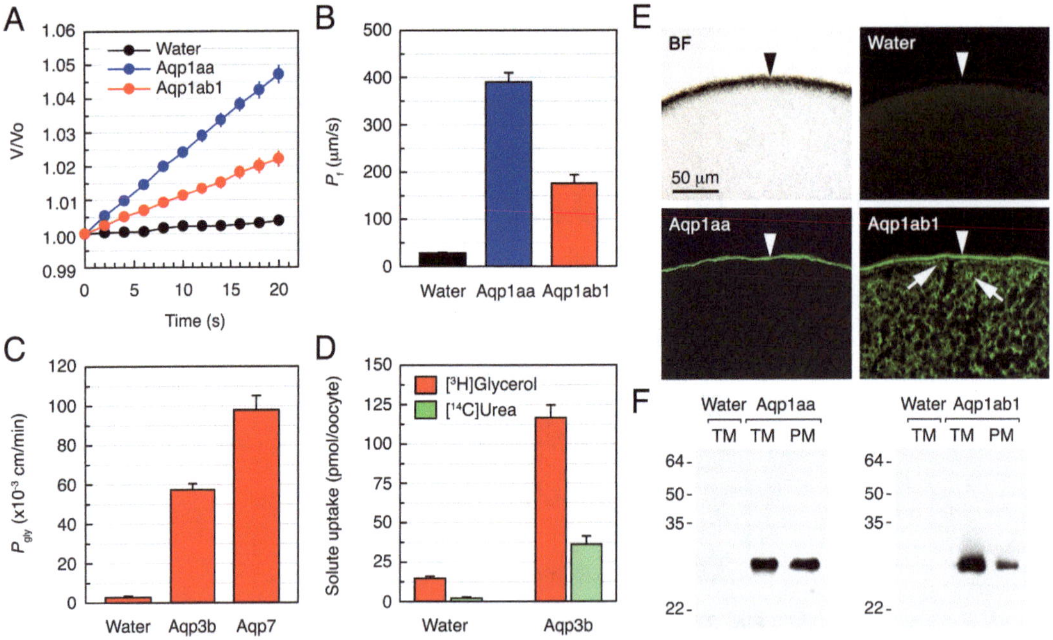

Fig. 4 Examples of the functional characterization of fish aquaporins in *Xenopus* oocytes. (**a**) Changes in volume of oocytes injected with water (control, black) or 1 ng of cRNA encoding piscine Aqp1aa (blue) or Aqp1ab1 (red). (**b**) P_f values of oocytes shown in (**a**). (**c**) Glycerol permeability (P_{gly}) of control oocytes and oocytes injected with 1 ng cRNA encoding zebrafish aquaglyceroporins Aqp3b or Aqp7. (**d**) Radiolabeled glycerol and urea uptake during 10 min of control and Aqp3b-injected oocytes. In (**a**)–(**d**), data are the mean ± SEM of ten oocytes. (**e**) Immunolocalization of seabream Aqp1aa and Aqp1ab1 in oocytes using specific antibodies. The arrowhead points to the plasma membrane, whereas the arrows indicate partial cytoplasmic retention of the Aqp1ab1 channel. (**f**) Immunoblotting analysis of Aqp1aa and Aqp1ab1 on total and plasma membrane (TM and PM, respectively) proteins extracted from the oocytes shown in (**e**). Note that the relative ratio of Aqp1ab1 in the PM to TM is lower than the same ratio of Aqp1aa

relative oocyte volume ($d(V/V_0)/dt$), the molar volume of water ($V_W = 18$ cm^3/ml), and the oocyte surface area (S), using the formula $V_0(d(V/V_0)/dt)/(SV_W(Osm_{in} - Osm_{out}))$. The surface area of the oocyte is considered to be nine times the apparent area because of membrane folding [25]. The P_{gly} and P_{urea} are calculated from oocyte swelling using the formula $[d(V/V_0)dt]/(S/V_0)$ [26]. Typical P_f and P_{gly} data obtained from fish aquaporins and aquaglyceroporins are shown in Fig. 4a–c (*see* **Note 11**).

3.3.2 Solute Uptake Assays Using Radiolabeled Compounds

1. Transfer ten oocytes to a 30-mm Petri dish, and incubate with 200 μl of isotonic MBS containing 5 μM (20 μCi) of the radiolabeled solute (e.g., [1,2,3-^{3}H]glycerol or [^{14}C]urea) and 1 mM of cold (non-labeled) solute for 10 min at room temperature (*see* **Note 12**).
2. Wash oocytes three times rapidly with ice-cold MBS + 1 mM cold solute.

3. For determination of time zero (necessary to subtract the signal from externally bound solute and calculate the uptake rate, *see* below), oocytes are treated as above except that they are exposed to the radiolabeled solutes for a few seconds and immediately washed.
4. Transfer individual oocytes to a scintillation vial containing 400 μl 10% SDS, and digest for 1 h at room temperature.
5. Add 4 ml of scintillation liquid, vortex, and measure radioactivity in a scintillation counter.
6. The rate of solute uptake (pmol total solute/oocyte/10 min) is calculated using the formula (((TEST − TO)/POS)/(CONC))/((RAD/VOL)/COLD)) × 1000, where POS are the counts per minute (cpm) of 1 μCi of solute; TEST and T0 are the cpm values of each oocyte exposed to solute during 10 min and at time 0, respectively; CONC is the concentration of the radioactive solute (in Ci/mmol); RAD is the amount of radioactive solute in the MBS; VOL is the volume of the MBS (0.2 ml); and COLD is the concentration of cold solute (1 mM). Typical data for glycerol and urea uptake by oocytes expressing zebrafish Aqp3b are shown in Fig. 4d.

3.4 Assessment of Protein Levels of Expressed Aquaporins in Xenopus Oocytes

To draw conclusions on the differences in permeability between aquaporins, the detection of expressed channels at the plasma membrane of *Xenopus* oocytes and the determination of their subcellular distribution are very important (Fig. 4e, f). The examination of channel trafficking to the plasma membrane is critical, for instance, when the permeability of wild-type and mutated aquaporins are to be compared. In this section, very simple biochemical procedures for the extraction of total and plasma membrane proteins from oocytes [23], as well as fluorescent methods for the immunolocalization of the proteins, will be described. For the extraction of proteins, we will review only the sample preparation in this section, since immunoprecipitation and western blotting are standard techniques. Finally, we will also outline an easy protocol for Co-IP which can be very useful to examine interactions of aquaporins with aquaporin splice forms or mutants and investigate dominant-negative regulatory mechanisms [27, 28].

3.4.1 Total Membrane Protein Extraction

1. Homogenize ten oocytes in a 1.5-ml Eppendorf tube containing 200 μl of ice-cold HbA buffer using an automatic pipette until no particles (dark granules) are visible (e.g., about 20 times using a Gilson™ PIPETMAN P200).
2. Centrifuge at 200 × *g* for 5 min at 4 °C and transfer the supernatant to a clean 1.5-ml Eppendorf tube. Repeat this step once.
3. Centrifuge supernatant at 16,873 × *g* for 20 min at 4 °C.

4. Resuspend the pellet in 60 μl of 1× Laemmli sample buffer (one oocyte equivalent per 6 μl). Heat the sample at 95 °C for 10 min and cool down at room temperature.
5. Aliquot the sample, snap freeze with liquid nitrogen, and store at −80 °C.
6. Proceed with standard protocol for immunoblotting.

3.4.2 Plasma Membrane Protein Extraction

1. Wash ten oocytes two times with ice-cold MBSS.
2. Rotate oocytes immersed in MBSS with 1% Ludox in shaker for 1 h at 4 °C, and wash three times in cold MBSS.
3. Rotate again the oocytes in MBSS with 0.1% polyacrylic acid for 1 h at 4 °C, and wash three times in cold MBS.
4. Homogenize oocytes in 500 μl ice-cold HbA as described above and add 700 μl of HbA to the homogenate.
5. Centrifuge at 27 × *g* and 4 °C for 20 s.
6. Discard 1 ml from the top of the sample and add 1 ml of fresh HbA. Repeat this procedure four times, with the third centrifugation for 30 s and the last centrifugation at 38 × *g* for 30 s.
7. Add 1 ml of HbA. At this point large leaflets should be visible when tubes are filled. Spin down plasma membrane for 30 min at 16,873 × *g* and 4 °C.
8. Resuspend pellet in 50 μl of 1× Laemmli sample buffer (plasma membrane of one oocyte per 5 μl), and boil at 95 °C for 10 min. Snap freeze in liquid nitrogen and store at −80 °C.
9. Proceed with standard protocol for immunoblotting.

3.4.3 Co-IP Assay

1. Homogenize the pellet of total membrane from 30 oocytes expressing 2 variants of the same aquaporin in 600 μl of RIPA buffer.
2. Collect 50 μl and mix with 2× Laemmli sample buffer as the "input" control sample of the IP.
3. Activate Dynabeads™ (the day before): Wash 1× 1 mg of beads with buffer B, and add 15 μg of the antibody specific for one of the aquaporin variants, or of rabbit IgG (negative control), in a final volume of 150 μl in buffer B. Vortex and add 100 μl of buffer C. Rotate the mixture overnight at room temperature. Next day wash 1× with buffer D for 1 h at room temperature and twice with buffer E (*see* **Note 13**).
4. Resuspend the beads with the rest of the oocyte homogenate (550 μl) with the activated Dynabeads, and rotate as above overnight at room temperature.
5. Next day wash the beads three times with PBST, and elute the antibody-protein complexes with 50 μl of 1× Laemmli sample buffer during 10 min in a vortex. Heat at 90 °C for 5 min.
6. Proceed with standard protocol for immunoblotting (usually 20 μl of the elution per lane is loaded for western blot).

3.5 Immunofluorescence Microscopy

1. Fix ten oocytes with 4% PFA for 6 h at room temperature.
2. Wash two times with PBS for 30 min under gentle agitation. Store in 70% ethanol at 4 °C overnight.
3. Dehydrate oocytes through sequential passages in 100% ethanol (two times for 1 h), 50%/50% dilution of 100% ethanol/xylene (vol:vol) for 1 h, and finally 100% xylene for 1 h. Keep oocytes in a fresh xylene solution overnight at room temperature under constant agitation.
4. Transfer oocytes to histology tissue cassettes and place them into a beaker with Paraplast at 60 °C for 3 h, change to another fresh bath of Paraplast for 3 h, and to a third bath overnight. Transfer oocytes with a pipette into an embedding mold containing fresh Paraplast at 60 °C and a cassette on top. Let to solidify at room temperature overnight and store the block at 4 °C.
5. Make 7-μm-thick sections with a microtome. Dispense the sections on UltraStick™/UltraFrost™ adhesion microscope slides and dry for 48 h.
6. Deparaffinize and rehydrate tissue sections: Remove the Paraplast from the sections and rehydrate through a sequential ethanol gradient as follows:

Solution	Time (min)	Number of changes
100% xylene	20	2
100% ethanol	5	2
96% ethanol	5	1
70% ethanol	5	1
40% ethanol	5	1
1× PBS	5	2

7. Immerse the slides in PBS containing 0.2% Triton-X100 for 15 min for antigen retrieval, and wash three times the slides for 5 min each with PBS.
8. Incubate the sections with blocking solution for 60 min.
9. Incubate the sections overnight at 4 °C with the primary antibodies diluted in PBS (dilution depends on each antibody).
10. Wash the slides three times 5 min each with PBS, and incubate with the diluted secondary antibody for 1 h at room temperature.
11. Wash the slides three times 5 min each with PBS, and mount with Fluoromount™ mounting medium squeezing out any air

bubbles. Allow to dry overnight at room temperature. Store slides at 4 °C protected from light.

4 Notes

1. The pT7Ts is a vector closely related to pSP64T [29] which bears three cloning sites (*Bgl*II, *Eco*RV, and *Spe*I) flanked by 5′ and 3′ untranslated regions (UTRs) of *Xenopus* β-globin mRNA, followed by a 30 mer poly-A and a 30 mer poly-C, for efficient translation and stability of the cRNA. Other expression vectors in which aquaporin constructs are also flanked with *Xenopus* β-globin 5′ and 3′ UTRs, such as pGEMHE [30], are often used. Dual-expression vectors such as pXOOM [31] may be of practical interest since they also allow the transfection of the cloned gene into mammalian cells [32].
2. Ovarian lobes are typically obtained from anesthetized frogs by surgical laparotomy. Detailed protocols for this procedure are available [33], but they need to be approved by local ethical or animal experimentation committees. If accredited animal facilities with approved procedures for laparotomy are not available, oocytes can be obtained from commercial sources, although the viability of these oocytes is usually lower than that of freshly collected oocytes and the cost is high.
3. The remaining collagenase stock should be stored at −20 °C sealed with Parafilm to avoid hydration.
4. The solution of DTT should be prepared fresh for maximum activity. As an alternative agent to reduce protein disulfide bonds of proteins, β-mercaptoethanol can be used, although it is more volatile and toxic than DTT.
5. If desired, human influenza hemagglutinin (HA) or Flag™ epitope tags can be introduced in the N- or C-terminus of the aquaporin sequence, using standard molecular biology techniques, for further immunological detection of the protein product. It is important to consider that the anti-HA or anti-Flag antibodies to be used should not cross-react with any endogenous protein of *Xenopus* oocytes.
6. Discard follicles with oocytes showing black spots in the vegetal or animal pole, discolored oocytes, and those showing a white spot in the animal pole indicative of germinal vesicle breakdown (GVBD). These latter oocytes underwent meiotic maturation, in which meiosis is progressing to or already reached metaphase II [34]. At this stage and also when GVBD in the oocyte occurs spontaneously during the incubation periods (possibly because oocytes have been hormonally stimulated

prior to their collection), we have observed that aquaporin trafficking to the oocyte plasma membrane is impaired. Therefore, if ~30% or more of the follicles in the ovary show oocytes at the GVBD stage, it is better to discard the ovary and use another female for oocyte collection.

7. For experiments in which the dominant-negative effect by a mutant aquaporin or a splice form on the wild-type (WT) channel needs to be investigated, the cRNAs of the WT and variant are co-expressed in oocytes. The amount of cRNA for the WT should be the minimum to measure a significant increase in oocyte permeability with respect to the control oocytes (see below), so the effect of the expression of increasing amounts of the variant (e.g., 0.01, 0.1, 1, or 10 times with respect the WT) can be detected. The cRNAs should be concentrated accordingly to inject the desired amount of the WT and variant in a volume of 50 nl.

8. Control oocytes should be at the same developmental stage, from the same frog and isolated at the same time as those used for aquaporin expression. For aquaporin research, controls injected with water are suitable, possibly because frog oocytes express endogenous channels and transporters at a low level and there is little endogenous transport activity in the oocyte plasma membrane [20, 35]. In our experience, we have also observed that water-injected oocytes and uninjected oocytes show the same permeability, despite the potential damage to the plasma membrane during the microinjection. However, caution should be taken to choose the appropriate controls when electrophysiological recordings are carried out, since it is known that the activity of endogenous ion channels may be unspecifically stimulated by the expression of foreign proteins [35], or they can form heteromultimers with the proteins translated from injected RNA [20].

9. In some laboratories, the P_f of aquaporin-expressing oocytes is estimated in hyperosmotic rather than hypoosmotic MBS by adding sucrose. In this case, oocyte shrinkage is measured, which have the advantage of preventing the dilution of the intracellular contents and of ion transport across the oocyte membrane [36]. However, to ensure that the P_f is not affected by ion dilution during oocyte volume increase under external hyposmolarity, the NaCl concentration in the MBS (88 mM) can be reduced to 78 mM, 50 mM, or 25 mM, and sucrose is added to a concentration of 20 mM, 56 mM, and 126 mM, respectively, to obtain isotonic solutions as determined with an osmometer. Oocytes are equilibrated in these solutions, and subsequently oocyte swelling is measured in the same solutions, which are made hyposmotic by removing sucrose, and thus the ion concentrations remain constant.

10. Other compounds can be tested for aquaporin permeability using volumetric assays, such as polyols (ethylene glycol, propylene glycol, 1,2-propanediol, 1,2-butandiol), DMSO, purines, pyrimidines, and monocarboxylates such as β-hydroxybutyrate [37–40].
11. As a way to confirm that oocyte water and solute permeability is mediated by the expressed aquaporins, mercury ($HgCl_2$) is often used as a typical inhibitor of most aquaporins, as it forms disulfide bonds with cysteines in the pore region and collapses channel conductance [41]. To test mercurial inhibition, oocytes are incubated in MBS containing 100–300 μM $HgCl_2$ for 15 min before and during the swelling assays. The reversibility of the mercury inhibition is determined by rinsing the same oocytes three times with fresh MBS and further incubation with 5 mM of the reducing agent β-mercaptoethanol (which reduces disulfide bonds) for 15 min prior to the swelling assays.
12. Radiolabeled compounds indicated in **Note 10** can also be tested.
13. For the Co-IP, two different antibodies specific for each aquaporin variant are necessary. One of them is used for the IP and the other for the immunoblotting. This can be solved by expressing the two aquaporin forms with different epitope tags.

Acknowledgments

Work in our laboratory is supported by grants from the Spanish Ministry of Economy, Industry and Competition (MINECO) (Grant nos. AGL2010-15597, AGL2013-41196-R, AGL2016-76802-R) and Catalan Government (2009 SGR 1050, 2014 SGR 1351, 2017 SGR 1042). F.C. and A.F. are recipients of a "Ramón y Cajal" contract (RYC-2015-17103) and a predoctoral grant (BES-2014-068745), respectively, from Spanish MINECO.

References

1. Preston GM, Carroll TP, Guggino WB, Agre P (1992) Appearance of water channels in *Xenopus* oocytes expressing red cell CHIP28 protein. Science 256:385–387
2. King LS, Kozono D, Agre P (2004) From structure to disease: the evolving tale of aquaporin biology. Nat Rev Mol Cell Biol 5:687–698
3. Finn RN, Chauvigné F, Hlidberg JB, Cutler CP, Cerdà J (2014) The lineage-specific evolution of aquaporin gene clusters facilitated tetrapod terrestrial adaptation. PLoS One 9: e113686
4. Finn RN, Cerdà J (2015) Evolution and functional diversity of aquaporins. Biol Bull 229:6–23
5. Finn RN, Cerdà J (2018) Aquaporin. In: Choi S (ed) Encyclopedia of signaling molecules, 2nd edn. Springer, New York, pp 1–18

6. Wu B, Beitz E (2007) Aquaporins with selectivity for unconventional permeants. Cell Mol Life Sci 64:2413–2421
7. Maurel C, Boursiac Y, Luu DT, Santoni V, Shahzad Z, Verdoucq L (2015) Aquaporins in plants. Physiol Rev 95:1321–1358
8. Rojek A, Praetorius J, Frøkiaer J, Nielsen S, Fenton RA (2008) A current view of the mammalian aquaglyceroporins. Annu Rev Physiol 70:301–327
9. Verkman AS, Anderson MO, Papadopoulos MC (2014) Aquaporins: important but elusive drug targets. Nat Rev Drug Discov 13:259–277
10. Madsen SS, Engelund MB, Cutler CP (2015) Water transport and functional dynamics of aquaporins in osmoregulatory organs of fishes. Biol Bull 229:70–92
11. Talbot K, Kwong RW, Gilmour KM, Perry SF (2015) The water channel aquaporin-1a1 facilitates movement of CO_2 and ammonia in zebrafish (*Danio rerio*) larvae. J Exp Biol 218:3931–3940
12. Cerdà J, Chauvigné F, Finn RN (2017) The physiological role and regulation of aquaporins in teleost germ cells. Adv Exp Med Biol 969:149–171
13. Eskova A, Chauvigné F, Maischein HM, Ammelburg M, Cerdà J, Nüsslein-Volhard C, Irion U (2017) Gain-of-function mutations in Aqp3a influence zebrafish pigment pattern formation through the tissue environment. Development 144:2059–2069
14. Vorontsova I, Gehring I, Hall JE, Schilling TF (2018) Aqp0a regulates suture stability in the zebrafish lens. Invest Ophthalmol Vis Sci 59:2869–2879
15. Tingaud-Sequeira A, Calusinska M, Finn RN, Chauvigné F, Lozano J, Cerdà J (2010) The zebrafish genome encodes the largest vertebrate repertoire of functional aquaporins with dual paralogy and substrate specificities similar to mammals. BMC Evol Biol 10:38
16. Gurdon JB, Lane CD, Woodland HR, Marbaix G (1971) Use of frog eggs and oocytes for the study of messenger RNA and its translation in living cells. Nature 233:177–182
17. Liu K, Nagase H, Huang CG, Calamita G, Agre P (2006) Purification and functional characterization of aquaporin-8. Biol Cell 98:153–161
18. Kai L, Kaldenhoff R, Lian J, Zhu X, Dötsch V, Bernhard F, Cen P, Xu Z (2010) Preparative scale production of functional mouse aquaporin 4 using different cell-free expression modes. PLoS One 5:e12972
19. Solenov EI, Baturina GS, Katkova LE, Zarogiannis SG (2017) Methods to measure water permeability. Adv Exp Med Biol 969:263–276
20. Bianchi L, Driscoll M (2006) Heterologous expression of *C. elegans* ion channels in *Xenopus* oocytes. WormBook 1:1–16
21. Li C, Wang W (2017) Molecular biology of aquaporins. Adv Exp Med Biol 969:1–34
22. Deen PM, Verdijk MA, Knoers NV, Wieringa B, Monnens LA, van Os CH, van Oost BA (1994) Requirement of human renal water channel aquaporin-2 for vasopressin-dependent concentration of urine. Science 264:92–95
23. Kamsteeg EJ, Deen PM (2001) Detection of aquaporin-2 in the plasma membranes of oocytes: a novel isolation method with improved yield and purity. Biochem Biophys Res Commun 282:683–690
24. Laemmli UK (1970) Cleavage of structural proteins during the assembly of the head of bacteriophage T4. Nature 227:680–685
25. Zampighi GA, Kreman M, Boorer KJ, Loo DDF, Bezanilla F, Chandy G, Hall JE, Wright EM (1995) A method for determining the unitary functional capacity of cloned channels and transporters expressed in *Xenopus laevis* oocytes. J Membr Biol 148:65–78
26. Verkman AS, Ives HE (1986) Water permeability and fluidity of renal basolateral membranes. Am J Phys 250:F633–F643
27. Francis P, Chung JJ, Yasui M, Berry V, Moore A, Wyatt MK, Wistow G, Bhattacharya SS, Agre P (2000) Functional impairment of lens aquaporin in two families with dominantly inherited cataracts. Hum Mol Genet 9:2329–2334
28. De Bellis M, Pisani F, Mola MG, Basco D, Catalano F, Nicchia GP, Svelto M, Frigeri A (2014) A novel human aquaporin-4 splice variant exhibits a dominant-negative activity: a new mechanism to regulate water permeability. Mol Biol Cell 25:470–480
29. Krieg PA, Melton DA (1984) Functional messenger RNAs are produced by SP6 in vitro transcription of cloned cDNAs. Nucleic Acids Res 12:7057–7070
30. Liman ER, Tytgat J, Hess P (1992) Subunit stoichiometry of a mammalian K+ channel determined by construction of multimeric cDNAs. Neuron 9:861–871
31. Jespersen T, Grunnet M, Angelo K, Klaerke DA, Olesen SP (2002) Dual-function vector for protein expression in both mammalian cells and *Xenopus laevis* oocytes. BioTechniques 32:536–538

32. Vivaudou M, Todorov Z, Reyes-Mejia GC, Moreau C (2017) Ion channels as reporters of membrane receptor function: automated analysis in *Xenopus* oocytes. Methods Mol Biol 1635:283–301
33. NIH Intramural Research Program: Animal Research Advisory Committee Guidelines (2016) Guidelines for egg and oocyte harvesting in *Xenopus laevis*. https://oacu.oir.nih.gov/sites/default/files/uploads/arac-guidelines/oocyte_harvest.pdf
34. JL M, Gautier J, Langan TA, Lohka MJ, Shenoy S, Shalloway D, Nurse P (1989) Maturation-promoting factor and the regulation of the cell cycle. J Cell Sci Suppl 12:53–63
35. Miller AJ, Zhou JJ (2000) *Xenopus* oocytes as an expression system for plant transporters. Biochim Biophys Acta 1465:343–358
36. Søgaard R, Zeuthen T (2008) Test of blockers of AQP1 water permeability by a high-resolution method: no effects of tetraethylammonium ions or acetazolamide. Pflugers Arch 456:285–292
37. Tsukaguchi H, Shayakul C, Berger UV, Mackenzie B, Devidas S, Guggino WB, van Hoek AN, Hediger MA (1998) Molecular characterization of a broad selectivity neutral solute channel. J Biol Chem 273:24737–24743
38. Zeuthen T, Klaerke DA (1999) Transport of water and glycerol in aquaporin 3 is gated by H (+). J Biol Chem 274:21631–21636
39. Yamaji Y, Valdez DM Jr, Seki S, Yazawa K, Urakawa C, Jin B, Kasai M, Kleinhans FW, Edashige K (2006) Cryoprotectant permeability of aquaporin-3 expressed in *Xenopus* oocytes. Cryobiology 53:258–267
40. Chauvigné F, Lubzens E, Cerdà J (2011) Design and characterization of genetically engineered zebrafish aquaporin-3 mutants highly permeable to the cryoprotectant ethylene glycol. BMC Biotechnol 11:34
41. Hirano Y, Okimoto N, Kadohira I, Suematsu M, Yasuoka K, Yasui M (2010) Molecular mechanisms of how mercury inhibits water permeation through aquaporin-1: understanding by molecular dynamics simulation. Biophys J 98:1512–1519

Chapter 3

Computer-Assisted Sperm Analysis to Test Environmental Toxicants

Tímea Kollár, Ákos Horváth, and Zsolt Csenki-Bakos

Abstract

Fish sperm show many measurable parameters which react sensitively in a dose- and time-dependent way to toxic exposure. Fish sperm is therefore used as an in vitro toxicology test system. One of the most sensitive and easily detectable parameters is progressive motility which can be measured by a computer-assisted sperm analysis (CASA) system. Here we describe a simple protocol to test the effect of environmental toxicants by using zebrafish (*Danio rerio*) sperm.

Key words CASA, Zebrafish sperm, Progressive motility

1 Introduction

In vitro test systems—due to their time- and cost-efficiency as well as their practical and ethical advantages—are becoming more and more widespread in toxicological assessments. Fish sperm has many advantages compared to cell and tissue cultures used generally in in vitro tests which make it a suitable model for toxicological assessments. In this context, one has to disregard the fact that the main function of sperm is to fertilize the eggs: sperm is considered a single cell type which has the following advantages over cell cultures:

1. Fresh sperm can be collected from the donor anytime.
2. It can be collected by stripping which is considered noninvasive.
3. No need for the time- and labor-intensive long-term maintenance of cells.
4. No DNA transcription in mature spermatozoa which could repair the damages.

Roland Dosch (ed.), *Germline Development in the Zebrafish: Methods and Protocols*, Methods in Molecular Biology, vol. 2218, https://doi.org/10.1007/978-1-0716-0970-5_3, © Springer Science+Business Media, LLC, part of Springer Nature 2021

5. Many measurable parameters (e.g., viability, motility, enzyme activity) are sensitive and can be detected easily, fast, and objectively.

These parameters are able to react with a dose and time response to a toxic exposure; for their investigation many techniques are available (e.g., staining methods, measurement of enzyme activity, motility assessment) [1]. The simplest and fastest method to detect the quality of fish sperm is motility assessment, which can be carried out by a computer-assisted sperm analysis (CASA) system [2, 3]. The system is based on a camera connected to a microscope equipped with a negative phase contrast objective of 10× or 20× magnification which is able to record short (typically 0.5 s) video sequences. The records are analyzed with a software developed specifically for sperm analysis which is able to detect spermatozoa based on their size and can calculate various motility parameters [4]. Due to this, the toxicology-aimed motility assessment of fish sperm is gaining popularity [5–13]. Nonetheless, this method is not standardized in case of fish; there are many differences among the performed experiments (e.g., handling of sperm, applied extenders, exposure duration); thus, the comparison of the results can be problematic. Because of this, we developed a simple and standardized toxicology test system based on zebrafish sperm analysis which was successfully tested on several heavy metals [12].

2 Materials

1. MS 222 (tricaine methanesulfonate) stock solution for anesthesia [14]: Measure 400 mg ethyl 3-aminobenzoate methanesulfonate and dissolve it in 97.9 mL distilled water. Adjust the pH to 7.0–7.4 by adding approx. 2.1 mL of 1 M Tris. Aliquot into 4.2 mL stock solutions and store at −20 °C until use.
2. Cyprinid immobilizing solution (200 mM KCl, 30 mM Tris) [15]: Dissolve 3.728 g KCl and 0.90855 g Tris in 200 mL distilled water. Adjust the pH with HCl to 8.0 and add up to 250 mL with distilled water. Store the solution at 4 °C until use.
3. Test solution: Dissolve the tested substance in cyprinid immobilizing solution to prepare a stock solution. Check the pH of the stock solution and keep it between 7.8 and 8.2 to avoid that pH fluctuation (and not the toxic effect) cause motility reduction. Make a double-concentrated dilution series (maximum four concentrations at the same time) by using cyprinid immobilizing solution (*see* **Notes 1** and **2**).

For sperm stripping:

4. Sponge.
5. Paper towels.
6. Forceps.
7. Calibrated glass capillary.
8. Microcentrifuge tube (1.5-mL Eppendorf or 0.2-mL PCR tube).
9. Stereomicroscope at 10× magnification.

For the motility assessment:

10. Computer-assisted sperm analysis (CASA) system: SpermVision CASA system (Minitüb, Tiefenbach, Germany) consisting of an Olympus BX41 phase contrast microscope with a 20× negative phase contrast objective, coupled to a JAI CV-A10CL camera (frame rate, 60 frames/s) and connected to a computer running the CASA software SpermVision version 3.7.4 (Minitube of America, Venture Court Verona, USA). However, with the settings given later (*see* **Note 3**), any CASA system should be suitable for the test.
11. Makler-type cell counting chamber.
12. System water (in which zebrafish are held).
13. Activating solution: 0.1% bovine serum albumin (BSA) in system water. Keep on melting ice.

3 Methods

3.1 Stripping

1. Prepare fresh anesthesia solution. Dilute 4.2 mL MS 222 in 100 mL system water.
2. Place the fish into the anesthesia solution, and wait until they stop voluntary movement and turn upside down (*see* **Note 4**).
3. Rinse the fish in system water to prevent contamination of sperm with the anesthesia solution [16].
4. Fix the fish on a wet sponge with the abdomen facing upward, and dry the urogenital papilla with a paper towel to prevent activation of sperm by water.
5. Apply gentle pressure to the abdominal wall by forceps under a stereomicroscope at 10× magnification. Collect the outflowing sperm with a calibrated glass capillary to evaluate its quantity (Fig. 1; *see* **Notes 5–11**).
6. Pool the sperm in a microcentrifuge tube in 50 μL cyprinid immobilizing solution until a total 10 μL sperm is collected (approx. 10–15 males).
7. Store the sample on melting ice.

Fig. 1 Stripping of male zebrafish with forceps and collection of the outflowing sperm with a calibrated glass capillary

8. Place the fish back into clear system water to regain consciousness and keep them under normal housing conditions. Strip the same males after at least 1 week of rest.

3.2 Assessment of Fresh Sperm Motility

1. Activate 0.5 μL sperm in 2.0–2.5 μL of activation solution in a Makler chamber.
2. Use the CASA system to record motility in two different fields before the motility of sperm slows down.
3. Repeat the activation at least once (*see* **Notes 12** and **13**).
4. Use the sample for toxicant testing if the progressive motility is higher than 60%.

3.3 Sperm Dilution and Exposure

1. Create five aliquots from the pooled sperm sample: measure 10 μL samples into microcentrifuge tubes.
2. Dilute four subgroups with 10 μL of the double-concentrated solutions (50%/50% (v/v) dilution rate) to reach the final test concentration of the toxicant.
3. As a control, dilute the fifth subgroup with 10 μL cyprinid immobilizing solution (Fig. 2).
4. Store the samples on melting ice during the exposure.

3.4 Assessment of Sperm Motility

1. Thirty minutes after the start of the toxic exposure, activate 1 μL of each exposed sample in 2.0–2.5 μL BSA containing system water in the Makler chamber.
2. Assess motility according to **steps 3** and **4** of Subheading 3.3.
3. Repeat the measurement at 120 and 240 min.

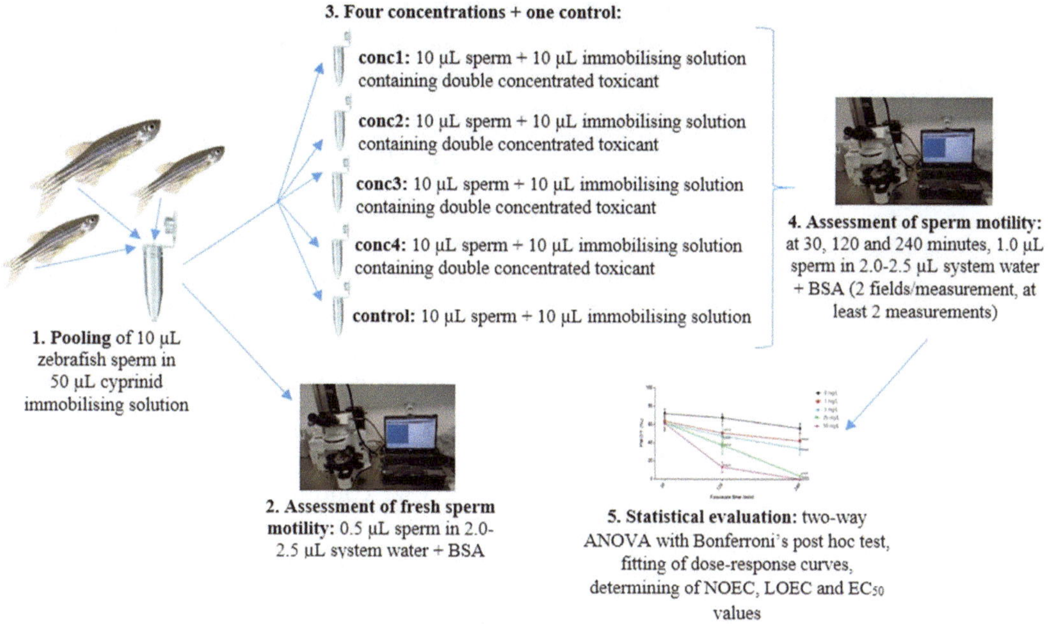

Fig. 2 Flowchart highlighting the steps to assess toxicants with zebrafish sperm and computer-assisted sperm analysis (CASA)

3.5 Statistical Analysis

1. Homogeneity of variances and normality of distribution is verified prior to statistical tests (*see* **Note 14**).
2. Two-way ANOVA is used to determine which concentrations had a major effect on progressive motility at different measurement points in time (significance level: $p = 0.05$).
3. By the help of Bonferroni's post-hoc test, no observed effect concentration (NOEC) and lowest observed effect concentration (LOEC) values can be detected at different exposure durations.
4. At those points of time, when the values of measured variables reduce beyond half of the control's, median effective concentrations (EC_{50} values) with standard deviation are calculated by fitting dose-response curves [17].

4 Notes

1. The remaining cyprinid immobilizing solution can be stored on −20 °C and can be reused for approximately 1 month.
2. If the tested chemical has poor solubility in water, sonication or the use of a solvent (dimethyl sulfoxide—DMSO—at a maximum concentration of 0.1%) is allowed for improved solubility. In the latter case, a solvent control group must also be assessed included to test the effect of the solvent on sperm motility.

3. Apply the following settings for CASA measurements:
   (a) Threshold limit of the PMOT: straight-line distance between the initial and the end point of movement DSL > 5 μm.
   (b) Pixel/μm ratio: 151/100.
   (c) Size of head area 1–100 μm^2.
4. Wild-type zebrafish from the AB line are used in the experiments which have been bred and raised from proprietary broodstock for several generations.
5. Sexually mature (at least 6 months old) males are selected and kept separately in 3-L polycarbonate tanks (maximum 25 fish in 3 L), in a recirculating zebrafish housing system.
6. In the recirculating system, water quality parameters (25 ± 2 °C; pH 7.0 ± 0.2; conductivity 525 ± 50 μS) and photoperiod (14-h light:10-h dark cycle) are constant.
7. Zebrafish are fed two times a day with Zebrafeed by Sparos (Olhão, Portugal) supplemented with live *Artemia salina* nauplii every second day.
8. It is advisable not to feed the fish on the morning of stripping in order to avoid the contamination of sperm with feces during the stripping.
9. During the stripping process, anesthetize a maximum of two to three fish at the same time.
10. Replace the anesthetic solution with fresh solution after ~20 min to prevent depletion of oxygen to prevent the death of fish.
11. During the stripping, urine can also leave the urogenital papilla. It can be distinguished based on the color: sperm is opalescent, while urine is transparent. Avoid collecting urine with sperm in order to prevent contamination and premature activation of sperm.
12. If there is a substantial difference in the progressive motility between the two measurements of the same sample (more than 10%), repeat it one more time and discard the records from the outlying replicate.
13. Clean the Makler chamber, and always change the pipette tip between measurements in order to prevent cross-contamination of the samples.
14. For a proper statistical evaluation, at least five independent samples are needed.

Acknowledgments

This work was supported by the EFOP-3.6.3-VEKOP-16-2017-00008 project co-financed by the European Union and the European Social Fund.

References

1. Cabrita E, Martínez-Páramo S, Gavaia PJ, Riesco MF, Valcarce DG, Sarasquete C, Herráez MP, Robles V (2014) Factors enhancing fish sperm quality and emerging tools for sperm analysis. Aquaculture 432:389–401
2. Kime DE, Van Look KJW, McAllister BG, Huyskens G, Rurangwa E, Ollevier F (2001) Computer-assisted sperm analysis (CASA) as a tool for monitoring sperm quality in fish. Comp Biochem Physiol Part C 130:425–433
3. Fauvel C, Suquet M, Cosson J (2010) Evaluation of fish sperm quality. J Appl Ichthyol 26:636–643
4. Rurangwa E, Kime DE, Ollevier F, Nash JP (2004) The measurement of sperm motility and factors affecting sperm quality in cultured fish. Aquaculture 234:1–28
5. Ciereszko A, Dabrowski K (2000) In vitro effect of gossypol acetate on yellow perch (Perca flavescens) spermatozoa. Aquat Toxicol 49:181–187
6. Roesty M, Ordonez FJ, Rosety-Rodríguez M, Rosety JM, Rosety I (2003) In vitro acute toxicity of anionic surfactant linear alkylbenzene sulphonate (LAS) on the motility of gilthead (Sparus aurata L.) sperm. Histol Histopathol 18:475–478
7. Van Look KJW, Kime DE (2003) Automated sperm morphology analysis in fishes: the effect of mercury on goldfish sperm. J Fish Biol 63:1020–1033
8. Abascal FJ, Cosson J, Fauvel C (2007) Characterization of sperm motility in sea bass: the effect of heavy metals and physicochemical variables on sperm motility. J Fish Biol 70:509A–522A
9. Hatef A, Alavi SMH, Golshan M, Linhart O (2013) Toxicity of environmental contaminants to fish spermatozoa function in vitro—a review. Aquat Toxicol 140-141:134–144
10. Hayati A, Giarti K, Winarsih Y, Amin MHF (2017) The effect of cadmium on sperm quality and fertilization of Cyprinus carpio L. J Trop Biodiv Biotechnol 2:45–50
11. Rocha S, Streit Junior DP, Marques LS, Varela Junior AS, Corcini CD, Hoshiba MA (2017) Toxic effects of mercury chloride on silver catfish (Rhamdia quelen) spermatozoa. Aquac Res. https://doi.org/10.1111/are.13543
12. Kollár T, Kása E, Ferincz Á, Urbányi B, Csenki-Bakos Z, Horváth Á (2018) Development of an in vitro toxicological test system based on zebrafish (Danio rerio) sperm analysis. Environ Sci Pollut Res 25(15):14426–14436
13. Kollár T, Kása E, Csorbai B, Urbányi B, Csenki-Bakos Z, Horváth Á (2018) In vitro toxicology test system based on common carp (Cyprinus carpio) sperm analysis. Fish Physiol Biochem 44(6):1577–1589
14. Meyers JR (2018) Zebrafish: development of a vertebrate model organism. Curr Protoc Essent Lab Tech e19. https://doi.org/10.1002/cpet.19
15. Saad A, Billard R (1987) Spermatozoa production and volume of semen collected after hormonal stimulation in the carp, Cyprinus carpio. Aquaculture 65:67–77
16. Bromage N (1992) Propagation and stock improvement. In: Shepherd J, Bromage N (eds) Intensive fish farming. Balckwell Scientific Publications, Oxford, pp 103–153
17. Ritz C, Baty F, Streibig JC, Gerhard D (2015) Dose-response analysis using R. PLoS One 10 (12):e0146021

Chapter 4

Cryopreservation and Transplantation of Spermatogonial Stem Cells

Zoran Marinović, Jelena Lujić, Qian Li, Yoshiko Iwasaki, Béla Urbányi, Goro Yoshizaki, and Ákos Horváth

Abstract

Cryopreservation as a method that enables long-term storage of biological material has long been used for the conservation of valuable zebrafish genetic resources. However, currently, only spermatozoa of zebrafish can be successfully cryopreserved, while protocols for cryopreservation of eggs and embryos have not yet been fully developed. Transplantation of germline stem cells (GSCs) has risen as a favorable method that can bypass the current problem in cryopreservation of female genetic resources and can lead to reconstitution of fish species and lines through surrogate production. Here, we describe essential steps needed for the cryopreservation of spermatogonial stem cells (SSCs) and their utilization in the conservation of zebrafish genetic resources through SSC transplantation and surrogate production.

Key words Microinjection, Freezing, Biobanking, Conservation, Surrogate production

1 Introduction

As a method that enables long-term storage of biological material, cryopreservation has been used for storage of zebrafish genetic resources [1]. However, currently, only spermatozoa of zebrafish can be successfully cryopreserved [2, 3], while protocols for cryopreservation of zebrafish eggs and embryos have not yet been fully developed. Even though some progress has been made in cryopreservation of zebrafish embryos [4], these methods still fail to produce functional embryos which would be able to reconstitute a whole zebrafish line. Additionally, even though the application of sperm cryopreservation has been proven in practice [5], the line needs to be recovered using wild-type eggs. In such case, the F1 generation will be 100% heterozygous, while in the F2 generation, only 25% of the fish are homozygous making recovery of the given line laborious.

Recently, transplantation of spermatogonial stem cells (SSCs) has risen as a favorable method that can bypass the current problem

Roland Dosch (ed.), *Germline Development in the Zebrafish: Methods and Protocols*, Methods in Molecular Biology, vol. 2218, https://doi.org/10.1007/978-1-0716-0970-5_4, © Springer Science+Business Media, LLC, part of Springer Nature 2021

in cryopreservation of female genetic resources and can lead to reconstitution of fish species and lines through surrogate production [6]. Until now, its application has been demonstrated in several fish species [6–9]. The method entails isolation of SSCs which represent the baseline cells of spermatogenesis from a donor individual and their subsequent transplantation into a recipient individual. SSCs are most commonly transplanted into the body cavity of recipient larvae, and due to their limited migratory ability, they can colonize the recipient gonads, incorporate into it, proliferate, differentiate, and give rise to functional gametes. Additionally, as demonstrated in previous studies [10–12], these cells can also transdifferentiate into oogonial stem cells in female recipients and thus give rise to functional eggs and subsequently donor-derived offspring after mating with recipient males. Through cryopreservation, SSCs can be stored for an indefinite time. Furthermore, these cells are commonly functional after thawing and retain their ability to colonize recipient gonads after transplantation, proliferate, and give rise to functional gametes [8, 11, 13, 14].

To ensure that recipients will give rise only to donor-derived offspring, autologous recipient germ cells need to be eliminated. Even though there are several strategies to conduct sterilization, in the present study, microinjection of morpholino oligonucleotides (MOs) against the *dead end* gene was used. MOs are small synthetic oligonucleotides that bind to mRNA and prevent translation [15]. The *dead end* gene is involved in maintaining the germline fate of primordial germ cells (PGCs) and their migration to the genital ridge [16]. Knockdown or knockout of this gene cause mis-migration of PGCs and subsequently their apoptosis or transdifferentiation into other cell types [16–18], thus leaving the gonads consisting solely of supportive somatic cells.

The only possible downside to the application of SSC transplantation into sterilized zebrafish larvae is that the knockdown of the *dead end* gene induces an all-male population [17, 18]. However, part of SSCs can be transplanted into sterilized individuals, while a part can be transplanted into a non-sterilized batch of wild-type larvae in which the females will carry donor-derived genetic material. When spawning males from the sterilized batch with females from the non-sterilized batch, one part of the progeny will be homozygous for the given trait. Additionally, preliminary results obtained by Saito et al. (cited from Goto and Saito [19]) display that eggs can be obtained from sterilized zebrafish recipients after hormonal treatment where in such case, the production of 100% donor-derived eggs and subsequently 100% donor-derived and homozygous offspring would be possible within the F1 generation.

The method displayed in this study can be used complementary to sperm cryopreservation. A major difference between the two methods is that SSC cryopreservation and transplantation require sacrificing the donor individual, while sperm cryopreservation enables a long-term utilization of a single male. However, when taking into account that ~50% of SSCs survive the cryopreservation procedure and that these cells incorporate into 30–50% of the recipients, approximately 10–15 recipients producing donor-derived gametes can be produced from a single donor fish. Therefore, for conservation of a specific line, part of the males can be kept as part of the breeding colony and for milt production, while one part can be sacrificed for SSC cryopreservation. In the case that maintaining the breeding colony is not possible, only SSC manipulation strategies can be employed.

2 Materials

2.1 Equipment

2.1.1 Dissection

1. Stereomicroscope.
2. Fluorescent stereomicroscope.
3. Dissection mat.
4. Spring scissors.
5. Microscissors.
6. Tweezers.
7. Acupuncture or dissection needles.
8. 30 μm filters.

2.1.2 Cryopreservation

1. −80 °C freezer.
2. CoolCell box (e.g., BioCision).
3. Alternatively, a controlled-rate freezer.
4. Liquid nitrogen canisters.

2.1.3 Microinjection and Transplantation

1. Glass micropipette puller (e.g., Narishige PC-10).
2. Micropipette grinder (e.g., Narishige EG-401).
3. Microinjector (e.g., Narishige IM-9A).
4. Glass capillaries (1 mm inner diameter, 0.6 mm inner diameter, without internal filament).
5. Hemocytometer.
6. Depression microscope slides.

2.2 Chemicals and Buffers

1. 200 mg/l MS-222 solution—prepare a 200 mg/ml stock solution by diluting 200 mg of MS-222 into 1 ml ddH_2O. For the final dilution, add 1 ml of the stock solution into 1 l of system water.
2. 0.03% 2-phenoxyethanol—for the final dilution, add 300 μl of 2-phenoxyethanol to 1 l of system water.

3. Leibovitz L-15 medium (with or without glutamine).
4. Fetal bovine serum (FBS).
5. PBS: prepare a 10× stock solution. For 1 l of 10× stock solution, mix 80 g NaCl, 2 g KCl, 14.4 g Na_2HPO_4, 2.4 g KH_2PO_4, sterile filter, or autoclave. For making 1× PBS, dilute the stock solution 10× and adjust the pH to 7.4.
6. 20 mg/ml collagenase stock solution in PBS (*see* **Note 1**).
7. 15 mg/ml trypsin stock solution in PBS (*see* **Note 1**).
8. 1 mg/ml DNase I stock solution in PBS (*see* **Note 1**).
9. Dissociation solution containing L-15, 2 mg/ml collagenase, 1.5 mg/ml trypsin, and 30 μg/ml DNase I.
10. PKH-26 linker dye kit for general membrane labeling.
11. Morpholino oligonucleotide against the *dead end* gene (MO1-*dnd*; 5′- GCTGGGCATCCATGTCTCCGACCAT -3′ [20])—upon arrival, resuspend the obtained lyophilized solid in an appropriate volume of ddH_2O to obtain a 1 mM stock solution according to manufacturer's instructions (*see* **Note 2**).
12. Extender for cryopreservation containing 55.27 mM HEPES, 375.48 mM NaCl, 7.28 mM KCl, 23.10 mM KH_2PO_4, 3.82 mM Na_2HPO_4, 3.64 mM sodium pyruvate, 2.6 mM $CaCl_2{\cdot}2H_2O$, and 1.4 mM $MgCl_2{\cdot}6H_2O$, pH 7.8.
13. Cryomedium containing 35.2% extender, 1.3 M DMSO, 0.1 M trehalose, and 1.5% BSA.
14. Medium for short-term hypothermic storage (optional) containing L-15, 10% FBS, and 20 mM HEPES, pH 7.4.
15. Trypan blue.
16. Mineral oil.
17. Methylene blue.

2.3 Zebrafish

Wild-type or any other preferred line.

3 Methods

3.1 Anesthesia, Dissection, and Testis Collection

1. Set up the dissection area by firstly sterilizing the working environment with 70% EtOH; setting up the stereomicroscope, dissection mat, and acupuncture needles; preparing the dissection tools needed; and preparing the anesthesia bath.
2. Transfer a selected number of zebrafish males into a small breeding tank.
3. Euthanize the fish by placing them in a dish containing 200 mg/l MS-222 (*see* **Note 3**).

4. Shortly transfer the fish into a dish containing 70% EtOH to sterilize the skin, then dry the fish on a sterile paper towel, and place it on a dissection mat on its dorsal side (*see* **Note 4**).
5. Cut off the pelvic and pectoral fins to facilitate the dissection.
6. Snip the skin and the underlying muscle between the pectoral fins, and cut the skin and underlying muscle along the belly until the anal fin.
7. Open the body cavity by pinning the left and right body walls onto the dissection mat with acupuncture or dissection needles.
8. To remove the testes, either detach the intestines from the peritoneum near the anus and lift the whole gastrointestinal tract toward the pericardial cavity of the fish, thus exposing the two testes which are located on the sides of the swimming bladder, or move the gastrointestinal tract to one side of the body cavity and excise the testis from the opposing side (*see* **Note 5**).
9. Sterilize the testes either in 70% EtOH for few seconds [21] or in 0.1% commercial bleach for 2 min [22].
10. Transfer the testes into a 96-well plate filled with L-15 medium and keep on ice until further work.

3.2 Freezing of Testes

1. Freshly prepare the cryomedium as described above. Label the cryotubes and add 1 ml of cryomedium into each tube.
2. Place the CoolCell boxes into a refrigerator (4 °C) or cover by ice to pre-cool.
3. Transfer the testes of each individual male into a separate cryotube.
4. Incubate the cryotubes for 20 min on ice (or at 4 °C).
5. Transfer the cryotubes into a CoolCell box and insert the box into a −80 °C deep freezer (or alternatively into dry ice). This procedure will enable the cooling of samples at a rate of ~−1 °C/min. After ~100 min, take the box out of the deep freezer and plunge the cryotubes into liquid nitrogen.
6. Alternatively to the CoolCell box procedure, insert the cryotubes into a controlled-rate freezer. Set the cooling profile to cool from 4 °C to −80 °C at a rate of −1 °C/min, with a 15-min holding period at −80 °C. After the cooling is complete, plunge the cryotubes directly into liquid nitrogen.
7. Place the cryotubes onto metal cane holders under the surface of liquid nitrogen, and store them in a storage canister dewar until further use.

3.3 Tissue Dissociation

1. Freshly prepare enzymatic solutions prior to tissue dissociation.
2. Add fresh or cryopreserved testes from one male into 500 μl of the dissociation solution in 2 ml Eppendorf tubes.
3. Mince the testicular tissue into small fragments and incubate on a shaking plate for 90 min at 25 °C.
4. Stop the enzymatic reaction by adding 100 μl of FBS (10% v/v FBS) and 400 μl of L-15, and gently shake the solution.
5. Filter the obtained solution through 30 μm filters into a 1.5 ml Eppendorf tube.
6. Centrifuge the solution at 200 × *g* for 10 min at room temperature (RT).
7. Remove the supernatant and resuspend the pellet containing spermatogonia in the medium for short-term hypothermic storage. Manually shake the tube until cells in the pellet are completely resuspended. Keep at 4 °C until further use (*see* **Note 6**).

3.4 Preparation of Needles

Needle preparation encompasses two main actions: (1) pulling of glass capillaries to create the needles and (2) grinding the tip of the needles to create a sharp edge that is able to penetrate zebrafish larvae. In the current study, the glass capillaries were pulled on a Narishige PC-10 one-stage glass micropipette puller and grind the needles on a Narishige EG-401 micropipette grinder; however, other equipment for pulling and grinding glass needles can be used.

3.4.1 Pulling of Glass Needles

1. Open the lid and place the glass capillary into the holder. Screw both knobs tight so that the capillary does not fall off. The capillary should be placed in such a way that the center of the capillary is directly inside the heater coil. Close the lid.
2. Press the start button to initiate heating (*see* **Note 7**). The heating coil will start to heat up which will in turn melt the center of the capillary. Under the gravitational pull of the weights, the lower part of the capillary will drop, thus creating two glass needles with sharp edges.
3. Remove both needles and place them into a Petri dish.

3.4.2 Grinding the Tip of the Needle

1. Turn on the micropipette grinder and wait until the grinding stone starts revolving.
2. Add water into the grinding stone through a syringe so that the surface of the stone is wet.
3. Mount a needle onto the pipette holder and turn the knob tightly so that the needle does not move.
4. Lower the needle to the stone and start grinding (the angle between the needle and the grinding stone is 40°). During

grinding, pay special attention to the diameter of the needle opening through the microscope; diameter of the needle should be around 5 μm for the injection of morpholino oligonucleotides and around 30 μm for the transplantation of spermatogonia.

5. When the desired diameter is reached, lift the needle from the stone and release it from the holder.
6. Mount the needle on a syringe and wash it out with ddH_2O five to ten times. Place the sharpened and washed needles in a separate Petri dish.

3.5 Microinjection of Embryos with MO1-dnd

1. In the late afternoon the day before injecting MO into embryos, set up breeding pairs by placing male and female zebrafish into a breeding tank with separators to separate males and females.
2. On the morning of injection, remove the divider, and leave the male and female fish together to spawn. Check every 15 min for freshly laid eggs. Collect the eggs and wash them two times in fresh system water.
3. Prior to injection, prepare a Petri dish covered with 2% agar. Additionally, place a microscope slide into the Petri dish which will aid during the injection.
4. Load the needle with approximately 10 μl of MO1-*dnd*. Place the embryos into the previously prepared Petri dish in a line along the microscopic slide. The slide will act as a "wall" on which the embryos will lean on and will be immobilized during the injection.
5. Inject approximately 3 ng of MO into the yolk of each embryo. Embryos should not be more advanced than four-cell stage during injection.
6. After injection, transfer the embryos into a new Petri dish containing fresh system water and 0.0002% methylene blue. Incubate the embryos at 26 °C. Change the water every day and remove dead embryos and the chorions after hatching.

3.6 Labeling Cells with PKH-26 Linker Dye

In the case that donor germ cells do not originate from a line in which germ cells are fluorescently labeled, they need to be labeled prior to transplantation to ensure proper visualization. The most commonly used is the PKH-26 fluorescent linker dye that intercalates into the lipid bilayer of cell membranes. However, this dye enables visualization of only several daughter generations after cells start to proliferate [23] which may be restrictive for this method.

1. If working with fresh spermatogonia, continue from **step** 7 of Subheading 3.3.

2. If working with frozen spermatogonia, continue from **step** 7 of Subheading 3.2. Take the cryotubes out of the storage canister into a separate styrofoam box filled with liquid nitrogen. Thaw the samples in a 25 °C water bath until they are completely thawed. Transfer the testes into a 96-well plate filled with L-15 medium. Wash off the cryoprotectants two times in L-15, and after the last wash, continue with the dissociation procedure (**step 1**, Subheading 3.3).
3. Wash the cells 3× in PBS by centrifuging at 200 × *g* for 10 min at RT.
4. Dilute a small fraction of the washed cell suspensions with an equal volume of 0.4% trypan blue dye, mix well, and incubate for 1 min. Transfer the obtained solution into a hemocytometer and count the total number of cells.
5. Centrifuge the cell suspension one more time at 200 × *g* for 10 min at RT and dispose the supernatant. Resuspend the pellet in Diluent C obtained in the PKH-26 linker dye kit by adding 50 μl of Diluent C to each one million cells in the suspension (e.g., for three million cells, use 150 μl Diluent C).
6. Immediately after resuspending the pellet, add 50 μl Diluent C mixed with 3 μl of the PKH-26 dye per each one million cells in the suspension (e.g., for three million cells, add 150 μl Diluent C and 3 μl of the PKH-26 dye).
7. Incubate for 5 min at RT.
8. Stop the reaction by adding L-15 supplemented with 10% FBS up to 1 ml, and centrifuge at 200 × *g* for 10 min at RT.
9. Wash the suspension two more times with L-15 supplemented with 10% FBS. Keep the suspension at 4 °C until further use (*see* **Note 6**).

3.7 Spermatogonia Transplantation

1. If transplanting spermatogonia of a line whose germ cells are not fluorescently labeled, continue from **step 9** of Subheading 3.6.
2. If transplanting spermatogonia from a line whose germ cells are fluorescently labeled, for fresh spermatogonia continue from **step** 7 of Subheading 3.3, while for frozen spermatogonia do the **step 2** of Subheading 3.6.
3. Dilute a small fraction of the obtained cell suspensions in an equal volume of 0.4% trypan blue dye, mix well, and incubate for 1 min. Transfer the obtained solution into a hemocytometer and count the number of viable spermatogonia.
4. Centrifuge the cell suspensions at 200 × *g* for 10 min at RT, and resuspend the pellet in the storage medium (or simple L-15 medium) so that the final concentration will count approximately 15,000 spermatogonia/μl.

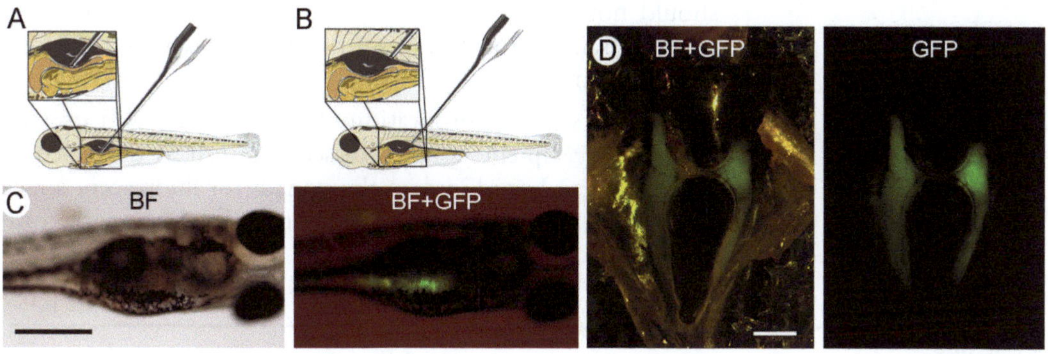

Fig. 1 Transplantation of zebrafish spermatogonial stem cells (SSCs). Points of injection during SSC transplantation are (**a**) between the swimming bladder and the intestines or (**b**) behind the swimming bladder. (**c**) Visualization of GFP-labeled SSCs after injection into wild-type recipients. (**d**) MO-sterilized recipients of YFP-labeled SSCs displayed both developed testes. The fluorescent signal within the wild-type recipient indicates that the germ cells within the testes are of donor-derived origin. Scale bars: (**c**) 0.5 mm; (**d**) 1 mm

5. Pipette the cell suspension into a depression microscope slide, and cover the suspension with mineral oil so that the medium of the suspension does not evaporate.
6. Insert a previously prepared needle into the microinjector.
7. Move the needle with the micromanipulator into a dish filled with mineral oil, and start loading the oil into the microinjector.
8. Once sufficient oil is added, lift the micromanipulator, and remove the dish with the mineral oil.
9. Place prepared cell suspension and load the cells into the needle.
10. In the meantime, anesthetize the 7 dpf MO-treated zebrafish larvae in 0.03% 2-phenoxyethanol for few minutes until they stop moving.
11. Transfer the anesthetized fish into a Petri dish covered with 2% agar with a Pasteur pipette. Wipe off the excess water with a sterile paper towel.
12. Place the larvae under the stereomicroscope. By using the micromanipulator, penetrate the body wall of the larvae with the needle, and inject approximately 200 nl of the cell suspension into the abdominal cavity of each larvae. The tip of the needle should be inserted into the body cavity at one of two possible entry points: (1) between the swimming bladder and the intestines (Fig. 1a) or (2) behind the swimming bladder (Fig. 1b). After the injection, pull back the needle and inject the subsequent individual (Fig. 1c).
13. After injection, place the injected larvae into a Petri dish containing fresh system water in an incubator at 26 °C. Larvae

should not be fed at least 1 day after the procedure. After 2 days, transfer the larvae into the holding tanks and rear them until maturity.

14. In order to verify incorporation of the donor-derived spermatogonia in recipient gonads, dissect the fish as described in Subheading 3.1, and inspect the gonads for fluorescent signal originating from the fluorescent tag or PKH-26 under a fluorescent stereomicroscope (Fig. 1d).

4 Notes

1. After the enzyme is fully dissolved in PBS, stock solutions should be filtered through 0.2 μm filters, aliquoted into smaller volumes (e.g., 50 μl), and stored at −20 °C.
2. The resuspended stock solution should be kept at room temperature. After prolonged storage, autoclaving is recommended to restore potentially lost activity of the oligonucleotide.
3. Before proceeding with the dissection, make sure that fish have been properly euthanized by verifying that the gills have stopped moving. Additionally, to ensure euthanization, fish may be decapitated at this point.
4. All surgical material needs to be sterilized by immersion into 70% EtOH prior to dissection of each individual fish.
5. Make sure that the intestines are not punctured during the dissociation as it can lead to severe contamination. When excising the testes, pull the testes with tweezers while cutting the underlying connective peritoneum with microscissors to ensure that the testes remain intact.
6. Cells resuspended in the storage medium can be stored at 4 °C for at least 2 days without significantly losing viability. However, we recommend to use the cells shortly after dissociation and store them hypothermically for prolonged periods of time.
7. Temperature is dependent on the humidity of the room, as well as the lengths of the needle tip needed. Temperature should be optimized before starting the procedure.

References

1. Matthews JL, Murphy JM, Carmichael C et al (2018) Changes to extender, cryoprotective medium, and in vitro fertilization improve zebrafish sperm cryopreservation. Zebrafish 15:279–290. https://doi.org/10.1089/zeb.2017.1521
2. Carmichael C, Westerfield M, Varga ZM (2009) Cryopreservation and in vitro fertilization at the Zebrafish International Resource Center. In: Lieschke GJ, Oates AC, Kawakami K (eds) Zebrafish: methods and protocols. Humana Press, New York, pp 45–65

3. Morris JP IV, Berghmans S, Zahrieh D et al (2003) Zebrafish sperm cryopreservation with N,N-dimethylacetamide. BioTechniques 35:956–968. https://doi.org/10.1016/S0091-679X(04)77034-X
4. Khosla K, Wang Y, Hagedorn M et al (2017) Gold nanorod induced warming of embryos from the cryogenic state enhances viability. ACS Nano 11:7869–7878. https://doi.org/10.1021/acsnano.7b02216
5. Murray KN, Varga ZM, Kent ML (2016) Biosecurity and health monitoring at the Zebrafish International Resource Center. Zebrafish 13: S30–S38. https://doi.org/10.1089/zeb.2015.1206
6. Okutsu T, Shikina S, Kanno M et al (2007) Production of trout offspring from triploid salmon parents. Science 317:1517. https://doi.org/10.1126/science.1145626
7. Lacerda SMSN, Batlouni SR, Silva SBG et al (2006) Germ cells transplantation in fish: the Nile-tilapia model. Anim Reprod 3:146–159
8. Franěk R, Marinović Z, Lujić J et al (2019) Cryopreservation and transplantation of common carp spermatogonia. PLoS One 14: e0205481. https://doi.org/10.1371/journal.pone.0205481
9. Yoshizaki G, Yazawa R (2019) Application of surrogate broodstock technology in aquaculture. Fish Sci 83:429–437. https://doi.org/10.1007/s12562-019-01299-y
10. Lee S, Iwasaki Y, Shikina S, Yoshizaki G (2013) Generation of functional eggs and sperm from cryopreserved whole testes. Proc Natl Acad Sci U S A 110:1640–1645. https://doi.org/10.1073/pnas.1218468110
11. Yoshizaki G, Fujinuma K, Iwasaki Y et al (2011) Spermatogonial transplantation in fish: a novel method for the preservation of genetic resources. Comp Biochem Physiol Part D Genomics Proteomics 6:55–61. https://doi.org/10.1016/j.cbd.2010.05.003
12. Okutsu T, Suzuki K, Takeuchi Y et al (2006) Testicular germ cells can colonize sexually undifferentiated embryonic gonad and produce functional eggs in fish. Proc Natl Acad Sci 103:2725–2729
13. Lee S, Yoshizaki G (2016) Successful cryopreservation of spermatogonia in critically endangered Manchurian trout (Brachymystax lenok). Cryobiology 72:165–168. https://doi.org/10.1016/j.cryobiol.2016.01.004
14. Yoshizaki G, Lee S (2018) Production of live fish derived from frozen germ cells via germ cell transplantation. Stem Cell Res 29:103–110. https://doi.org/10.1016/j.scr.2018.03.015
15. Škugor A, Tveiten H, Krasnov A, Andersen Ø (2014) Knockdown of the germ cell factor Dead end induces multiple transcriptional changes in Atlantic cod (Gadus morhua) hatchlings. Anim Reprod Sci 144:129–137. https://doi.org/10.1016/j.anireprosci.2013.12.010
16. Gross-Thebing T, Yigit S, Pfeiffer J et al (2017) The vertebrate protein dead end maintains primordial germ cell fate by inhibiting somatic differentiation. Dev Cell 43:704–715. https://doi.org/10.1016/j.devcel.2017.11.019
17. Li Q, Fujii W, Naito K, Yoshizaki G (2017) Application of dead end-knockout zebrafish as recipients of germ cell transplantation. Mol Reprod Dev 84:1100–1111. https://doi.org/10.1002/mrd.22870
18. Slanchev K, Stebler J, de la Cueva-Mendez G, Raz E (2005) Development without germ cells: the role of the germ line in zebrafish sex differentiation. Proc Natl Acad Sci 102:4074–4079. https://doi.org/10.1073/pnas.0407475102
19. Goto R, Saito T (2019) A state-of-the-art review of surrogate propagation in fish. Theriogenology 133:216–227. https://doi.org/10.1016/j.theriogenology.2019.03.032
20. Ciruna B, Weidinger G, Knaut H et al (2002) Production of maternal-zygotic mutant zebrafish by germ-line replacement. Proc Natl Acad Sci U S A 99:14919–14924. https://doi.org/10.1073/pnas.222459999
21. Marinović Z, Lujić J, Kása E et al (2018) Cryopreservation of zebrafish spermatogonia by whole testes needle immersed ultra-rapid cooling J Vis Exp:e56118. https://doi.org/10.3791/56118
22. Sakai N (2006) In vitro male germ cell cultures of zebrafish. Methods 39:239–245. https://doi.org/10.1016/j.ymeth.2005.12.008
23. Wallace PK, Tario JD Jr, Fisher JL et al (2008) Tracking antigen-driven responses by flow cytometry: monitoring proliferation by dye dilution. Cytom A 73:1019–1034. https://doi.org/10.1002/cyto.a.20619

Chapter 5

Masculinization of Zebrafish Through Partial Depletion of Primordial Germ Cells by Injecting Diluted Morpholino Oligonucleotides into Embryos

László Orbán, Jolly M. Saju, Keh-Weei Tzung, and Woei Chang Liew

Abstract

The regulation of reproduction in zebrafish, the prime model of fish research, is not fully understood. An efficient tool to gain a better understanding of this complicated process is utilization of severely sex-biased families or groups. Here, we describe a method for partial depletion of primordial germ cells (PGCs) that leads to eventual masculinization of zebrafish. The technique is based on injecting early embryos with diluted morpholino oligonucleotides that temporarily interfere with the production of Dead end (Dnd), an RNA-binding protein essential for PGC survival. In addition, we also propose the use of eviscerated trunk, as a suitable alternative for examining gonadal expression in juvenile zebrafish.

Key words Sex bias, Testis development, Isolation of early gonads, Gonadal transcriptome, Primordial germ cells

1 Introduction

Although the zebrafish (*Danio rerio*) has become a well-established model for many aspects of biological research [1–4], regulation of its sex determination and gonad differentiation are not well understood [5–7]. Recent data indicate that wild populations have a ZW/ZZ-type chromosomal sex determination [8, 9] that may have been lost repeatedly during domestication, resulting in polygenic sex determination [10]. In addition, all zebrafish juveniles start to develop a "juvenile ovary" during their early development [11] that will be later converted into a testis in the males in a process called "gonadal transformation" [12, 13]. In addition to the genetic process, environmental factors, including temperature,

The methodologies described in this chapter were developed at Temasek Life Sciences Laboratory (TLL), Singapore, during the period between 2006 and 2017. All four authors were affiliated with the Reproductive Genomics Group of TLL during the above time period. L.O. and K.W.T. are no longer with TLL, whereas J.M.S. and L.W.C. have moved to different research groups within TLL.

Roland Dosch (ed.), *Germline Development in the Zebrafish: Methods and Protocols*, Methods in Molecular Biology, vol. 2218, https://doi.org/10.1007/978-1-0716-0970-5_5, © Springer Science+Business Media, LLC, part of Springer Nature 2021

density, and availability of food, can also have an effect on the decision of sexual phenotypes [14, 15].

The essential role of germline stem cells or primordial germ cells (PGCs) in the formation and differentiation of (zebra)fish gonad has been described earlier [16, 17]. Complete elimination of PGCs in zebrafish led to the formation of sterile males with "testicular shells [18]." Later, it has been demonstrated by our team and our collaborators that reduction of PGC count by injecting early embryos with diluted morpholino oligonucleotides that temporarily interfere with the production of a protein essential for their survival results in a drastic shift of the sex ratio of a treated stock toward the males [19]. Alternative approaches to produce severely sex-biased lines include early treatment with hormone inhibitors [20], heat treatment [15, 21], multigenerational selection against the "other sex" [10], or genome manipulation [22].

Here, we describe the procedure of PGC depletion through morpholino injection into embryos of a transgenic zebrafish line [23], the sorting of injected embryos, and generation of an eviscerated trunk that could be used as an alternative for a gonad in gene expression analysis during the phases of early development.

2 Materials

2.1 Preparation of the Microinjection Needle

1. Dual-stage glass micropipette puller (*see* **Note 1**).
2. Borosilicate capillary glass with filament (length 10 cm), appropriate outside diameter and inside diameter (*see* **Note 2**).

2.2 Microinjection of Embryos with Diluted dnd*-MO*

1. Transgenic reporter zebrafish line with PGC-enhanced gene expression (*see* **Note 3**).
2. Stereomicroscope.
3. Plastic microinjection mold (*see* **Note 4**).
4. Injection plate with long parallel slots molded into 1% agarose (*see* **Note 5**).
5. Incubator set to 28 °C for embryo development.
6. Microinjector device (*see* **Note 6**).
7. Bottled nitrogen gas.
8. Micromanipulator with magnetic stand and a cast-iron base (optional; *see* **Notes 7** and **8**).
9. Extra-long pipette tips (optional).
10. Plastic Petri dishes (10 cm diameter).
11. Embryo medium (E3): dissolve 0.03% ocean salt in double distilled water.
12. 1% methylene blue.

13. Blue E3 medium: add 100 μL 1% methylene blue to 1 L of E3 medium.
14. 1% agarose dissolved in E3 medium.
15. Anti-*dead end (dnd)* morpholino (from here *dnd*-MO) solution (100 μM; 5′-GCTGGGCATCCATGTCTCCGACCAT-3′) (*see* **Note 9**).
16. Food dye (optional).

2.3 Dechorionation

1. Incubator set to 28 °C.
2. E3 medium and blue E3 medium.
3. Pronase (Roche) 1 mg/mL.
4. Glass Pasteur pipettes with tips blunted by fire-polishing.
5. Three 1 L beakers filled with E3 medium.

2.4 Sorting the Embryos According to Their PGC Number

1. Widefield fluorescent microscope.
2. Glass or plastic Petri dishes (10 cm diameter).
3. Microscopic glass slides.
4. Glass Pasteur pipettes with tips blunted by fire-polishing.

2.5 Dissection of Larvae

1. Stereomicroscope with fluorescent viewing capabilities (*see* **Note 10**).
2. Microscopic glass slide.
3. Tungsten ultrafine dissection needle (0.125 mm; *see* **Note 11**).
4. Stainless steel dissecting needle (0.5 mm, 45°; *see* **Note 11**).
5. Microdissecting needle holder.
6. #11 surgical blade.
7. #5 pointed forceps (*see* **Note 12**).
8. 1.5 mL Eppendorf tubes.
9. Liquid nitrogen.

3 Methods

3.1 Preparation for Microinjection

1. Insert a glass capillary tube (with filament) into the dual-stage glass micropipette puller, and pull it to generate two symmetrical needles (*see* **Notes 13** and **14**).
2. Generate an injection plate with parallel channels by partially immersing a custom-engineered plastic mold into a Petri dish containing warm 1% agarose dissolved in E3 medium (ca. 5–6 mm depth), and allow to gel.
3. Collect fertilized zebrafish eggs from a spawning tank containing zebrafish brooders that are homozygous transgenic to a

reporter gene resulting in PGC-enhanced gene expression in every offspring individual (*see* **Note 15**).

4. Wash the one-cell-stage zebrafish embryos in E3 medium and transfer them into a Petri dish.

3.2 Microinjection of Embryos with Diluted dnd-MO

1. Arrange the one-cell-stage embryos in a single line into the linear slots of the injection plate (*see* **Note 16**).
2. Choose the *dnd*-MO with the required concentration selected earlier in a pilot experiment (*see* **Note 17**). Add 10 μL of 1% methylene blue or food dye to the solution.
3. Insert the blunt end of the injection needle into the holder of microinjector.
4. Hold the needle with its holder in your hand and place its tip carefully onto the agarose surface. Gently break the tip of the injection needle using a fine forceps or the edge of a scalpel (*see* **Note 18**).
5. Load a microinjection needle with the diluted *dnd*-MO solution selected by titration (*see* **Note 19**).
6. Place an injection plate filled with embryos onto the stage of a stereomicroscope.
7. Inject the diluted *dnd*-MO solution into the top of yolk cell right below the embryonic pole of each embryo (*see* **Note 20**).
8. Carefully rinse the injected embryos into a Petri dish containing blue E3 medium.
9. Keep the embryos in an incubator set at 28 °C.

3.3 Dechorionation of Microinjected Embryos

Two different procedures have been used for this purpose in our lab. Either of them can be used to remove the chorion from microinjected embryos at 24 h postfertilization with high survival rates. They are as follows:

Procedure A:

- Remove the Petri dish containing the injected embryos from the incubator, and replace the blue E3 medium with E3 medium containing 1 mg/mL pronase.
- Place the dish containing the embryos under a stereomicroscope and watch them closely.
- When the first chorion breaks, an 8-shaped structure will form as the internal pressure will squeeze part the embryo through the hole (*see* **Note 21**).
- Lift the Petri dish with the embryos and carefully pour its content into a 1 L beaker containing E3 medium.

- Decant most of the solution, and pour the remaining solution with the embryos into another 1 L beaker containing E3 medium.
- Repeat the previous step again.
- Finally, decant most of the solution from the third beaker, and transfer the remaining E3 medium with dechorionated embryos to a new Petri dish.

Procedure B:

- Remove the Petri dish containing the injected embryos from the incubator, and replace the blue E3 medium with that containing 1 mg/mL pronase.
- Place the dish containing the embryos under a stereomicroscope and watch them closely.
- Check the condition of chorions constantly under a stereomicroscope by gently pressing them every minute with a forceps. The treatment is complete when the shape of chorions won't return to a sphere.
- Wash the embryos thoroughly in blue E3 medium.
- Pipette the embryos back and forth in blue E3 medium using a glass Pasteur pipette with tip blunted by fire-polishing in order to break open the chorions, and release the embryos (*see* **Note 22**).

3.4 Sorting the Embryos According to Their PGC Number

- Using a glass Pasteur pipette, add one drop of embryo medium onto a glass slide, and place one embryo into the droplet.
- Observe the embryo under a widefield fluorescence microscope, and rotate it with a blunt needle or a hair loop to have a clear lateral view (*see* **Note 23**).
- Count the number of fluorescently labeled PGCs in the gonadal ridge (Fig. 1; *see* **Note 24**).
- Sort embryos into three different groups: group I (1–7 PGCs), group II (8–15 PGCs), and group III (>15 PGCs; *see* **Note 25**). Uninjected controls can be used to determine the sex ratio of the batch, whereas embryos with no PGC can serve as "PGC-less controls" (*see* **Notes 26–28**).
- Grow sorted embryos in the respective groups for downstream analysis.

3.5 Dissection of Larvae

- Euthanize the larvae (7–21 dpf) in icy water.
- Transfer a single larva onto a clean microscope glass slide and position it along its lateral side, and then remove excess water from the glass slide by blotting.

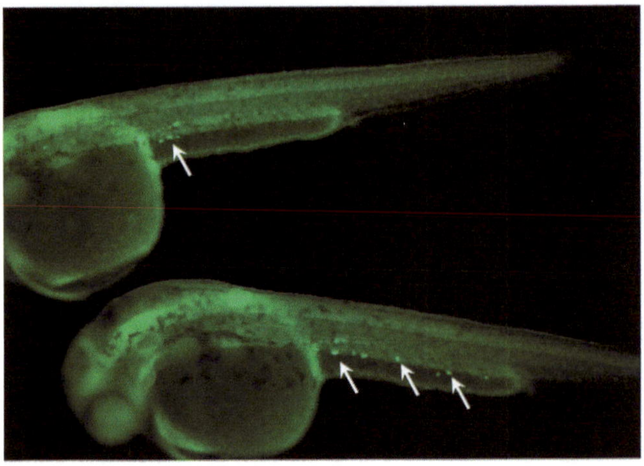

Fig. 1 Counting primordial germ cells in embryos transgenic with *Tg*(*vasa:vasa-EGFP*) reporter construct at the age of 24–28 hours post-fertilization (hpf). The larvae were injected with diluted anti-dead end (Dnd) morpholino oligos. Two larvae with a reduced number of primordial germ cells (PGCs, some labeled with arrows) are shown from the group containing seven to ten PGCs

- The following operations should be performed under the vision of a stereomicroscope with an EGFP filter. Using Tungsten dissecting needle (1 μm tip) or the pointed edge of a 26G needle, carefully make an incision on the ventral edge of larval belly from the cloaca to the gill operculum while positioning and holding the head of larva with a superfine dissecting needle (45° angle) simultaneously (Fig. 2; *see* **Note 29**).
- Gently eviscerate the larvae using a pointed #5 tweezer by pulling out the intestine from the cloaca toward the operculum (*see* **Note 30**). Remove the mass of internal organs, including the swim bladder. The gonads should still remain attached to the internal cavity.
- Using a #11 surgical blade, carefully make a vertical cut at the posterior end of gill operculum and at the cloaca (Fig. 2). Transfer the trunk into a sterile 1.5 mL centrifuge tube and snap freeze in liquid nitrogen (*see* **Note 31**).

3.6 Gene Expression Profiling

- Perform total RNA extraction according to the manufacturer's protocol.
- Gene expression profiling can be performed by qPCR or high throughput sequencing (*see* **Note 32**).

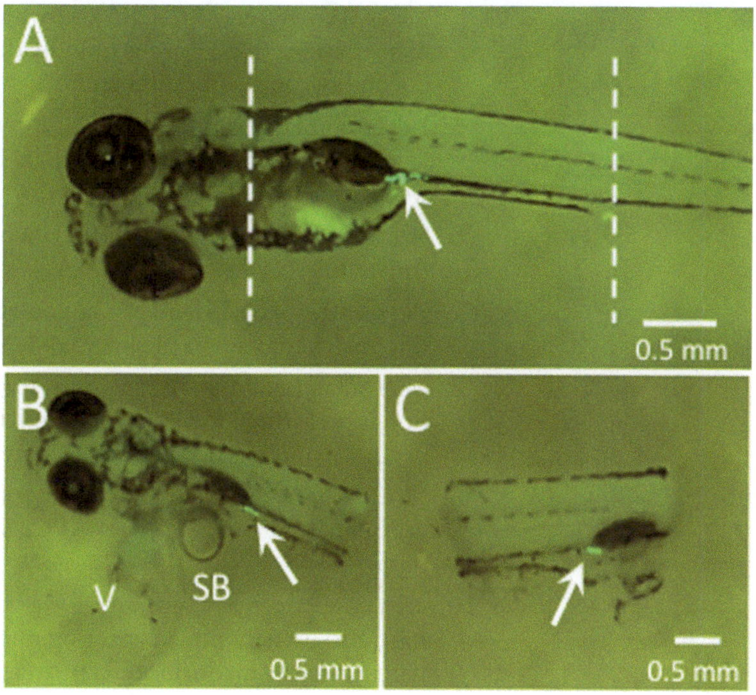

Fig. 2 Dissection of a *Tg(vasa:vasa-EGFP)* transgenic larva at 7 days post-fertilization (dpf) of age. (**A**) The whole larva with primordial germ cells (PGCs) showing high Egfp intensity (arrows). Dotted lines indicate vertical cuts at the posterior end of the operculum and cloaca, respectively, to isolate the trunk. (**B**) Eviscerated trunk with the gonadal ridge containing the PGCs still attached to the internal cavity. The swim bladder (SB) and viscera (V) were gently removed. (**C**) Eviscerated trunk with the gonadal ridge and PGCs, but without the head. Arrows label the PGCs on all three panels

4 Notes

1. We used P97 from Sutter Instrument (Novato, CA, USA).
2. We used borosilicate capillary glasses of 10 cm with filaments from Sutter Instrument.
3. For our experiments, we used homozygous transgenic individuals from the *Tg(vasa:vasa-EGFP)* zebrafish line (AB-based) for investigating the role of PGCs during gonadal development. The transgenic line was developed by the team of Lisbeth Olsen at the SARS Center, whereas the homozygous transgenic lineage was generated in our lab at TLL. Injection of GFP-*nos3* 3'UTR mRNA is an alternative method to identify the PGCs [24].
4. Typically, this plastic mold contains four to six parallel ridges that form wedge-shaped linear slots in hot agarose when partially immersed. We have used TU-1 from Adaptive Science

Tools (Worcester, MA, USA). However, several other custom-made designs (e.g., dozens of individual wells) can also be used depending on the requirements.

5. Injection plates should be filled up with E3 medium and covered. The plates can be stored in a refrigerator for several weeks. They can be used repeatedly, until damaged or dried.
6. We recommend the PLI-100 Pico-Injector from Harvard Apparatus (Holliston, MA, USA) with double foot pedals.
7. We use the MN-151 micromanipulator from Narishige Scientific Instruments Lab (Tokyo, Japan).
8. Micromanipulators are necessary for those cases, when injection is performed with them. For microinjection with handheld needle, a much more affordable simple holder is suitable.
9. Morpholinos designed against any other gene, whose product is essential for early survival of PGCs, are also suitable for this purpose.
10. As a cost-effective alternative for a fluorescent microscope, we have also used a simple dissecting microscope (Leica) equipped with a custom-made fluorescent attachment (MAA-03/B; BLS Ltd., Budapest, Hungary).
11. For the Tungsten ultrafine dissection needle, we use RS-6063 with a 1 micron tip, whereas for the stainless steel dissecting needle with a 10 micron tip. As more affordable alternatives, 26–30G needles can also be used., All of these needles are from Roboz Surgical Instrument Co. (Gaithersburg, MD, USA). To protect the needles when not in use, we use a needle holder (RS-6061, Roboz).
12. We recommend Dumostar Tweezers from World Precision Instruments (Sarasota, FL, USA).
13. Optimizing the settings on the puller is important for producing useful needles. Needles produced with different settings should be pretested on smaller batches of fertilized eggs to find the best combination.
14. Once the right setting is found, it is worth generating several needles, and store them by sticking them carefully in parallel onto a strip of BlueTac pushed into a Petri dish. The top of the dish should be labeled to indicate the status of needles (e.g., "pulled," "reusable," etc.).
15. For these experiments, we recommend using zebrafish pairs that earlier yielded offspring with (severe) female bias (typically more than 70% females), to make sure that the difference between the sex ratios of individuals developing from the group with severely depleted PGCs and those from the uninjected ones will be substantial.

16. When injecting with micromanipulators, the embryos must be carefully arranged with their animal pole facing toward 10–11 o'clock (for right-handed users) or toward 1–2 o'clock (for left-handed users). When microinjecting with handheld needle, eggs can be carefully rotated with the needle tip.
17. We recommend titrating the MO amount injected per cell to arrive to the range ideal for the purpose. We typically started to inject 30–40 pg per embryo and gradually increased the amount in two to three steps up to 100–120 pg per embryo, until we arrived to a level of depletion that yielded enough "low PGC" individuals in a repeatable and reliable manner. Under the conditions used by us, typically 60–80 pg MO per embryo was suitable. It is easier to inject different concentrations than different volumes of the same solution, as estimating the amount based on the diameter of the colored ball injected is rather difficult.
18. Make sure to break the tip evenly without creating a sharp end and removing too much. Liquid oozing out rapidly from the end of the tip is an indication that a new capillary will be needed. Abrupt change of the injected volume during the process indicates a broken tip. At that point the capillary must be replaced.
19. The needle can be loaded from the front by carefully placing an open Eppendorf tube with the diluted MO solution under a stereomicroscope, inserting the tip of the needle into the solution and carefully sucking up the required volume. Alternatively, it can be also loaded from the blunt end by a micropipette using long, fine "Microloader"-type tips.
20. At the end of the injection process, intact, calibrated injection needles should be rinsed clean with RO water and saved until the next use by sticking them carefully onto a strip of BlueTac pushed into a Petri dish.
21. Avoid extensive delays at this step as they might lead to complete loss of the embryos due to over-digestion.
22. Use glass Pasteur pipettes with tips blunted by fire-polishing for handling dechorionated embryos in order to reduce their damage.
23. As a potential alternative, digital quantification of the total GFP signal intensity within the gonadal region by widefield fluorescence microscopy using a 16-bit cooled CCD camera (CoolSNAP HQ, Teledyne Photometrics, Tucson, AZ, USA) was tested. The assignment of fish based on the two approaches was in good agreement (87% for the "low PGC" group and 100% of the time for the "high PGC" group). However, it was observed that individuals scored by digital GFP quantitation showed delayed growth and had a much higher mortality rate

than those scored manually, likely due to phototoxicity resulting from longer exposure to the ultraviolet (UV) component of the mercury lamp. Thus, this should not be considered as an alternative for counting of the PGCs under a microscope.

24. We recommend carefully rotating the embryo and viewing from at least two angles to make sure that all PGCs are accounted for.
25. Manual sorting of the individuals into groups based on their actual PGC count is a suitable approach for the reduction of diversity. Earlier, we have successfully used a similar strategy for narrowing down the wide variation in the extent and timing of gonadal transformation of transgenic zebrafish males by selecting those that showed minimal level of GFP expression to minimize the chance of accidentally selecting female individuals [25].
26. Group I embryos should yield adults with severe male bias, whereas adults developing from Group III embryos show female bias.
27. According to the literature, PGC-less controls should yield exclusively sterile males.
28. It is a well-known fact that MO technology is suffering from problems, including efficacy, off-target effects, and reproducibility [26, 27]. The MO-based PGC depletion is an artificial method that allows us to push the sex ratio of a set of individuals to extreme male bias and analyze differences between (mostly) males and controls at the same genetic background. In our experiments we observed variable PGC counts that could be the result of the inherent variability of the MO-injection technique or the consequence of the differential response from the two sexes. It is possible that the "additional males" that appear in the MO-depleted batch compared to controls are actually genetic females that had a lower number of PGCs within the female range.
29. Take care not to puncture the intestine.
30. Although we mostly initiated the removal of the intestine from the direction of the cloaca, this procedure can also be performed by starting from the anterior end.
31. The head of larva can be saved for transcriptional analysis, whereas the tail portion can be used for genotyping, if necessary.
32. qPCR analysis of 70 genes with potentially sex-associated function indicated that <10% of the genes show >25% residual activity in the gonad-less trunks compared to the gonad-containing (full) trunks (Table S7 of [19]), proving that the presence of other tissues did not have a substantial effect on the gonadal expression level of most genes.

Acknowledgments

The work described in the MS was funded by the National Research Foundation, Prime Minister's Office, Singapore, under its Competitive Research Programme (Award No: NRF-CRP7-2010-001) and by internal research grants from Temasek Life Sciences Laboratory. During the writing of this review, L.O. was supported by the Frontline Research Grant of the National Research, Development and Innovation Office of Hungary (KKP126764). We would like to thank Lisbeth Olsen for the transgenic line and Lydia Tan for helping with the figures.

References

1. Ota S, Kawahara A (2014) Zebrafish: a model vertebrate suitable for the analysis of human genetic disorders. Congenit Anom 54:8–11
2. Ablain J, Zon LI (2013) Of fish and men: using zebrafish to fight human diseases. Trends Cell Biol 23:584–586
3. Ribas L, Piferrer F (2014) The zebrafish (*Danio rerio*) as a model organism, with emphasis on applications for finfish aquaculture research. Rev Aqua 6:209–240
4. Ulloa PE, Medrano JF, Feijoo CG (2014) Zebrafish as animal model for aquaculture nutrition research. Front Genet 5:313
5. Liew WC, Orbán L (2014) Zebrafish sex: a complicated affair. Brief Funct Genomics 13:172–187
6. Kossack ME, Draper BW (2019) Genetic regulation of sex determination and maintenance in zebrafish (*Danio rerio*). In: Capel B (ed) Current topics in developmental biology, vol 134. Academic, New York, pp 119–149
7. Nagabhushana A, Mishra RK (2016) Finding clues to the riddle of sex determination in zebrafish. J Biosci 41:145–155
8. Anderson JL, Rodriguez Mari A, Braasch I, Amores A, Hohenlohe P, Batzel P, Postlethwait JH (2012) Multiple sex-associated regions and a putative sex chromosome in zebrafish revealed by RAD mapping and population genomics. PLoS One 7:e40701
9. Wilson CA, High SK, McCluskey BM, Amores A, Y-l Y, Titus TA, Anderson JL, Batzel P, Carvan MJ, Schartl M, Postlethwait JH (2014) Wild sex in zebrafish: loss of the natural sex determinant in domesticated strains. Genetics 198:1291–1308
10. Liew WC, Bartfai R, Lim Z, Sreenivasan R, Siegfried KR, Orban L (2012) Polygenic sex determination system in zebrafish. PLoS One 7:e34397
11. Takahashi H (1977) Juvenile hermaphroditism in the zebrafish, *Brachydanio rerio*. Bull Fac Fish Hokkaido Univ 28:57–65
12. Wang XG, Bartfai R, Sleptsova-Freidrich I, Orban L (2007) The timing and extent of 'juvenile ovary' phase are highly variable during zebrafish testis differentiation. J Fish Biol 70:33–44
13. Hsiao CD, Tsai HJ (2003) Transgenic zebrafish with fluorescent germ cell: a useful tool to visualize germ cell proliferation and juvenile hermaphroditism *in vivo*. Dev Biol 262:313–323
14. Santos D, Luzio A, Coimbra AM (2017) Zebrafish sex differentiation and gonad development: a review on the impact of environmental factors. Aquat Toxicol 191:141–163
15. Ribas L, Liew WC, Díaz N, Sreenivasan R, Orbán L, Piferrer F (2017) Heat-induced masculinization in domesticated zebrafish is family-specific and yields a set of different gonadal transcriptomes. Proc Natl Acad Sci USA 114: E941–E950
16. Dranow DB, Tucker RP, Draper BW (2013) Germ cells are required to maintain a stable sexual phenotype in adult zebrafish. Dev Biol 376:43–50
17. Siegfried KR, Nusslein-Volhard C (2008) Germ line control of female sex determination in zebrafish. Dev Biol 324:277–287
18. Slanchev K, Stebler J, de la Cueva-Méndez G, Raz E (2005) Development without germ cells: the role of the germ line in zebrafish sex differentiation. Proc Natl Acad Sci USA 102:4074–4079
19. Tzung K-W, Goto R, Saju Jolly M, Sreenivasan R, Saito T, Arai K, Yamaha E, Hossain Mohammad S, Calvert Meredith EK, Orbán L (2015) Early depletion of primordial

germ cells in zebrafish promotes testis formation. Stem Cell Rep 4:61–73

20. Luzio A, Monteiro SM, Garcia-Santos S, Rocha E, Fontaínhas-Fernandes AA, Coimbra AM (2015) Zebrafish sex differentiation and gonad development after exposure to 17-α-ethinylestradiol, fadrozole and their binary mixture: a stereological study. Aquat Toxicol 166:83–95
21. Abozaid H, Wessels S, Horstgen-Schwark G (2012) Elevated temperature applied during gonadal transformation leads to male bias in zebrafish (*Danio rerio*). Sex Dev 6:201–209
22. Lau ES-W, Zhang Z, Qin M, Ge W (2016) Knockout of zebrafish ovarian aromatase gene (*cyp19a1a*) by TALEN and CRISPR /Cas9 leads to all-male offspring due to failed ovarian differentiation. Sci Rep 6:37357
23. Krøvel AV, Olsen LC (2002) Expression of a *vas::EGFP* transgene in primordial germ cells of the zebrafish. Mech Dev 116:141–150
24. Saito T, Fujimoto T, Maegawa S, Inoue K, Tanaka M, Arai K, Yamaha E (2006) Visualization of primordial germ cells in vivo using GFP-nos1 3'UTR mRNA. Int J Dev Biol 50:691–699
25. Hossain MS (2010) Molecular analyses of gonad differentation in zebrafish. PhD Thesis, NUS, Singapore, 1–171.
26. Eisen JS, Smith JC (2008) Controlling morpholino experiments: don't stop making antisense. Development 135:1735–1743
27. Stainier DYR, Raz E, Lawson ND, Ekker SC, Burdine RD, Eisen JS, et al. (2017) Guidelines for morpholino use in zebrafish. PLoS Genet 13(10): e1007000

Chapter 6

Chemical Genetics: Manipulating the Germline with Small Molecules

Youngnam N. Jin and Randall T. Peterson

Abstract

Primordial germ cells (PGCs) are the precursor cells that form during early embryogenesis and later differentiate into oocytes or spermatozoa. Abnormal development of PGCs is frequently a causative factor of infertility and germ cell tumors. However, our understanding of PGC development remains insufficient, and we have few pharmacological tools for manipulating PGC development for biological study or therapy. The zebrafish (*Danio rerio*) embryos provide an excellent in vivo animal model to study PGCs, because zebrafish embryos are transparent and develop outside the mother. Importantly, the model is also amenable to facile chemical manipulations, including scalable screening to discover novel compounds that alter PGC development. This chapter describes methodologies for manipulating the germline (i.e., PGCs) with small molecules and for monitoring PGC development. Utilizing the 3′UTR of PGC marker genes such as *nanos3* and *ddx4/vasa* is a key component of these methodologies, which consist of expressing fluorescent or luminescent proteins in PGCs, treatment with small molecules, and quantitative observation of PGC development.

Key words Zebrafish, Embryo, Primordial germ cells, Germline, Germline stem cells, *Nanos3*, *Ddx4/Vasa*, 3′UTR, Chemical genetics, Small-molecule screens

1 Introduction

The survival of species relies on the ability of germ cells to preserve its genome and transmit it to offspring. Primordial germ cells (PGCs) are precursors of germ cells that will give rise to gametes. Abnormal events during PGC development can lead to dreadful consequences such as infertility and germ cell tumors. Understanding how PGCs are developed and how germ cells are regulated is greatly important for improvement of human life. Chemical tools for manipulating PGCs offer potential to improve our understanding of PGC biology and perhaps intervene in pathologies associated with abnormal PGC development.

The specification of PGCs in zebrafish (*Danio rerio*) embryos occurs from the very beginning of embryogenesis, prior to the

Roland Dosch (ed.), *Germline Development in the Zebrafish: Methods and Protocols*, Methods in Molecular Biology, vol. 2218, https://doi.org/10.1007/978-1-0716-0970-5_6, © Springer Science+Business Media, LLC, part of Springer Nature 2021

onset of maternal-to-zygotic transition (MZT) [1] in which newly transcribed RNAs begin to be generated, a process called zygotic genome activation (ZGA). This implies that the early developmental processes including PGC development depend on maternally supplied RNAs and proteins and that RNA regulation including translation and localization is critical for early embryogenesis. Moreover, early PGCs remain transcriptionally repressed until a later stage, whether they are specified by germ plasm-dependent determinative or inductive mechanisms, while transcription in somatic cells has been activated [2–6]. This is thought to be important for preventing premature differentiation into meiotic cells and/or for ensuring suppression of somatic fate. Another important feature in zebrafish embryos is that maternally supplied mRNAs exhibit short poly(A) tails [7], which is thought to keep translation of mRNAs repressed until needed. Cytoplasmic polyadenylation (CPA) is a major process for translation activation during early embryogenesis by adding poly(A) tails. However, only selected maternal mRNAs undergo CPA [8], suggesting that early embryos very likely utilize alternative mechanisms of translation in addition to canonical translation. Given these unique features in early embryos, there is much we still need to learn about the fundamental biological processes occurring during PGC development, which will likely give rise to improved treatments for infertility and germ cell tumors.

Chemical genetics is an efficient way to discover the functions of proteins and to understand signaling pathways in cells or whole organisms. While classical forward genetic screening can be a powerful approach for identifying genes involved in a particular process, genetic screening often requires a tremendous amount of labor, time, and cost. This can be particularly pronounced when screening for early embryonic phenotypes because maternally deposited RNAs can often mask the effects of gene mutations in a traditional zygotic screen. Chemical screens are generally faster and more cost-effective than genetic screens. Furthermore, chemical screening can identify compounds that target both maternally and zygotically expressed proteins or compounds whose biological effects arise from their engagement of multiple targets. More importantly, it allows for the discovery of small-molecule inhibitors or activators for target proteins which can be directly applied for therapeutic strategies or simply used as biological tools. Because of their ability to discover completely novel targets or unexpected biological processes, phenotypic screening using cell lines or whole organisms is advantageous over target-based chemical screening. Whole-organism screening offers many advantageous features, including the ability to identify compounds that work in a tissue-specific manner, that are nontoxic to the organism, that circumvent drug metabolism, or that cross the blood-brain barrier. Rodents and other mammalian model organisms are generally not suitable for chemical screening due to high cost, large body size, low throughput, low fecundity, and ethical concerns. Invertebrate animal models such as a nematode worm (*C. elegans*) and the fruit fly (*D. melanogaster*) may not be able to model some human diseases or organs. In addition, chemical treatment for worms and fruit flies is rather a cumbersome task due to limitations of liquid medium. On the other hand, zebrafish embryos or larvae are readily used for whole-organism high-throughput chemical screening [9–12]. The zebrafish is evolutionary closer to mammals than worms and flies. High fecundity, low-cost maintenance, rapid embryo development outside of mother's body, embryo/larvae transparency, small body size of embryos/larvae, and simple liquid treatment highlight zebrafish as an ideal animal model for high-throughput chemical screening.

As outlined elsewhere in this volume, the zebrafish has emerged as an excellent in vivo genetic model organism for studying PGCs [13, 14]. Since the development of PGCs including specification, migration, and maintenance is complete 24 hour postfertilization (hpf), zebrafish embryos can be used for chemical library screening to identify small molecules that impact PGC development. Indeed, using a transgenic zebrafish line expressing EGFP in PGCs [15], we were able to discover two structurally similar small molecules, we named primordazine A and B, that specifically inhibit a noncanonical form of translation called PAINT (poly(A)-tail-independent noncanonical translation) during early embryogenesis and further demonstrated that PAINT is essential for PGC maintenance [16].

In this chapter, we provide detailed protocols for chemical library screening to identify small molecules that intervene in the developmental processes of PGCs. This assay allows for screening compounds in 96-well format and can be completed in about 2 days for chemical screening using a transgenic zebrafish line, *Tg (ddx4:EGFP)* [15]. We also described quantitative analytic methods to assess PGC development by counting GFP-positive PGCs or by measuring the luciferase activities after injection of luciferase reporter mRNAs.

2 Materials

2.1 Zebrafish Lines

1. Zebrafish TuAB or AB strain for wild-type strain.
2. Transgenic *Tg(ddx4/vasa:EGFP)* line [15].

2.2 In Vitro Transcription of the Reporter mRNA

1. Reporter plasmids: EGFP-*nanos3*–3′UTR [17], RLuc-*nanos3*–3′UTR [16], FLuc-SV40pA, or any other EGFP or luciferase reporter derivatives.
2. Linearized plasmid templates.

3. 10× digestion buffer and restriction enzymes for linearization of plasmids.
4. Thermocycler for in vitro transcription.
5. Gel extraction and PCR cleanup kits.
6. In vitro transcription kits for capped RNA synthesis (e.g., m MESSAGE mMACHINE T7 (T3 or SP6) Transcription Kit, Thermo Fisher Scientific) (*see* **Note 1**).
7. DNase I.
8. Poly(A) tailing kit.
9. RNA cleanup kit (e.g., RNA Clean & Concentrator-25, Zymo Research).
10. NanoDrop or an equivalent spectrophotometer.

2.3 Microinjection

1. Borosilicate glass capillary tube (OD 1.0 mm, ID 0.78 mm, length 100 mm).
2. Dual-stage glass micropipette puller.
3. Micromanipulator.
4. Microinjector (e.g., FemtoJet 4i, Eppendorf).
5. Stereomicroscope.
6. Eppendorf Microloader tips.
7. Stage micrometer.
8. Microinjection mold (Adaptive Science Tools, TU-1).

2.4 Small-Molecule Preparation

1. Small-molecule libraries (normally 10 mM in DMSO).
2. 100% dimethyl sulfoxide (DMSO, molecular grade).
3. Aluminum foil.

2.5 Assessment of PGC Development

1. Fluorescence stereomicroscope with a GFP filter.
2. Microplate reader capable of measuring luminescence.
3. Pronase stock solution: Make 10 mg/ml in E3 medium, aliquot 1 ml into 1.5 ml microtube, and store at −20 °C.
4. 25 × 75 × 1 mm microscope slides.
5. 22 × 22 mm coverslips.
6. Phosphate-buffered saline (PBS): 137 mM NaCl, 2.7 mM KCl, 10 mM Na_2HPO_4, 1.8 mM KH_2PO_4.
7. 1× passive lysis buffer (Promega).
8. Dual-Glo Luciferase Assay kit (Promega).

2.6 Common Materials

1. 60× E3 medium stock solution: 180 mM NaCl, 10 mM KCl, 20 mM $CaCl_2$, 20 mM $MgCl_2$.

2. E3 medium: Add 160 ml of 60× E3 medium stock solution and adjust the volume to 10 L with distilled water.
3. System water: Water used to maintain zebrafish made of 0.5–1 g/l of sea salt with a conductivity of 300–1500 μS and buffered with sodium bicarbonate between pH 6.8 and 8.0.
4. Dumont #5 or #55 forceps.
5. Disposable transfer pipettes.
6. Squeeze bottles with E3 medium: They are used for rinsing the collected embryos or for refilling petri dishes or plates that contain the embryos.
7. Petri dishes: 100 mm for collecting embryos and 35 mm dishes for treating embryos with chemicals.
8. 96-well plates: Transparent 96-well plates for chemical treatment, opaque 96-well plates for measurement of luciferase activities.
9. Incubator for embryos, 28.5 °C.

3 Methods

PGCs are one of the earliest specified cell types during embryogenesis. PGC development begins from the very early embryo stage and depends on maternally provided germ plasm. Specified PGCs can be observed at as early as 3 hpf. Therefore, in order to intervene in the developmental processes of PGCs, it is important to treat embryos with small molecules at early developmental time point.

3.1 Preparation of Zebrafish Embryos

1. Place one male and one to two female *Tg(ddx4:EGFP)* adult fish in a mating container, and separate them by sex using a divider (*see* **Note 2**). Keep them overnight at 28.5 °C. For injection of mRNAs into embryos, use wild-type zebrafish such as TuAB.
2. Remove a divider the next morning and wait for 5–10 min until spawning (*see* **Note 3**).
3. Collect the embryos by pouring the contents of mating containers through a mesh strainer. Rinse the embryos using a squeeze bottle with E3 medium.
4. Transfer eggs from the mesh strainer to 100 mm dishes using a squeeze bottle with E3 medium.
5. Remove unfertilized or deformed embryos, other debris, or contaminants using a transfer pipette. Replenish with fresh E3 medium.
6. Transfer five to ten embryos at one to two-cell stages per well with 200 μl of E3 medium into 96-well plates that are not coated (*see* **Note 4**).
7. Go to step in Subheading 3.4 for chemical treatment.

3.2 In Vitro Transcription of the Reporter mRNA

1. Digest the reporter plasmid(s) for the template(s) to be in vitro transcribed at 37 °C overnight (*see* **Note 5**).
2. Purify the digested template plasmid using a PCR cleanup kit. If multiple bands occur, purify the desired band using a Gel extraction kit. The final concentration of linearized plasmid is typically 80–150 ng/μl in 50 μl of nuclease-free water.
3. Set up the in vitro transcription reaction according to the manufacturer's protocol.
4. Mix and incubate for at least 3 h to overnight at 37 °C in a thermocycler.
5. Add 1 μl of RNase-free DNase I, mix, and incubate at 37 °C for 15 min (*see* **Note 6**).
6. Purify RNA using a RNA cleanup kit and elute RNA with 25 μl of nuclease-free water.
7. Measure the RNA concentration using NanoDrop. Typically, RNA concertation will be around 1000 ng/μl. Keep the RNA at −80 °C until use or go to the next step.
8. Dilute the RNA into working concentration for injection (*see* **Note 7**). Store the rest of RNA at −80 °C until use.

3.3 Microinjection

There are some cases in which EGFP or a luciferase mRNA fused with *nanos3*–3′UTR will be injected into embryos at one-cell stage, for example, if the *Tg(ddx4:EGFP)* fish line is not available, if the effect of mutation of *nanos3*–3′UTR on response to chemicals is to be investigated, if translation efficiency of the reporter mRNA rather than PGC number in response to chemicals is being evaluated, or if a more accurate quantitative method such as the luminescence measurement is required.

1. Make glass injection needles by pulling glass capillary tubes using a dual-stage glass micropipette puller.
2. Load 4–5 μl of 100–200 ng/μl of mRNA solution in nuclease-free water using an Eppendorf Microloader tip (*see* **Note 7**).
3. Mount the injection needle to the needle holder in the micromanipulator, and connect tubing from the needle holder to the microinjector.
4. Break the tip of the injection needle using forceps under a stereomicroscope, and adjust the injection volume to 1.5–2 nl using a stage micrometer by setting up the injection pressure.
5. Place one-cell-stage embryos in an injection mold using a transfer pipette (*see* **Note 8**). The injection mold is made of 2% agarose with E3 medium in a 100 mm petri dish.

6. Arrange embryos into the grooves in the injection plate under a stereomicroscope using a syringe needle, and rotate embryos to put a cell layer parallel to the wall of groove (*see* **Note 9**). Locate the first embryo in the center of the stereomicroscopic view.
7. Bring down the injection needle close to the first embryo using the micromanipulator.
8. Guide the injection needle into the cell using a micromanipulator, and dispense mRNA solution by introducing the injection pressure. Pull out the injection needle out of the cell using the micromanipulator.
9. Repeat **step 8** for additional injection (*see* **Note 10**).
10. Go to step in Subheading 3.4 for chemical treatment (*see* **Note 11**).
11. Assess PGC development by GFP fluorescence (go to the step in Subheading 3.6) or by luminescence (go to the step in Subheading 3.7), depending on mRNAs for injection.

3.4 Treatment of Embryos with Small-Molecule Libraries

1. Thaw a 96-well chemical library plate containing DMSO stock solutions in a desiccator to prevent water condensation during thawing.
2. For chemical screen, add compounds to the wells of a 96-well plate using a 0.2 μl pin tool (*see* **Note 12**). For testing individual chemicals of interest or different concentrations, prepare the working solution of small molecules in DMSO, and treat embryos with desired concentrations of small molecules (*see* **Note 11**).
3. Thoroughly mix compounds by shaking the plate on the bench or by pipetting up and down with a multichannel pipettor until no precipitation can be observed. Cover the plate with aluminum foil (*see* **Note 13**).
4. Incubate the plate at 28.5 °C.

3.5 Qualitative Assessment of PGCs by Imaging GFP Fluorescence

1. When embryos have developed to ~24 hpf, bring the plate out of the incubator, and place it on the fluorescence stereomicroscope.
2. Assess PGC development by visually observing cells which show bright GFP fluorescence (*see* **Note 14**).
3. Take a fluorescence image per well with consistent settings such as magnification, exposure time, and gain (Fig. 1).

3.6 Quantitative Assessment of PGC by Counting the GFP Fluorescence

To confirm positive hits or to quantify the number of PGCs, the number of PGCs can be counted under the fluorescence stereomicroscope (Figs. 2 and 3).

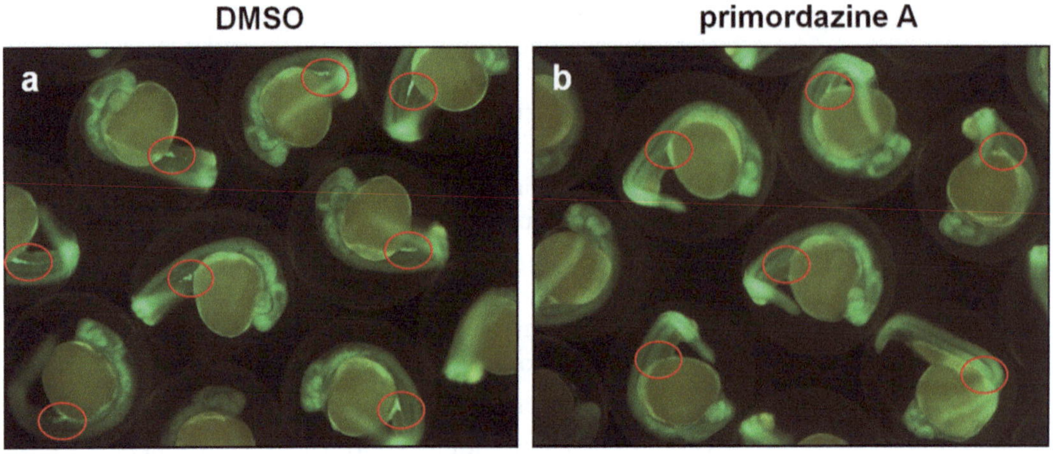

Fig. 1 Representative images of *Tg(ddx4:EGFP)* embryos at 24 hpf. (**a**) Embryos were treated with DMSO. (**b**) Embryos were treated with 10 μM primordazine A. Images were taken at 24 hpf using a fluorescence stereomicroscope with the same exposure time and magnification. Note that these transgenic embryos at 24 hpf show brighter GFP fluorescence in PGCs than in somatic cells. The GFP fluorescence in somatic cells will become weaker as embryos develop, while PGCs maintain the bright fluorescence for at least a week. The red circles indicate the area where PGCs are supposed to be observed

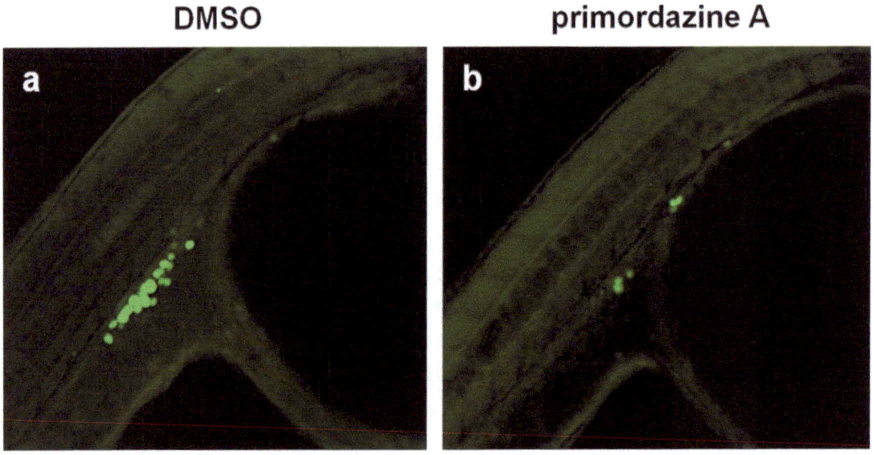

Fig. 2 Representative magnified images of *Tg(ddx4:EGFP)* embryos at 24 hpf. Embryos were treated with DMSO (**a**) or 10 μM primordazine A (**b**). Images were taken at 24 hpf using a fluorescence stereomicroscope with the greatest magnification. For the quantitative analysis, the number of PGCs which show the bright GFP fluorescence will be counted

1. Take the plate out of the incubator at 24 hpf.
2. Dechorinate embryos by adding 20 μl of pronase stock solution to a single well of 96-well plate (*see* **Note 15**).
3. Incubate the plate at room temperature and gently swirl the embryos until the chorions fall off from the embryos. It typically takes 10–15 min (*see* **Note 16**).

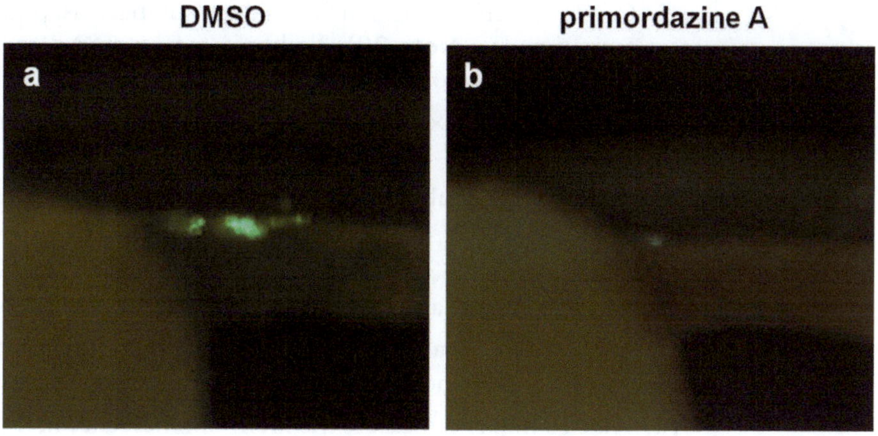

Fig. 3 Representative images of embryos injected with EGFP-*nanos3*–3′UTR reporter mRNA. Embryos were treated with DMSO (**a**) or 10 μM primordazine A (**b**) after injection with EGFP-*nanos3*–3′UTR mRNA. Images were taken at 24 hpf with a fluorescence stereomicroscope with the same exposure time and magnification. Note that GFP fluorescence of EGFP-*nanos3*–3′UTR in PGCs is much brighter, while GFP fluorescence in somatic cells is barely detectable. This is because *nanos3*–3′UTR facilitates mRNA decay via miRNA pathway only in somatic cells but not in PGCs

4. Wash at least three times with E3 medium.
5. Transfer one to two embryos onto a microscope slide and remove excess solution with a transfer pipette. Put a 22 × 22 mm coverslip gently on top of embryos by releasing the coverslip from one side to the other. This will flatten the embryo while keeping the basic organization of the embryo intact.
6. Place the microscope slide under the fluorescence stereomicroscope. Increase the magnification to find PGCs which should be recognized by their bright GFP fluorescence (*see* **Note 17**).
7. Count PGCs, make a note, and wipe the coverslip and the slide. Repeat **steps 5** and **6** until at least ten embryos are examined.

3.7 Quantitative Assessment of PGC by Measuring Luminescence

When embryos are injected with the luciferase reporter mRNA mixtures, embryos will be collected at earlier time point such as 6 hpf as compared to GFP reporter mRNA (*see* **Note 18**).

1. At 6 hpf, bring the plate or dishes of treated embryos from the incubator (*see* **Note 11**).
2. Transfer 20–30 healthy embryos to a 1.5 ml microtube using a P1000 or P200 pipette with the pipette tip being cut (*see* **Notes 4** and **19**).
3. Spin down for 20–30 s at ~20 × *g*, and discard the supernatant by aspiration with ~100 μl of residual solution being left.
4. Add 1 ml of PBS to wash embryos. Repeat **step 3**.

5. Carefully remove as much residual solution as possible by aspiration (*see* **Note 20**). Embryos can be stored at −80 °C until use or go to the next step.
6. Add 3 μl of 1× passive lysis buffer per one embryo (*see* **Note 21**).
7. Homogenize embryos with a plastic pestle using a motorized pestle mixer.
8. Centrifuge at 15,000 rcf for 1 min and transfer 20 μl of solution to a well of an opaque 96-well plate.
9. Add 70 μl of Dual-Glo Luciferase reagent to the wells containing the lysates and mix by gently tapping.
10. Wait 10 min and measure the firefly luminescence in the microplate reader.
11. Add 70 μl of Dual-Glo Stop & Glo reagent to the same well and mix by gently tapping.
12. Wait 10 min and measure the Renilla luminescence in the microplate reader.
13. Normalize the Renilla luminescence by the firefly luminescence since the firefly luminescence is supposedly unsusceptible to drug treatment and used as an indicator of the amount of injected mRNA. To obtain the relative response ratios, the normalized value of drug treatment condition will be divided by the normalized value of DMSO condition.

4 Notes

1. DNA templates for in vitro transcription need to have a T7, T3, or SP6 promoter sequence upstream of gene of interest. T7 and T3 RNA polymerases are more efficient than SP6 RNA polymerase. If a DNA template does not have any promoter sequence, it can be added by PCR with a forward primer containing one of the RNA promoter sequence.
2. Feeding fish 1 h prior to setting them up for mating enhances mating rates and egg quantity. One female fish may lay about 200 eggs. However, the number of eggs varies so prepare for enough fish for mating. Adult fish ideally need to rest at least 1 week before they can be used for mating again.
3. Replenishing the container with fresh system water and making a slope by raising one side of the container normally increases mating rates. After removing a divider, stay away a couple of meters from fish so they are not disturbed during mating.
4. To ensure 200 μl of total volume with embryos and E3 medium, cut 3–4 mm of a 1000 μl pipette tip to allow embryos

to go through. Set the volume of pipette to 200 μl and take five to ten embryos into the pipette tip. Transfer them into a well of 96-well plate so the final volume in a well will be 200 μl.

5. To linearize the template plasmid, digest 10 μg of plasmid DNA using a restriction enzyme that cuts the plasmid after the sequence of 3′UTR. The reaction mixture includes 5 μl of 10× CutSmart buffer (for NEB enzymes), 10 μg of plasmid DNA, 10–20 U of restriction enzyme, and nuclease-free water to make up 50 μl.
6. Although a poly(A) tail will enhance the translation efficiency of mRNA, it may affect the characteristics of mRNA via the 3′UTR such as the specific localization, RNA stability, or translational response to chemicals. Therefore, do not add a poly (A) tail for PGC reporter mRNAs. However, we add a poly (A) tail to the firefly luciferase mRNA using a poly(A) tailing kit, because we found a small molecule, primordazine, does not affect translation of mRNA with a poly(A) tail so this mRNA can be used as an internal control [16].
7. In general, 200 ng/μl of mRNA solution will be enough for maximum expression. However, the amount of mRNA can be adjusted in order to reduce the toxicity or to increase the expression level. For the luciferase assay, the mixture of 200 ng/μl of Renilla reporter mRNA with *nanos3*–3′UTR and 20 ng/μl of the firefly reporter, such as Luc2, mRNA with a poly(A) tail was used for the injection. In addition, to help visualize mRNA solution during injection, 0.02–0.05% phenol red can be used. Phenol red at this concentration does not show any negative effect on embryo development.
8. The groove of microinjection mold consists of a beveled bottom and a perpendicular wall. Place the perpendicular wall to the left side of grooves on the stage of stereomicroscope. Injection needle will approach the embryos from the right side.
9. Embryos tend to float, which will interfere with injection. To prevent embryos from floating, gently push them to the grooves with a needle.
10. The embryos will develop into the two-cell stage around 40–45 min postfertilization so injection should be finished within 40 min or so. Try to inject as many embryos as possible before embryos get to two-cell stage.
11. Because generally more than 10 embryos should be used for the quantitative data, 30–40 embryos will be placed in a 6-well plate or 35 mm dishes with 3 ml of E3 medium, and positive hits or compounds of interest will be added. Embryos injected with the reporter mRNAs will also be placed in a 6-well plate or 35 mm dishes and treated with chemicals accordingly. To transfer embryos after injection, rinse off injected embryos

with E3 medium from the injection mold to a 100 mm petri dish, and transfer them to a 6-well plate or 35 mm dishes using a P1000 or P200 pipette with an appropriate tip of which the end has been cut out. Make up to 3 ml with E3 medium.

12. Dip a pin tool into E3 medium up and down three times to ensure chemicals being completely released from the pins.
13. The development of PGCs begins from the very early embryonic stage. So it is very critical to treat embryos at early and consistent time point, for example, 1 hpf.
14. *Tg(ddx4:EGFP)* embryos exhibit GFP fluorescence throughout the whole body, while PGCs show brighter GFP fluorescence. Most of the embryos in normal (nontreated) or control (DMSO-treated) condition contain 25–40 PGCs.
15. Add 1/10 of pronase stock solution for dechorionation, for example, 20 μl of pronase stock to 200 μl of E3 medium.
16. Chorions can be torn off with forceps or syringes with a 27 G needle under a stereomicroscope. If you have only a few embryos to count, for example, 20–30 embryos, peeling off chorions may be a better choice than using pronase.
17. PGCs may be clustered and overlapping. To distinguish individual PGCs, gently push the coverslip to spread clustered PGCs. The number of PGCs will be unchanged from 1 day postfertilization (dpf) to 2 dpf so it is possible to count PGCs until 2 dpf. However, PGCs become more clustered and fragile as embryo develops.
18. GFP protein is known to be a more stable protein in cells as compared to luciferases. The half-life of GFP is approximately 24–26 h in mammalian cells [18], while the half-life of Renilla or firefly luciferase is about 4.5 h or 3 h, respectively [19, 20]. Once GFP protein is translated, it will stay in the cell even after chemical treatment inhibits level of its mRNA or translational efficiency of the reporter mRNA. Therefore, GFP is the better marker for ensuring the integrity of cells, while luciferases are the better indicator for monitoring the dynamics of cellular processes or translation efficiency. For example, to assess PGC development in response to small molecules, it is better to use GFP to ensure PGC existence, but measuring the luciferase activity is a better method to monitor changes in translation or transcription in PGCs.
19. Chemical treatment often delays zebrafish development. Keep an eye on the embryos until they reach "shield" stage, a developmental stage of ~6 hpf.
20. The strong vacuum suction may aspirate the embryos. To avoid this, put a 10 μl pipette tip at the vacuum line to reduce the suction power. Or reduce the vacuum power if possible.

21. The volume of 1× passive lysis buffer to be added will be determined by the number of embryos collected for the luciferase assay. Therefore, write down the number of collected embryos on each microtube.

References

1. Strome S, Updike D (2015) Specifying and protecting germ cell fate. Nat Rev Mol Cell Biol 16:406–416. https://doi.org/10.1038/nrm4009
2. Siddiqui NU, Li X, Luo H et al (2012) Genome-wide analysis of the maternal-to-zygotic transition in Drosophila primordial germ cells. Genome Biol 13:R11. https://doi.org/10.1186/gb-2012-13-2-r11
3. Lebedeva LA, Yakovlev KV, Kozlov EN et al (2018) Transcriptional quiescence in primordial germ cells. Crit Rev Biochem Mol Biol 53:579–595. https://doi.org/10.1080/10409238.2018.1506733
4. Seydoux G, Dunn MA (1997) Transcriptionally repressed germ cells lack a subpopulation of phosphorylated RNA polymerase II in early embryos of Caenorhabditis elegans and Drosophila melanogaster. Development 124:2191–2201
5. Yamaguchi S, Kimura H, Tada M et al (2005) Nanog expression in mouse germ cell development. Gene Expr Patterns 5:639–646. https://doi.org/10.1016/j.modgep.2005.03.001
6. Fang F, Angulo B, Xia N et al (2018) A PAX5--OCT4–PRDM1 developmental switch specifies human primordial germ cells. Nat Cell Biol 20:655. https://doi.org/10.1038/s41556-018-0094-3
7. Subtelny AO, Eichhorn SW, Chen GR et al (2014) Poly(A)-tail profiling reveals an embryonic switch in translational control. Nature 508:66–71. https://doi.org/10.1038/nature13007
8. Winata CL, Łapiński M, Pryszcz L et al (2018) Cytoplasmic polyadenylation-mediated translational control of maternal mRNAs directs maternal-to-zygotic transition. Development 145:dev159566. https://doi.org/10.1242/dev.159566
9. Kithcart AP, MacRae CA (2018) Zebrafish assay development for cardiovascular disease mechanism and drug discovery. Prog Biophys Mol Biol 138:126–131. https://doi.org/10.1016/j.pbiomolbio.2018.07.002
10. McCarroll MN, Gendelev L, Keiser MJ, Kokel D (2016) Leveraging large-scale behavioral profiling in zebrafish to explore neuroactive polypharmacology. ACS Chem Biol 11:842–849. https://doi.org/10.1021/acschembio.5b00800
11. Oikonomou G, Prober DA (2017) Attacking sleep from a new angle: contributions from zebrafish. Curr Opin Neurobiol 44:80–88. https://doi.org/10.1016/j.conb.2017.03.009
12. Rennekamp AJ, Peterson RT (2015) 15 years of zebrafish chemical screening. Curr Opin Chem Biol 24:58–70. https://doi.org/10.1016/j.cbpa.2014.10.025
13. Raz E (2003) Primordial germ-cell development: the zebrafish perspective. Nat Rev Genet 4:690. https://doi.org/10.1038/nrg1154
14. Paksa A, Raz E (2015) Zebrafish germ cells: motility and guided migration. Curr Opin Cell Biol 36:80–85. https://doi.org/10.1016/j.ceb.2015.07.007
15. Krøvel AV, Olsen LC (2002) Expression of a vas::EGFP transgene in primordial germ cells of the zebrafish. Mech Dev 116:141–150. https://doi.org/10.1016/S0925-4773(02)00154-5
16. Jin YN, Schlueter PJ, Jurisch-Yaksi N et al (2018) Noncanonical translation via deadenylated 3′ UTRs maintains primordial germ cells. Nat Chem Biol 14:844. https://doi.org/10.1038/s41589-018-0098-0
17. Mishima Y, Giraldez AJ, Takeda Y et al (2006) Differential regulation of germline mRNAs in soma and germ cells by zebrafish miR-430. Curr Biol 16:2135–2142. https://doi.org/10.1016/j.cub.2006.08.086
18. Corish P, Tyler-Smith C (1999) Attenuation of green fluorescent protein half-life in mammalian cells. Protein Eng Des Sel 12:1035–1040. https://doi.org/10.1093/protein/12.12.1035
19. Leclerc GM, Boockfor FR, Faught WJ, Frawley LS (2000) Development of a destabilized firefly luciferase enzyme for measurement of gene expression. BioTechniques 29:590–601. https://doi.org/10.2144/00293rr02
20. Thorne N, Inglese J, Auld DS (2010) Illuminating insights into firefly luciferase and other bioluminescent reporters used in chemical biology. Chem Biol 17:646–657. https://doi.org/10.1016/j.chembiol.2010.05.012

Chapter 7

In Vitro Induction of Teleost PGCs

Vanesa Robles, David G. Valcarce, and Marta F. Riesco

Abstract

Primordial germ cells (PGCs) are unique cells in an embryo. These cells contain all genetic information and therefore represent the best source to store maternal and paternal genomes until embryo cryopreservation is achieved. However, the number of these cells in an embryo is very low limiting their potential application in cryopreservation and surrogate production. However, it was assumed that the induction of fish PGCs in vitro is not possible because in vivo they inherit germ plasm. In this chapter, we describe a successful differentiation protocol explaining the crucial factors and steps for in vitro PGC generation.

Key words Embryonic cell, Primordial germ cell, Differentiation, Culture, Morpholino microinjection, Transplant

1 Introduction

Primordial germ cells (PGCs) are the precursors of gametes. This condition makes them unique cells within the embryo because their genetic and epigenetic information will be transmitted to the progeny. These cells have a great potential for conservation purposes in teleosts, particularly considering that fish embryos and oocytes cannot be cryopreserved. PGC conservation is considered a good alternative for preserving the maternal genome [1]. Moreover, PGCs are used in surrogate production with important benefits for aquaculture. Transplantation and xenotransplantation of these cells have been done in different species shown by successful colonization of the genital ridge of the recipient and production of fully differentiated gametes in the adult [2–4]. This technology is particularly beneficial for those species with reproductive problems or for those teleosts with a problematic size of the breeders.

One limitation of these applications is the reduced number of PGCs per embryo [5]. In this sense, the possibility of differentiating PGCs from a more abundant cell population is interesting. Traditionally, it was thought that PGC induction is not possible in fish due to the in vivo requirement of inheriting germ plasm to

Roland Dosch (ed.), *Germline Development in the Zebrafish: Methods and Protocols*, Methods in Molecular Biology, vol. 2218, https://doi.org/10.1007/978-1-0716-0970-5_7, © Springer Science+Business Media, LLC, part of Springer Nature 2021

become PGCs [5]. We recently demonstrated that this differentiation is possible in vitro, evaluating the "PGC condition" of the resulting cells using different approaches: gene expression, vasa fluorescence detection, cell cycle study, identification of populations by flow cytometry, and transplantation experiments [5]. In this chapter, we present a detailed protocol for in vitro differentiation of teleost PGCs, highlighting the crucial steps in the procedure.

2 Materials

2.1 Zebrafish Embryo Maintenance

1. Fluorescent protein (EGFP) zf45 strain, *tg fvas:egfpg* zebrafish transgenic line.
2. Crossing tanks (e.g., Aquatic Habitats).
3. Sterile 10-ml plastic plates.
4. System water.
5. 0.1% methylene blue in system water.
6. 28 °C incubator.
7. 70% ethanol in system water.
8. 0.5% bleach solution in system water.

2.2 Embryonic Cell Recovery

1. Flow hood.
2. Fine watchmaker's forceps.
3. Stereomicroscope.
4. Leibovitz medium (L-15) (Sigma): 80% L-15 supplemented with 5% fetal bovine serum, 500 μg/ml of ampicillin, 500 μg/ml of streptomycin, and 500 IU/ml of penicillin (should the preparation of this buffer be described in **Note 1**).
5. 10-ml syringes.
6. 0.2-μm Millipore filter.

2.3 Embryonic Cell Culture

1. L-15 medium (*see* **Note 1**).
2. Heat-inactivated 15% fetal bovine serum (FBSHI).
3. Testicular cell culture medium (TCCM).
4. 500 g/ml of ampicillin in L-15 medium.
5. 500 lg/ml of streptomycin in L-15 medium.
6. 500 IU/ml of penicillin in L-15 medium.
7. Flow hood with stereomicroscope (e.g., Nikon).
8. High cell-binding dish (Nunc; VWR).

2.4 Embryonic Cell Differentiation to Primordial Germ Cells (PGCs)

1. TTCM.
2. Sterile Milli-Q water.
3. DMSO.
4. Centrifuge.
5. Filtered 20 mM acetic acid.
6. High cell-binding dish (Nunc; VWR).
7. 500 ng/ml bone morphogenetic protein 4 (BMP4).
8. 5 mM retinoic acid (RA).
9. 50 ng/ml epidermal growth factor (EFG).

2.5 Analysis of Migration Capacity of In Vitro Generated Primordial Germ Cells

1. 1 μg/μl Dnd-1 morpholino (Gene Tools).
2. 0.06% phenol red solution.
3. Glass micropipette needle (0.9-mm diameter borosilicate glass capillaries).
4. 100-mm Petri dish.
5. Slides.
6. Stereomicroscope equipped with green fluorescence filter (excitation wavelength, 530 nm).

3 Methods

3.1 Zebrafish Embryo Maintenance

1. Maintain zebrafish at 14-h light/10-h dark cycle with standard conditions.
2. The day prior to embryonic cell isolation, transfer the adult zebrafish vasa enhanced green fluorescent protein (EGFP) zf45 strain, *tg fvas:egfpg* transgenic line to the crossing tanks at 28 °C with six fish per tank at a male/female ratio of 1:2. The fluorescent protein (EGFP) *zf45 strain, tg fvas:egfpg* transgenic line was generated by the Krovel and Olsen laboratory [6] and is crucial to allow PGC identification in the embryos and embryonic cell cultures.
3. Transfer the fertilized eggs to sterilized plates with system water containing 0.1% blue methylene to avoid microbial contamination, and incubate them at 28 °C. Remove unfertilized eggs and debris with a transfer pipette.
4. When the embryos reach the appropriate developmental stage (blastula, 3.5 hpf), split into groups of approximately 50 individuals each.
5. Rinse with 70% ethanol for 10 s and then treat with bleach solution (0.5%) for 2 min to avoid culture contamination.
6. Rinse several times with Leibovitz (L-15) medium.

3.2 Embryonic Cell Recovery

1. Manually dechorionate zebrafish embryos at blastula stage using sterile fine forceps to maintain their integrity (*see* **Note 2**).
2. Manually excise the blastoderms from blastula-stage embryos (3.5 h post-fertilization) using fine watchmaker's forceps under a microscope inside of a flow hood. Separate carefully the blastoderms from the yolk with sterile forceps (*see* **Note 3**).
3. Wash the blastomeres in modified and sterile L-15 supplemented with 5% fetal bovine serum, 500 g/ml ampicillin, 500 g/ml streptomycin, and 500 IU/ml penicillin at room temperature until further use (*see* **Note 4**).
4. Culture these embryonic zebrafish cells (50 blastoderms per well; 50,000 cells approximately) in a high cell-binding dish in TCCM medium [7] (*see* **Note 5**) supplemented with 15% FBSHI at 28 °C (*see* **Note 6**).
5. Incubate the cultures 1 h for somatic cell adhesion, whereas embryonic cells remain in suspension maintaining their viability (*see* **Note 7**).

3.3 Embryonic Cell Differentiation to Primordial Germ Cell (PGC)

1. One hour after plating the cells, add the differentiation factors combined in different treatments (DFTs) [8–10] (*see* **Note 8**):
   (a) DFT1: bone morphogenetic protein 4 (BMP4) (500 ng/ml; BioVision).
   (b) DFT2: BMP4 (500 ng/ml) and RA (5 mM; Acros Organics).
   (c) DFT3: BMP4 (500 ng/ml) and EGF (50 ng/ml; Sigma).
   (d) DFT4: BMP4 (500 ng/ml), RA (5 mM), and EGF (50 ng/ml).

These compounds should be prepared according to their manufacturer's instructions and following specific health and safety guidelines (*see* **Notes 8** and **9**).

2. Incubate cell plates at 28 °C.
3. After 24 h of in vitro cell differentiation, check PGC fluorescence. For this purpose, the transgenic line is crucial. Somatic cells should adhere and the PGCs remain in suspension. Their bright fluorescence confirms their viability.

3.4 Analysis of Migration Capacity of In Vitro Generated Primordial Germ Cell

3.4.1 Recipient Sterilization

1. Obtain the zebrafish embryos as described in Subheading 2.1. The use of *tg fvas:egfpg* transgenic line is mandatory to ensure PGC removal by the morpholino (Fig. 1).
2. To allow for maximum numbers of eggs to be produced carefully, adjust the timing of egg collection without letting them pass beyond the single-cell stage.
3. Place a microscope slide in the inverted lid of a 100-mm Petri dish. Prior to injection, transfer the embryos in the one-cell stage using a transfer pipette to line up the eggs against the side of the slide forming a single column. Remove excess egg water from the Petri dish.
4. Microinject the recipient embryos in the one-cell stage with *dnd-1* morpholino (Gene Tools) at 1 μg/μl in 0.06% phenol red solution to allow the visualization of microinjected solution.
5. Backfill the needle (*see* **Note 10**) with mineral oil with a micropipette tip. Once the needle is installed, lower and submerge the needle into the in vitro generated PGC suspension.
6. Press the "Fill button" on the Nanoliter 2000 controller to pull the sample into the needle.
7. Determine the sample volume (4.6 nl) for microinjection using the DIP switches and press the Inject button.
8. After injecting a row of eggs, rinse them with a gentle stream of system water by a transfer pipette, and transfer them into a Petri dish with clean system water.
9. Replace the egg water in the dish periodically removing dead embryos, and at the same time, record the number of injected embryos.
10. Twenty-four hours post microinjection, check microinjected *tg fvas:egfpg* transgenic line zebrafish embryos. Successful morpholino-*dead-end* microinjected embryos lost fluorescent PGCs in their genital ridges (Fig. 1).
11. Collect the selected embryos without PGCs and put them in a Petri dish with clean system water. Check and replace the medium periodically until 7 dpf.

3.4.2 In Vitro Generated PGC Transplant

1. Select the PGC-like cells obtained from the differentiation treatments under a fluorescence stereomicroscope, and pick up them using a glass micropipette needle.
2. Inject the PGCs into the abdominal cavity of each previously sterilized larva at 7 dpf using a glass micropipette needle (0.9-mm diameter borosilicate glass capillaries) in a fluorescence stereomicroscope connected to a microinjector.

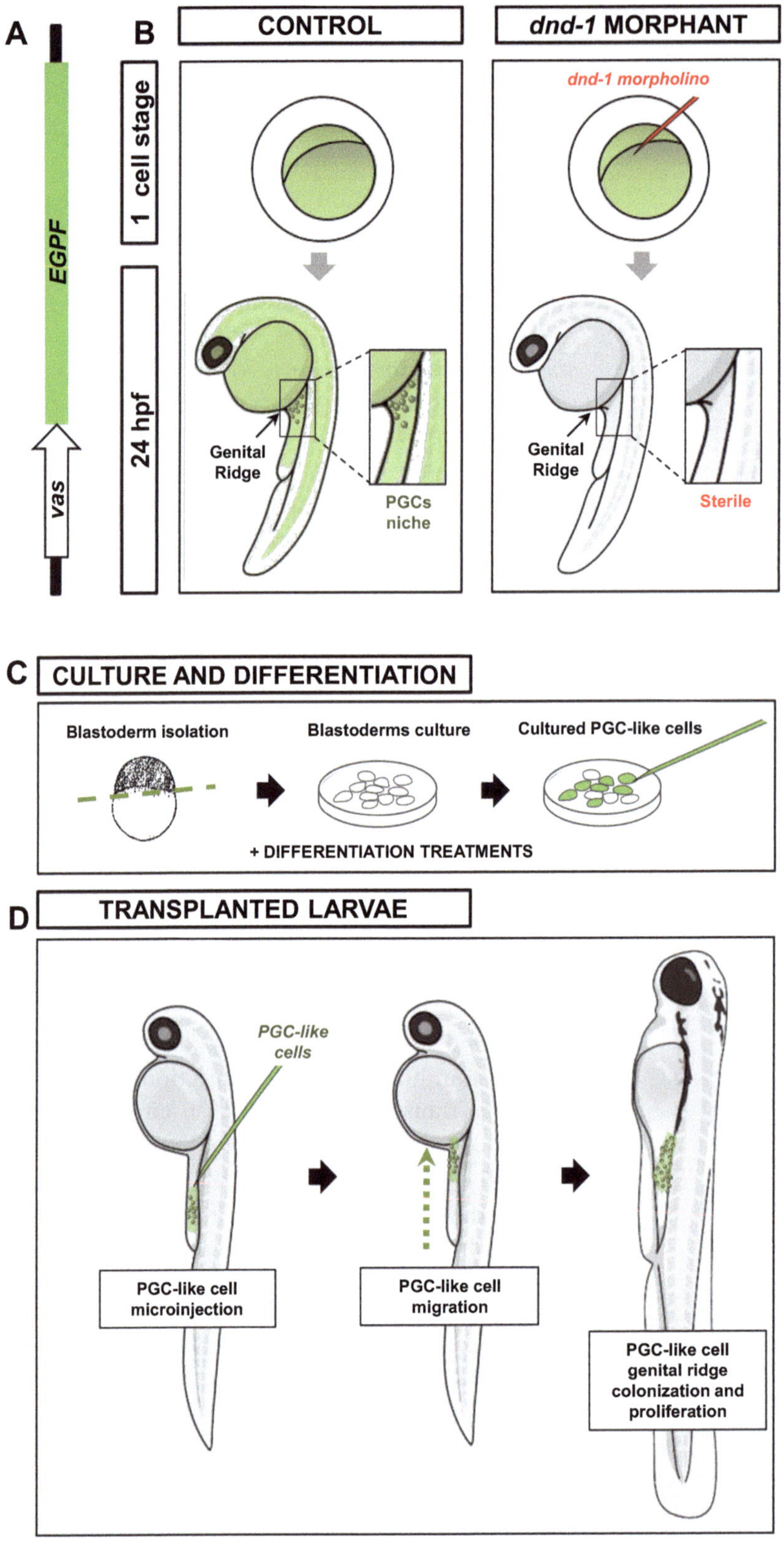

Fig. 1 Sterilization and transplant experiments in zebrafish. (**a**) Construction of zebrafish *tg fvas:egfpg* transgenic line. (**b**) Zebrafish recipient sterilization by *dead-end* morpholino microinjection in *tg fvas:egfpg* transgenic line. (**c**) Primordial germ cell (PGC) differentiation and transplant analyses. (**d**) Drawings were obtained and modified from Mind the Graph database

3. After transplantation, identify germline chimeras based on the presence and maintenance of PGC-like green fluorescent protein (GFP)-positive cells in the gonad region.
4. Confirm the ability of in vitro differentiated cells to migrate and colonize the genital ridge (Fig. 1), and take photographs of chimeric larvae with a fluorescence microscope equipped with a digital camera.
5. Examine periodically all the recipients by fluorescence microscopy (1, 2, and 3 days after transplantation).
6. Calculate the percentage of transplantation success (the number of embryos with genital ridge colonization divided by the number of embryos successfully transplanted).

4 Notes

1. The final concentration of L-15 medium is reduced from 100% to 80% by the addition of fetal bovine serum. All supplements were initially dissolved in 80% L-15 medium.
2. We do not recommend pronase treatment for embryo dechorionation at this early development stage. Manual dechorionation reduces embryo damage after a specific training and improves their survival.
3. It is important to avoid yolk contaminations (maintaining yolk integrity) and completely remove it with sterile fine forceps from the isolated blastoderms, because embryonic yolk is the major source of bacterial contamination at this stage. This step is crucial for the transient culture of blastoderm.
4. All antibiotic solutions (500 μg/ml ampicillin, 500 μg/ml streptomycin, and 500 IU/ml penicillin) should be prepared in sterile Milli-Q water and filtered using a 0.2-μm filter to avoid contaminations. We highly recommend the use of normocin for the transient culture. Normocin is an innovative antibiotic formulation designed to prevent cell lines from mycoplasma, bacterial, and fungal contaminations. We routinely add normocin to the cell culture medium in combination with penicillin/streptomycin to improve the antibacterial spectrum.
5. TCCM: this medium is slightly modified for the experiment. It was supplemented with 15% fetal bovine serum (as described previously [11] to maintain the zebrafish cell line from zebrafish blastula-stage embryos), 500 μg/ml of ampicillin, 500 μg/ml of streptomycin, and 500 IU/ml of penicillin. The concentration of L-15 medium has to be reduced from 100% to 80% to add fetal bovine serum (15%). All supplements are initially dissolved in 80% L-15.

6. Heat-inactivated serum facilitates the workflow, because it eliminates the need to aliquot and reduces the risk of contamination, while it maintains serum quality. Thaw the serum at 37 °C and mix. Transfer the serum bottle to a water bath at 56 °C for 30 min, mixing periodically every 10 min to ensure uniform heating and to avoid protein coagulation. After that, chill the serum in an ice bath.
7. Somatic cell adhesion is crucial for culture maintenance; in contrast, germ cells remain in suspension. Instead it is possible to use pretreated plates with different substrates such as Matrigel or laminin. Matrigel is a complex protein mixture required for optimal growth of cell culture, because it promotes cell adhesion and proliferation of embryonic cell cultures [12]. Laminin has been widely employed to induce neuronal differentiation, and we also used it successfully for blastoderm culture.
8. Different types, combinations, and concentrations of differentiation factors should be tested. These compounds are photosensitive and toxic (retinoic acid). Extreme caution and specific health and safety guidelines must be followed.
9. RA and BMP4 need to be reconstituted in DMSO or 20 mM acetic acid, respectively, instead of sterile water. RA stock solutions are insoluble in water and it is critical to pay attention to their maximum solubility.
10. Needle pulling, loading, and preparation: Pull a 1.0-mm OD glass capillary into two needles using a micropipette puller, and store in a 150-mm Petri dish by laying over silly putty ramps. Needles can be pulled in advance.

Acknowledgments

Authors would like to acknowledge project AGL2015 68330-C2-1-R and REPROSTRESS PID2019-108509RB-I00 (MINECO/FEDER) and PTA2016-11987-I (MINECO/FEDER) contract.

References

1. Robles V, Riesco MF, Psenicka M et al (2017) Biology of teleost primordial germ cells (PGCs) and spermatogonia: biotechnological applications. Aquaculture 472:4–20
2. Saito T, Goto-Kazeto R, Arai K et al (2008) Xenogenesis in teleost fish through generation of germ-line chimeras by single primordial germ cell transplantation. Biol Reprod 78:159–166
3. Saito T, Goto-Kazeto R, Fujimoto T et al (2010) Inter-species transplantation and migration of primordial germ cells in cyprinid fish. Int J Dev Biol 54:1479–1484
4. Lacerda SMSN, Batlouni SR, Costa GMJ et al (2010) A new and fast technique to generate offspring after germ cells transplantation in adult fish: the Nile tilapia (Oreochromis niloticus) model. PLoS One 5:e10740

5. Riesco MF, Valcarce DG, Alfonso J et al (2014) In vitro generation of zebrafish PGC-like cells. Biol Reprod 91:114
6. Krøvel AV, Olsen LC (2002) Expression of a vas::EGFP transgene in primordial germ cells of the zebrafish. Mech Dev 116:141–150
7. Sakai N (2002) Transmeiotic differentiation of zebrafish germ cells into functional sperm in culture. Development 129:3359–3365
8. Geijsen N, Horoschak M, Kim K et al (2004) Derivation of embryonic germ cells and male gametes from embryonic stem cells. Nature 427:148–154
9. Nayernia K, Nolte J, Michelmann HW et al (2006) In vitro-differentiated embryonic stem cells give rise to male gametes that can generate offspring mice. Dev Cell 11:125–132
10. Ge C, Yu M, Petitte JN et al (2009) Epidermal growth factor-induced proliferation of chicken primordial germ cells: involvement of calcium/protein kinase C and NFKB11. Biol Reprod 80:528–536
11. Christen B, Robles V, Raya M et al (2010) Regeneration and reprogramming compared. BMC Biol 8:5
12. Robles V, Martí M, Izpisua Belmonte JC (2011) Study of pluripotency markers in zebrafish embryos and transient embryonic stem cell cultures. Zebrafish 8:57–63

Chapter 8

Applying Rho Pathway Inhibitors to Investigate Germ Plasm Localization

Jerónimo Miranda and Denhí Schnabel

Abstract

The correct assembly, migration, and segregation of the mRNAs of the germ plasm during the first cell divisions are intimately connected to the cytoskeleton and cytokinesis.

RhoA is a key regulator of germ plasm localization during the first two cell division cycles in zebrafish embryos. Pharmacological inhibition of RhoA and his effector ROCK affected the correct assembly of microtubules in the cleavage furrow with the concomitant abnormal localization of germ plasm mRNAs. The inhibition of RhoA/ROCK pathway caused a significant decrease in the germ cell population later in development.

Key words Zebrafish, Cytoskeleton, Germ line, mRNA localization, Rho GTPases, RhoA, ROCK

1 Introduction

Germ plasm (GP) is a collection of specific mRNAs and proteins encoded by the maternal genome and deposited into a specific location in the oocyte (reviewed in [1]). Segregation of GP is essential during determination of the germ line in *Danio rerio* (zebrafish). After fertilization mRNAs of the GP, such as *vasa*, *nanos*, and dead end move from the egg yolk to the animal pole. Later during the first two cell divisions, GP mRNAs are recruited to the distal ends of the cleavage furrows; the mechanisms of assembly and transport of the different mRNAs are regulated in a spatiotemporal manner.

There is a clear interaction between the GP and the cytoskeleton during the first cell divisions; GP RNAs are associated with actin filaments at one-cell-stage embryos and are moved to the periphery by astral microtubules during the first cell divisions [2]. Furrow microtubules promote the compaction to the external part of the furrow [3].

Rho-GTPases are a family of small proteins that participate as molecular switches during the regulation of multiple cellular

Roland Dosch (ed.), *Germline Development in the Zebrafish: Methods and Protocols*, Methods in Molecular Biology, vol. 2218,
https://doi.org/10.1007/978-1-0716-0970-5_8,

© Springer Science+Business Media, LLC, part of Springer Nature 2021

processes, mainly by regulating cytoskeletal dynamics. Rho-GTPases cycle between an inactive state bound to GDP and an active state bound to GTP [4]. In the active state, Rho-GTPases can interact with various effectors to generate a cellular response. Rho-GTPases effectors are proteins with diverse functions such as kinases, scaffold proteins, or proteins that help cytoskeleton polymerization (reviewed by [5] and [6]).

There are 32 Rho-GTPase genes in the zebrafish genome [7]; nevertheless Rac1, Cdc42, and RhoA have been more studied, mainly because these are the Rho-GTPases that are present from yeast to chordates [8]. In particular, RhoA positively regulates cytokinesis in several cell types (reviewed by [9]). RhoA participates through mainly by three effectors, the family of the formins, citron kinase, and ROCK. Via the formins, RhoA promotes the polymerization of actin filaments [10]; the citron kinase participates in the last steps of cytokinesis [11]; and the activation of ROCK causes the phosphorylation in turn of the light chain of myosin II promoting the assembly of the contractile ring in conjunction with actin [12]. RhoA has also been implicated in the subcellular localization of various mRNAs. Disruption of RhoA interferes with β-actin mRNA localization in fibroblasts in vitro [13]; meanwhile, activation of the RhoA/ROCK/myosin II pathway regulates the targeting of mRNA to particular cell domains [14], and the RhoA/ROCK pathway actively participates in the formation of mRNA containing stress granules [15].

In zebrafish five homologous genes of RhoA have been described: *rhoaa*, *rhoab*, *rhoac*, *rhoad*, and *rhoae* [7]. It is not known if these proteins act redundantly during development. The generation of simultaneously mutant fish in all five homologues would be impractical. In these cases, the use of pharmacological inhibitors is an adequate approach (Fig. 1).

In zebrafish early embryogenesis, RhoA is distributed dynamically at the cell division furrows (Fig. 2); meanwhile RhoA effector ROCK is localized to the furrows but is confined to the distal ends, resembling GP components localization [16]. Specific inhibition of RhoA/ROCK pathway causes the incorrect localization of the germ plasm mRNAs (Fig. 2), by affecting microtubules organization at the cleavage furrow (Fig. 3) [16].

The use of inhibitors of RhoA/ROCK pathway allowed to understand the function of these proteins during the segregation of the GP with a concomitant decrease in the germ cell population later in development (Fig. 4) [16]. We anticipate that these techniques will be helpful for studying other maternally controlled processes in the early zebrafish embryo, such as cytokinesis or dorsoventral determination.

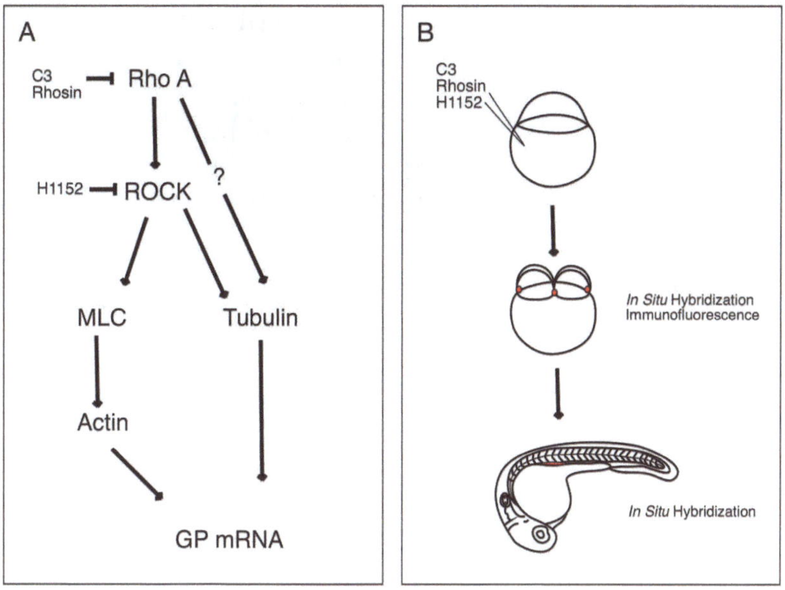

Fig. 1 Pharmacological inhibitors using in the adequate approach

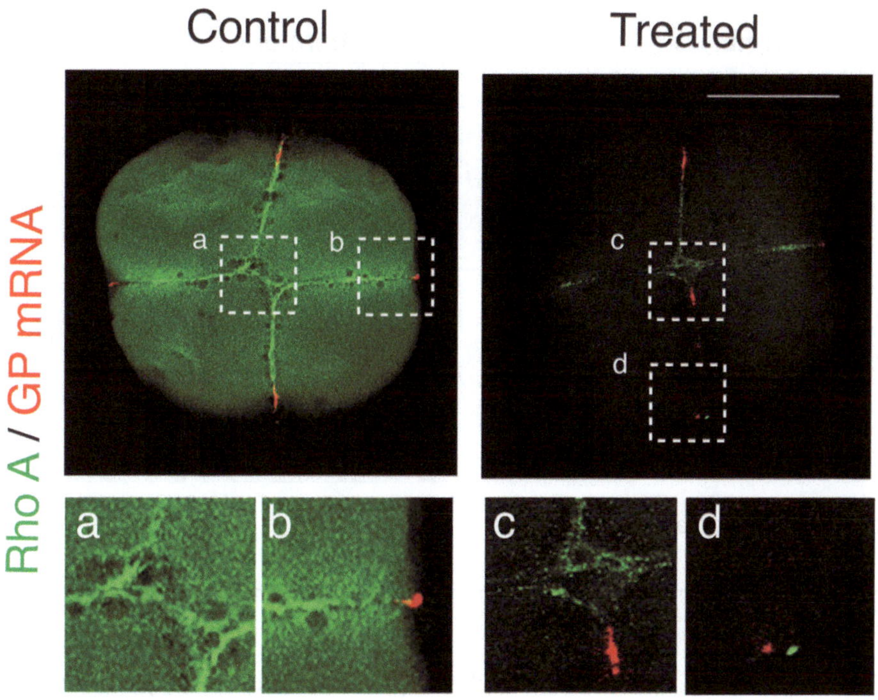

Fig. 2 Cell division furrows

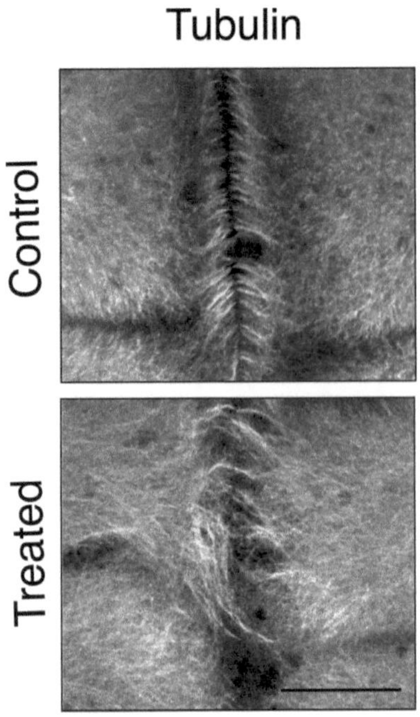

Fig. 3 Cleavage furrow

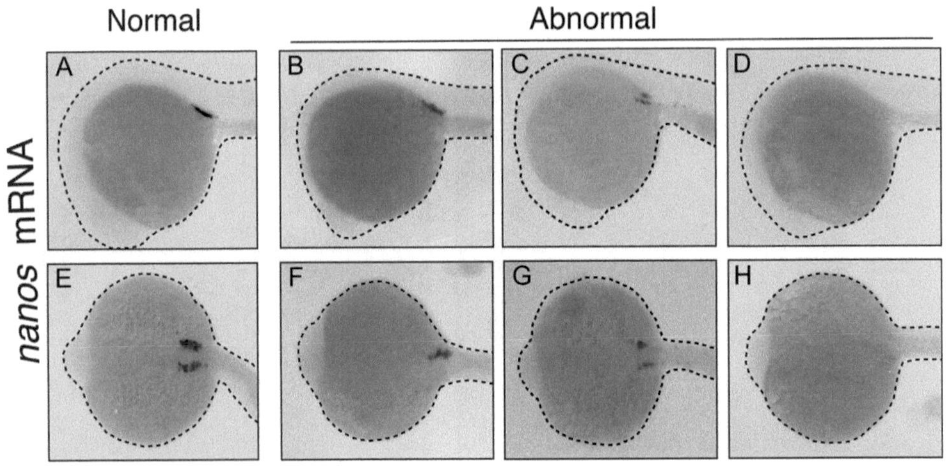

Fig. 4 Germ cell population later in development

2 Materials

2.1 Injections

1. Borosilicate glass capillaries needles for injection (1.0 mm ED × 0.78 mm ID) were pulled in a horizontal puller.
2. 4.16 μM C3 exoenzyme stock solution: Dilute 20 μg of Rho Inhibitor I (e.g., Cytoskeleton Inc.) with 200 μl of sterile water.

3. 25 mM Rhosin stock solution: Dilute 25 mg of Rhosin Rho inhibitor (e.g., Calbiochem) in 2.8 ml of DMSO.
4. 25 mM H1152 stock solution: Dilute 1 mg H-1152P dihydrochloride (e.g., Enzo Life Sciences) in 100 μl DMSO.
5. Injection buffer: 68.5 mM NaCl, 1.35 mM KCl, 5 mM Na_2HPO_4, 1 mM KH_2PO_4; sterilize by filtering.
6. Plastic transfer pipette.

2.2 Immunofluorescence Staining

1. 10× PBS (pH 7.5): 10.8 g Na_2HPO_4, 65 g NaH_2PO_4, 80 g NaCl, 2 g KCl; adjust pH 7.5 and bring to 1 l with water.
2. 4% PFA: 450 ml of H_2O is heated to 60 °C, and 20 grams of paraformaldehyde is added, under constant agitation. Add drops of 2 N NaOH until the solution gets clear. Cool down and 50 ml of 10× PBS is added. Adjust pH to 7.2 using HCl. Filter, aliquot, and store at −20 °C.
3. Tubulin fixation: 4% PFA, 0.25% glutaraldehyde, 5 mM EGTA, 0.2% Triton X-100.
4. 10% trichloroacetic acid (w/v) in distilled water.
5. PBST: 1× PBS, 1% (vol/vol) Triton X-100.
6. Fine forceps.
7. Molecular grade methanol.
8. Blocking solution: 1× PBS, 1% (vol/vol) Triton X-100, 0.1% (w/v) BSA.
9. 0.5 mg/ml $NaBH_4$ in 1× PBS.
10. 1% low-melting agarose in 1× PBS.
11. Anti-RhoA antibody, rabbit IgG (e.g., Santa Cruz Biotechnology (119, sc-179)).
12. Anti-RhoA-GTP antibody, mouse IgG2b (e.g., NewEast Biosciences (26904)).
13. Anti-ROCK 2α antibody, mouse IgG2b, IgM (e.g., AnaSpec (CT, Z-FISH; AS-55431s)).
14. Anti-phospho-myosin light chain 2 (Ser19) antibody, rabbit IgG (e.g., Cell Signaling Technology (3671)).
15. Anti-α-tubulin antibody, mouse IgG1 (e.g., Sigma (T9026)).
16. Secondary antibody anti-rabbit Alexa Fluor647 (e.g., Molecular Probes, Invitrogen (A-21244)).
17. Secondary antibody goat anti-mouse IgG Alexa Fluor488 (e.g., Molecular Probes, Invitrogen (A-11001)).
18. 0.5 mg/ml $NaBH_4$ in 1× PBS.

2.2.1 Nuclei Staining

1. 0.4 mg/ml RNase in PBS 1×.
2. 2.5 μg/ml Hoechst in blocking buffer.

2.3 In Situ Hybridization

1. Oligos for probe amplification:
   (a) *nanos3 Forward:* acaacggcgagactgagg,
   (b) *nanos3 Reverse:* cctgtagcgagcctgctg,
   (c) *vasa Forward:* gcgtgtccacctgctacc,
   (d) *vasa Reverse*: ttcatcacgggagccact,
2. Oligos used to amplify *nanos* and *vasa* produce a PCR fragment of 452 and 402 bp, respectively.
3. PBS 10×, pH 5.5: 10.8 g Na_2HPO_4, 65 g NaH_2PO_4, 80 g NaCl, 2 g KCl, dissolved in DEPC-treated water; adjust pH to 5.5; bring to 1 l with DEPC-treated water.
4. 4% PFA: 450 ml of DEPC-treated H_2O is heated to 60 °C, and 20 g of paraformaldehyde is added, under constant agitation. Add drops of 2 N NaOH until the solution gets clear. Cool down and 50 ml of 10× PBS is added. Adjust pH to 7.2 using HCl. Filter, aliquot, and store at −20 °C.
5. PBST: 1× PBS; 0.1% Tween 20 (vol/vol).
6. Fine forceps.
7. Gloves.
8. Molecular grade methanol.
9. Plastic transfer pipette.
10. 20 mg/ml proteinase K stock solution in water.
11. Hybridization solution: 50 mg/ml heparin, 500 mg/ml tRNA, 50% deionized formamide, 5× SSC, 0.1% Tween 20, pH 6 adjusted by adding 460 μl 1 M citric acid.
12. Hybridization medium (HM): 50% deionized formamide, 5× SSC, 0.1% Tween 20, pH 6 adjusted by adding 460 μl 1 M citric acid.
13. 20× SSC: 3 M NaCl, 300 mM trisodium citrate.
14. Blocking solution: 1× PBS, 2% sheep serum (vol/vol), 2 mg/ml BSA.
15. AP^-: 100 mM NaCl, 0.1% Tween 20, 100 mM Tris-HCl (pH 9.5 if the substrate is NBT, pH 8.2 if the substrate is *Fast-Red*).
16. AP^+: 50 mM $MgCl_2$ in AP^- solution.
17. Stop solution: 1× PBS, 1 mM EDTA, 0.1% Tween 20.
18. Glycerol.
19. 3% methylcellulose: Add 1.5 g of methylcellulose to 30 ml of 1× PBS in a Falcon 50 ml tube, and stir vigorously with a spatula until the methylcellulose is dissolved. Add 1× PBS until 50 ml. Allow to stand at 4 °C for more than 12 h. Make aliquots in 1 ml Eppendorf tubes and centrifuge maximum velocity in a benchtop centrifuge. Store at 4 °C until use.

20. *Fast-Red* staining solution: Dilute 250 μg/ml *Fast-Red* TR, 500 μg/ml naphthol-AS-MX-phosphate in AP^+ prepared with Tris-HCl pH 8.2 [17, 18].

3 Methods

3.1 Inhibitor Injections

Inhibition can be done by immersion in cell culture plates; nevertheless with this procedure, a lot of inhibitors should be used, and the amount that enters to each embryo cannot be controlled. On the other hand, injection of inhibitors directly in the yolk of one-cell-stage embryos allows to have equivalent effects in most embryos (Fig. 1b).

1. Zebrafish (*Danio rerio*) embryos were obtained from natural crosses and raised at 28 °C based on standard procedures [19].
2. Embryo stages were determined by morphological criteria according to Kimmel [20].
3. Zebrafish were handled in compliance with local animal welfare regulations.
4. Dilute inhibitor stock solutions to the following concentrations in injection buffer. C3, 520 nM; Rhosin, 5 mM; H-1152, 50 μM.
5. Remove the barrier separating male from female zebrafish. Set a 10-min timer as soon as the first embryos start falling into the tank floor.
6. Backload 4 μl of the diluted inhibitor solutions into a glass needle. Carefully break the tip of the needle with the help of a micromanipulator.
7. Deposit a drop of mineral oil on a micrometric ruler slide. To approximately calculate the volume of injection, introduce the tip of the needle into the mineral oil, and inject a bubble. Measure the diameter of the bubble. A bubble of 200 μm corresponds to approximately 4 nl (*see* **Note 1**). When the 10-min timer rings, collect the embryos that have lifted the chorion (*see* **Note 2**).
8. After all embryos from a groove are injected, embryos are removed with a transfer pipette and transferred to a 10 mm petri dish with embryo water.
9. Embryos are incubated at 28 °C until they reach the four- and eight-cell stage or until 24 h postfertilization. Along the incubation, embryos are monitored to remove dead embryos or embryos that do not develop properly.
10. When the embryos reach the desired step of development, the embryos are either fixed or dechorionated and fixed to follow the next steps for the immunodetection or in situ.

3.1.1 Immunofluorescence

1. Fix embryos at the desired stage with 4% PFA for 12–24 h at 4 °C.
2. Wash embryos 3 times 10 min each with PBST. Embryos should be dechorionated in this step with fine forceps.
3. Incubate embryos with methanol at −20 °C for 12 h (*see* **Note 3**).
4. Incubate embryos during 4 h in blocking solution at room temperature.
5. Incubate with primary antibody diluted 1:100 in blocking solution at 4 °C from 12 to 16 h with horizontal agitation at 100 rpm.
6. Wash embryos with blocking solution 3 times for 10 min each with horizontal agitation at room temperature.
7. Incubate for 2 h with the secondary antibody (Alexa 488 or 647) diluted 1:100 in blocking solution with horizontal agitation at room temperature avoiding the light from this moment.
8. Wash embryos 3 times with PBST for 10 min each with horizontal agitation at room temperature.
9. Mount embryos in 1% low-melting agarose in PBS.
10. Images were acquired in a confocal microscope objective 10× PlanApo numeric aperture 0.3.
11. Images were processed and analyzed using FIJI [21].

3.1.2 Tubulin Immunofluorescence

RhoA/ROCK pathway is typically associated with regulation of actin cytoskeleton but has been also implicated in regulating cell polarity and microtubules dynamics (Fig. 3). In order to detect tubulin is crucial to preserve the filaments, commonly fixation is achieved at 4 °C, but this causes the tubulin filaments quick depolymerization.

Modified from [2].

1. Embryos are dechorionated before fixation.
2. Fix embryos with tubulin fixative for 6 h at room temperature and then 12 h at 4 °C.
3. Wash with 1× PBS 10 min.
4. Incubate with $NaBH_4$/PBS for 30 min at room temperature (*see* **Note 4**).
5. Wash embryos 3 times with PBST 10 min at room temperature.
6. Incubate embryos with methanol at −20 °C for 12 h.
7. Block during 4 h in blocking solution.
8. Incubate with primary antibody anti-α-tubulin 1:500 in blocking solution at 4 °C for 12 h.

9. Wash embryos with blocking solution 3 times at room temperature.
10. Incubate for 2 h with secondary antibody Alexa 488 anti-mouse diluted 1:100 in blocking solution.
11. Wash 3 times for 10 min with PBST.
12. Embryos are mounted in 1% low-melting point agarose in PBS.
13. Images were acquired in a confocal microscope. Objective 20× and 40× numeric aperture 0.75.
14. Images were processed and analyzed using FIJI [21].

3.1.3 Active RhoA Immunofluorescence

Adapted from [22].

1. Fix embryos at the desired stage of development in trichloroacetic acid (TCA) 12 h 4 °C.
2. Wash once with 1× PBS.
3. Wash twice with 1× PBST 10 min each time.
4. Dechorionate with fine forceps.
5. Block with blocking solution for 4 h at room temperature with gentle shaking.
6. Incubate with primary antibody anti-RhoA-GTP diluted 1:100 in blocking solution at 4 °C for 12 h with gentle agitation.
7. Wash with blocking solution 3 times 10 min each at room temperature.
8. Incubate 2 h with secondary antibody diluted 1:100 in blocking solution.
9. Wash 3 times for 10 min with 1× PBST.
10. Mount embryos in 1% low-melting point agarose to observe in confocal microscope.
11. Images were acquired in a confocal microscope objective 10× numerical aperture 0.3.
12. Images were processed and analyzed using FIJI [21].

3.1.4 Nuclei Staining

(*See* **Note 5**)

1. At the end of the immunodetection, incubate embryos with RNase 1 h at room temperature.
2. Wash 3 times embryos with PBST for 10 min each with horizontal agitation at room temperature.
3. Incubate embryos with Hoechst 1 h at room temperature.
4. Wash 3 times 10 min each with 1× PBST.

3.2 In Situ Hybridization

1. Syntheses of probes for *nanos* and *vasa* were obtained by purifying total RNA from adult zebrafish ovaries. RNA was extracted from whole zebrafish ovaries. cDNA was synthetized with transcriptase reverse.
2. PCR products were obtained using the oligos described in Subheading 2, and the corresponding PCR products were cloned.
3. Digoxigenin antisense probes for *nanos* and *vasa* were synthesized using T7 and SP6 RNA polymerase; probes were purified, quantified, and stored at −20 °C.

   In situ hybridization was modified from [23].

1. Fifteen to 20 treated or control embryos were fixed with 4% PFA at the four-cell stage or 24 h postfertilization (hpf) during 12 h at 4 °C.
2. 24 hpf embryos were dechorionated before fixation.
3. Wash the embryos 3 times with PBST for 10 min each.
4. Four-cell-stage embryos were dechorionated after washed with PBST.
5. Embryos are transferred gently to glass vials with a transfer pipette.
6. Incubate embryos in 100% methanol at room temperature for 20 min.
7. Change methanol solution and incubate at −20 °C for 12 h.
8. Rehydrate embryos with serial dilutions of PBST/methanol 10 min each (PBST/methanol 25%; PBST/methanol 50%; PBST/methanol 75%).
9. Wash 4 times 5 min each with PBST.
10. Permeabilize the embryos with proteinase K for 30 s (four- and eight-cell embryos) or 10 min (24 hpf embryos) at a final concentration of 10 μg/ml (*see* **Note 6**).
11. Incubate immediately for 20 min with 4% PFA.
12. Wash 4 times with PBST for 5 min each time.
13. Add 1 ml of hybridization solution and incubate at 70 °C for 5 h under gentle agitation in rocking oven.
14. Add 500 μl of hybridization solution with 100 ng of DIG-marked probe for 12 h at 70 °C on a rocking oven.
15. Wash briefly with hybridization media (HM).
16. Replace progressively HM with 2× SSC with washes HM:SSC 2× 25%, 50%, and 75% 10 min at 70 °C each.
17. Wash with 100% 2× SSC for 10 min at 70 °C.
18. Wash twice of 30 min each with 0.2× SSC at 70 °C.

19. Replace SSC for PBST at room temperature with washes of PBST:SSC 0.2× at 25%, 50%, 75%, and 100%, 10 min each.
20. Incubate embryos for 3–4 h in blocking solution at room temperature.
21. Incubate for 12 h at 4 °C with gentle shaking with anti-digoxigenin antibody at 1:10,000.
22. Wash 6 times with PBST at room temperature 15 min each with gentle shaking.
23. Wash embryos 5 min at room temperature with AP^-.
24. Transfer embryos to a 24-well plate with transfer pipette.
25. Wash embryos 5 times 5 min each with AP^- at room temperature with rocking shaking.
26. Wash 3 times for 5 min each with AP^+ at room temperature with rocking shaking.
27. Add staining solution, and protect from light; the development of the signal is followed under a stereomicroscope until a desired stain is observed.
28. Discard the staining solution, and wash with Stop solution 3 times 15 min each time (avoid undesired background protecting embryos from light).
29. Replace Stop solution with glycerol and leave at least 12 h at room temperature with gentle agitation.
30. Change glycerol, embryos can be conserved at 4 °C.
31. To take pictures glycerol was changed with PBST and photographed in a petri dish with agarose in PBST, and to help orient the embryos, methylcellulose was used.
32. Photographs were obtained with a stereomicroscope.

3.3 Fluorescent In Situ and Immunofluorescence

In situ and immunofluorescence in the same embryo allows to determine the colocalization between the GP RNAm and RhoA. We could also observe in the septum where RhoA is not present the correlation with the mislocalization of the GP RNAm (Fig. 2).

Adapted from [24, 25]; in situ protocol is similar to the previously described, but *Fast-Red* substrate is used instead of NBT and BCIP.

Follow the same procedure of the in situ as previously described in Subheading 3.2.

1. Remove the anti-DIG antibody and wash 6 times for 15 min with PBST at room temperature with gentle agitation.
2. Wash twice with AP^- pH 8.2 for 5 min each time at room temperature.
3. Incubate embryos twice, 5 min each in AP^+ pH 8.2.
4. Incubate embryos in *Fast-Red* solution. Protect embryos from light. Under the microscope monitor the development of the red color until the desired intensity is observed.

5. Wash twice 5 min each time with AP^-.
6. Wash 3 times 10 min each time with PBST (do not use Stop solution since EDTA may interfere with the immunodetection of proteins in the cleavage furrow).
7. Fix with 4% PFA during 12 h at 4 °C.
8. Continue with the immunodetection as described previously, but the steps with methanol are avoided.

4 Notes

1. The formula for the volume of a sphere is $V = 4\pi r^3/3$. The radius of a 200 μm diameter bubble is 0.1 mm. Since 1×10^{-3} $mm^3 = 1$ nl, the volume of a 200 μm sphere is $4\pi (0.1\ mm)^3/3 \approx 4.19 \times 10^{-3}\ mm^3 = 4.19$ nl.
2. Quickness and not quantity is key, since germ plasm localization occurs very early in embryogenesis.
3. Embryos can be kept in methanol at −20 °C for months, but in practice we never wait more than a week before continuing the staining procedures.
4. Extra care is needed in this step to prevent damaging the embryos. Bubbles may attach to the embryos and they tend to stick to the walls of the tube.
5. Staining of nuclei is necessary in early stages from one- to eight-cell embryos. We observed that morphologically equivalent embryos localize the GP mRNAs differently, if the embryo has just divided or is going to divide. The number of nuclei in an embryo allows to determine more precisely the stages of the embryo, and a comparison between treated and control embryos is more accurate.
6. The 30-s proteinase K treatment is optional and even detrimental if you plan to follow with immunofluorescence or phalloidin staining. Methanol and/or detergents permeabilize four-cell embryos enough for stainings to be successful.

Acknowledgments

This work was supported by grants given to D.S. supported by P APIIT-UNAM IN200618. We thank Dulce I. Pacheco-Benítez for technical support with the animal care. Confocal microscopy service provided by Laboratorio Nacional de Microscopía Avanzada, Universidad Nacional Autónoma de México. We thank Laura S. Ramirez-Angeles for the support with figure editing.

References

1. Wylie C (1999) Germ cells. Cell 96:165–174
2. Theusch EV, Brown KJ, Pelegri F (2006) Separate pathways of RNA recruitment lead to the compartmentalization of the zebrafish germ plasm. Dev Biol 292:129–141
3. Lindeman RE, Pelegri F (2010) Vertebrate maternal-effect genes: insights into fertilization, early cleavage divisions, and germ cell determinant localization from studies in the zebrafish. Mol Reprod Dev 77:299–313
4. Etienne-Manneville S, Hall A (2002) Rho GTPases in cell biology. Nature 420:629–635
5. Aelst L, D'Souza-Schorey C (1997) Rho GTPases and signaling networks. Genes Dev 11:2295–2322
6. Bishop AL, Hall A (2000) Rho GTPases and their effector proteins. Biochem J 348:241–255
7. Salas-Vidal E, Meijer AH, Cheng X, Spaink HP (2005) Genomic annotation and expression analysis of the zebrafish Rho small GTPase family during development and bacterial infection. Genomics 86:25–37
8. Boureux A, Vignal E, Faure S, Fort P (2006) Evolution of the Rho family of Ras-like GTPases in eukaryotes. Mol Biol Evol 24:203–216
9. Piekny A, Werner M, Glotzer M (2005) Cytokinesis: welcome to the Rho zone. Trends Cell Biol 15:651–658
10. Watanabe S, Okawa K, Miki T, Sakamoto S, Morinaga T, Segawa K, Arakawa T, Kinoshita M, Ishizaki T, Narumiya S (2010) Rho and anillin-dependent control of mDia2 localization and function in cytokinesis. Mol Biol Cell 21:3193–3204
11. Naim V, Imarisio S, Di Cunto F, Gatti M, Bonaccorsi S (2004) Drosophila citron kinase is required for the final steps of cytokinesis. Mol Biol Cell 15:5053–5063
12. Matsumura F (2005) Regulation of myosin II during cytokinesis in higher eukaryotes. Trends Cell Biol 15:371–377
13. Latham VM, Yu EH, Tullio AN, Adelstein RS, Singer RH (2001) A Rho-dependent signaling pathway operating through myosin localizes beta-actin mRNA in fibroblasts. Curr Biol 11:1010–1016
14. Stuart HC, Jia Z, Messenberg A, Joshi B, Underhill TM, Moukhles H, Nabi IR (2008) Localized Rho GTPase activation regulates RNA dynamics and compartmentalization in tumor cell protrusions. J Biol Chem 283:34785–34795
15. Tsai N-P, Wei L-N (2010) RhoA/ROCK1 signaling regulates stress granules formation and apoptosis. Cell Signal 22:668
16. Miranda-Rodriguez JR, Salas-Vidal E, Lomeli H, Zurita M, Schnabel D (2017) Rho/ROCK pathway activity is essential for the correct localization of the germ mRNAs in zebrafish embryos. Dev Biol 421:27–42
17. Hauptmann G (2001) One-, two-, and three-color whole-mount in situ hybridization to Drosophila embryos. Methods 23:359–372
18. Jowett T (2001) Double in situ hybridization techniques in zebrafish. Methods 23:345–358
19. Westerfield M (1994) Zebrafish book: a guide for the laboratory use of zebrafish. Inst of NeuroScience, Eugene, OR
20. Kimmel CB, Ballard WW, Kimmel SR, Ullmann B, Schilling TF (1995) Stages of embryonic development of the zebrafish. Dev Dyn 203:253–310
21. Schindelin J, Arganda-Carreras I, Frise E, Kaynig V, Longair M, Pietzsch T, Preibisch S, Rueden C, Saalfeld S, Schmid B, Tinevez J-Y, White DJ, Hartenstein V, Eliceiri K, Tomancak P, Cardona A (2012) Fiji: an open-source platform for biological-image analysis. Nat Methods 9:676–682
22. Nishimura Y, Yonemura S (2006) Centralspindlin regulates ECT2 and RhoA accumulation at the equatorial cortex during cytokinesis. J Cell Sci 119:104–114
23. Thisse C, Thisse B (2008) High-resolution in situ hybridization to whole-mount zebra- fish embryos. Nat Protoc 3:59–69
24. Murdoch A, Jenkinson EJ, Johnson GD, Owen JJ (1990) Alkaline phosphatase-fast red, a new fluorescent label. J Immunol Methods 132:45–49
25. Lauter G, Söll I, Hauptmann G (2011) Two-color fluorescent in situ hybridization in the embryonic zebrafish brain using differential detection systems. BMC Dev Biol 11:43

Chapter 9

Cryopreservation of Pooled Sperm Samples

Jennifer L. Matthews and Zoltan M. Varga

Abstract

Cryopreservation of sperm cells is currently the most efficient tool for managing large and small collections of valuable genetic resources. Cryopreservation minimizes expenses for animal and facility maintenance such as personnel, water, power, and space. It extends the time offspring can be produced from individual organisms, reduces the need to maintain live populations, provides flexibility for planning future experiments and research projects, and can prevent catastrophic loss of irreplaceable research lines. In this chapter, we present the sperm collection, dilution, cryopreservation, thawing, and in vitro fertilization procedures used at the Zebrafish International Resource Center (ZIRC).

Key words Cryopreservation, In vitro fertilization, Protocol, Quality control, Quality assessment, Cell density, Cell motility, Zebrafish (*Danio rerio*)

1 Introduction

The number of zebrafish lines has increased considerably since several large-scale mutagenic screens generated thousands of novel genetically modified lines [1–5]. With the advent of genome editing technologies [6–8], the ability to generate even more lines in efficient and targeted ways has now accelerated dramatically. In addition, transgenesis is a well-established method for tissue-, organ-, and cell-specific observation of reporter genes [9–12], adding even more lines that are actively studied in the research community.

Frequently, more fish lines are maintained in a laboratory than can be studied, and some may have to be kept alive for months to years before investigation. However, the maintenance of live stocks requires sufficiently large facilities with space for water filtration equipment, racks, and aquaria. Aquatic life support systems require routine cleaning, maintenance, and repair, and animals require appropriate feeding, health monitoring, water conditioning, and filtration with the ensuing expenses for electricity, chemicals, water,

Roland Dosch (ed.), *Germline Development in the Zebrafish: Methods and Protocols*, Methods in Molecular Biology, vol. 2218, https://doi.org/10.1007/978-1-0716-0970-5_9, © Springer Science+Business Media, LLC, part of Springer Nature 2021

and personnel. Indeed, personnel represents the largest expense to ensure animals are well cared for [13, 14].

Cryopreservation of sperm cells is still the most cost-effective method to manage fish strains, even though there are some upfront investments for equipment such as liquid nitrogen (LN_2) freezers, bulk LN_2 storage tanks, racks, and safety equipment [15]. There may also be construction costs associated with running vacuum-jacketed pipelines and stainless-steel manifolds, if several freezers are used. Additionally, a time investment is necessary to train personnel so that they become proficient with cryopreservation methods. Then, sperm has to be obtained and several samples frozen for each individual fish line. However, these initial investments for cryogenic storage, training, and line preservation are rather negligible compared to the long-term cost savings cryopreserved samples generate when compared with live husbandry expenses [15]. LN_2 costs are currently between $0.3 and $1.5 per liter, depending on quantity, container type, and the delivery distance between laboratory and supplier. In the past year, for example, ZIRC spent $11,617 on LN_2 at an average of $0.31/L, including delivery fees. ZIRC houses currently 89,834 cryogenic samples which contain 12,196 lines or 45,679 alleles. The current annual expense for liquid nitrogen is therefore $0.95 per fish line, $0.25 per allele, or $0.13 per sample.

The process and effort of reactivating a cryopreserved line is also minimal. Obtaining eggs from females is a straightforward, well-established process in virtually all zebrafish laboratories [16]. Thawing of sperm samples and in vitro fertilization takes but a few minutes as we will show below. In sum, the long-term cost and overall effort for frozen lines is far lower than live fish maintenance, and the short-term effort for preserving and recovering lines is negligible.

At ZIRC, cryopreservation and cryobanking are *the* resource management tools for maintaining and distributing the ever-growing number of zebrafish lines [17]. In this chapter, we present our current cryopreservation and recovery processes, which include recent modifications that further refine our previously described method [18]. This chapter provides a step-by-step guide for (1) obtaining pooled sperm by stripping males, (2) determining sperm cell density with a spectrophotometer and diluting sperm for optimal storage and recovery, (3) freezing of samples, (4) thawing of frozen samples, and (5) in vitro fertilization of eggs. Together, these methods will provide laboratories with the ability to better plan their research programs and save funds. Once novel lines have been generated or acquired, they can be cryopreserved and backed up for later reactivation as needed.

2 Materials

2.1 Sperm Collection

1. Dissecting microscope with incident lighting.
2. 10 μL calibrated microcapillary pipettes (Drummond #2-000-010).
3. Aspirator tube assembly (included in each pack of microcapillary pipettes).
4. Millipore forceps with rubber tips (Millipore #XX6200006P), rubber tips made from heat-shrink tubing.
5. Plastic slotted spoon for moving fish out of anesthesia.
6. Soft and absorbent paper towels.
7. Clear and yellow 0.6 mL microcentrifuge tubes.
8. Sponge fish holder (in 35 × 10 mm petri dish).
9. Drawer/shelf anti-slip liner cut into a ~2 in. square.
10. Adjustable micropipettes: 200, 100, 20, 10, and 2 μL size.
11. Tricaine (MS-222) stock solution: 4 g/L in dH_2O; adjust to ~7.0 pH with Tris–HCL, pH 9.0.
12. Tricaine pre-anesthesia (optional for males, 48 mg/L, 12 mL tricaine stock solution in 1000 mL fish water).
13. Tricaine anesthesia (168 mg/L, 4.2 mL tricaine stock solution in 100 mL fish water).
14. Isotonic PBS fish rinse: Phosphate-buffered saline, pH 7.4, powder packets (Sigma #P3813), dissolved in 870 mL dH_2O, final osmolality of approximately 315–325 mmol/kg.
15. Fish water for recovery.
16. *E400G sperm extender:* To make 1 L of E400G, combine in 800 mL dH_2O: 9.70 g KCl, 2.92 g NaCl, 2.0 mL 1.0 M $CaCl_2$ (or 0.29 g $CaCl_2 \cdot 2H_2O$), 1.0 mL 1.0 M $MgSO_4$ (or 0.25 g $MgSO_4 \cdot 7H_2O$), 1.8 g D-(+)-glucose, 7.15 g HEPES, and 1 g gelatin from cold water fish skin (Sigma #G7041):
    (a) Add dry components first, stir to dissolve, then add liquid components, and stir.
    (b) Adjust pH to 7.9 with 5 M KOH.
    (c) Bring final volume to 1000 mL with dH_2O.
    (d) Check osmolality (should be close to 400 mmol/kg).
    (e) Filter-sterilize.
    (f) Store at 4 °C.

    The resulting solution contains 130 mM KCl, 50 mM NaCl, 2 mM $CaCl_2$, 1 mM $MgSO_4$, 10 mM D-(+)-glucose, 30 mM HEPES-KOH (pH 7.9), and 0.1% w/v gelatin. The osmolality of the resulting solution should be close to ~400 mmol/kg.

2.2 Cryopreservation

1. Cryogenic vials (0.5 mL Matrix Screw Top Storage Tubes, Thermo Scientific, Item #3745-BR, or 2 mL Corning vials, Item #430488 (*see* Freeze Rate, Subheading 2.3)).
2. Vial color coders (colored caps or cap inserts specific for vial type).
3. Micropipetters and tips.
4. 15 mL conical tubes (Falcon 352096) with cryovial spacers (*see* Freeze Rate, Subheading 2.3).
5. Styrofoam container or cooler (≈12″ × 12″) for dry ice.
6. Styrofoam container for liquid nitrogen (LN_2) tray.
7. Fiberglass tray for LN_2.
8. Cryovial storage box (specific for vial make and type).
9. Liquid nitrogen Dewar flask.
10. Waterproof Cryo-Gloves.
11. E400G sperm extender.
12. *RMMB cryoprotective medium* (100 mL) (follow this order):
    (a) Combine 20.0 g D-(+)-Raffinose pentahydrate (Sigma R7630 or 83400) and 70 mL dH_2O in a 250 mL beaker.
    (b) Place beaker in an evaporating dish (Pyrex 3140) or large beaker containing hot water (~70 °C) on a stir plate.
    (c) Stir mixture until Raffinose is completely dissolved.
    (d) Add 2.5 g skim milk (Difco #232100), and stir until completely dissolved.
    (e) Cool to room temperature.
    (f) Add 3 mL 1 M Bicine-NaOH (8.0).
    (g) Add 6.67 mL absolute methanol (Acetone-Free, Absolute, Certified ACS Reagent Grade).
    (h) Transfer to 100 mL volumetric flask, adjust final volume to 100 mL with dH_2O, and mix by inversion 3–4×.
    (i) Transfer to two 50 mL conical tubes.
    (j) Centrifuge at 15,000 × *g* for 20 min at 25 °C.
    (k) Transfer cleared supernatant into clean beaker.
    (l) Aliquot into 1.5 mL microfuge tubes, 1 mL each or other convenient volume for daily use.
    (m) Store frozen at −20 or −80 °C until use.

    The resulting solution will contain 20% (w/v) D-(+)-Raffinose pentahydrate, 2.5% (w/v) Difco Skim Milk, 6.67% (v/v) methanol, and 30 mM Bicine-NaOH.

2.3 Thawing and IVF

1. Plastic slotted spoon for transferring fish.
2. 35 or 60 mm Petri dishes.
3. White Taklon round paint brush, size 2.
4. Two micropipetters 200 μL and tips.
5. Water bath at 38 °C.
6. Timer: 2-min countdown.
7. Embryo medium or fish water.
8. Tricaine (MS-222) pre-anesthesia (48 mg/L).
9. Tricaine anesthesia (168 mg/L).
10. Isotonic PBS rinse.
11. Recovery fish water.
12. Sperm solution SS300 (for thawing samples that were frozen without milk, add 2 mg/mL Difco Skim Milk to the SS300; *see* below).
13. dH_2O.
14. *Sperm solution SS300* (1 L): Combine in 800 mL dH_2O: 8.2 g NaCl, 5 mL 1 M KCl (or 0.37 g KCL), 1 mL 1 M $CaCl_2$ (or 0.15 g $CaCl_2$-$2H_2O$), 1 mL 1 M $MgSO_4$ (or 0.25 g $MgSO_4 \cdot 7H_2O$), 1.8 g D-(+)-glucose, and 20 mL 1 M Tris-Cl (pH 8.0):
    (a) Add dry ingredients first, and stir to dissolve.
    (b) Add liquid ingredients, and stir to mix.
    (c) Bring final volume to 1000 mL with dH_2O.
    (d) Check osmolality (should be close to 300 mmol/kg).
    (e) Filter-sterilize.
    (f) Store at 4 °C.

    The resulting solution will contain 140 mM NaCl, 5 mM KCl, 1 mM $CaCl_2$, 1 mM $MgSO_4$, 10 mM D-(+)-glucose, and 20 mM Tris-Cl (8.0). The osmolality of the resulting solution should be close to ~300 mmol/kg.
15. *Sperm solution SS300 with 2 mg/mL Difco Skim Milk* (SS300 + Milk; *see* **Note 1**):
    (a) Add 100 mg Difco Skim Milk to 50 mL SS300, and stir or vortex to dissolve.
    (b) Aliquot into microcentrifuge tubes and store frozen at −20 °C.
    (c) Thaw and use at room temperature.

3 Methods

3.1 Collecting and Pooling of Sperm

1. Prepare a 0.5 mL microcentrifuge tube with E400G for each fish line (*see* **Note 2** for E400G starting volume considerations). Use a separate microcentrifuge sperm collection tube for each fish line.
2. Keep E400G and the collected sperm at room temperature throughout the collection and pre-freezing procedures.
3. Sedate or pre-anesthetize males (48 mg/L MS-222 in fish water) as needed for at least 10 min before the procedure (*see* **Note 3**).
4. Anesthetize two to three males in tricaine solution (168 mg/L MS-222 in fish water).
5. Briefly rinse a male in PBS isotonic fish rinse, dry it by rolling on a paper towel (*see* **Note 4**), and place it belly-up in a dampened sponge holder (Fig. 1a).
6. Place the end of a borosilicate glass microcapillary on the urogenital opening. Use rubber-tipped Millipore forceps to apply *gentle* abdominal pressure to the sides of the male (Fig. 1b). Collect sperm into the microcapillary.
7. Expel sperm immediately into the E400G in the microcentrifuge collection tube.
8. Transfer fish into fresh system water for recovery from anesthesia.
9. Continue collecting sperm from all males from the same family/stock, and pool into the same E400G microcentrifuge tube. Repeat **steps 4–8** for each male.

3.2 Estimation of Sperm Cell Densities and Sample Dilution

At ZIRC, we use a NanoDrop® 2000 Spectrophotometer to determine cell density by light absorption at 400 nm (*see* **Note 5**). The worksheet we use to determine NanoDrop® cell counts is available in the Cryopreservation and In Vitro Fertilization (IVF) section on the ZIRC site: https://zebrafish.org/wiki/protocols/start (*see* **Note 6**).

1. Estimate the volume of pooled sperm in E400G in the microcentrifuge tube for each line using a Pipetman. Draw sperm into the tip and adjust the pipette volume until all of the solution just fills the tip. Record the estimated volume for later use. While measuring, gently pipet the sperm to mix completely.
2. Prepare a 1:10 dilution of the sperm solution for density measurement. If the collected sample appears less opaque, i.e., less concentrated, a 1:5 dilution can be used. Pipette 9 μL (or 4 μL

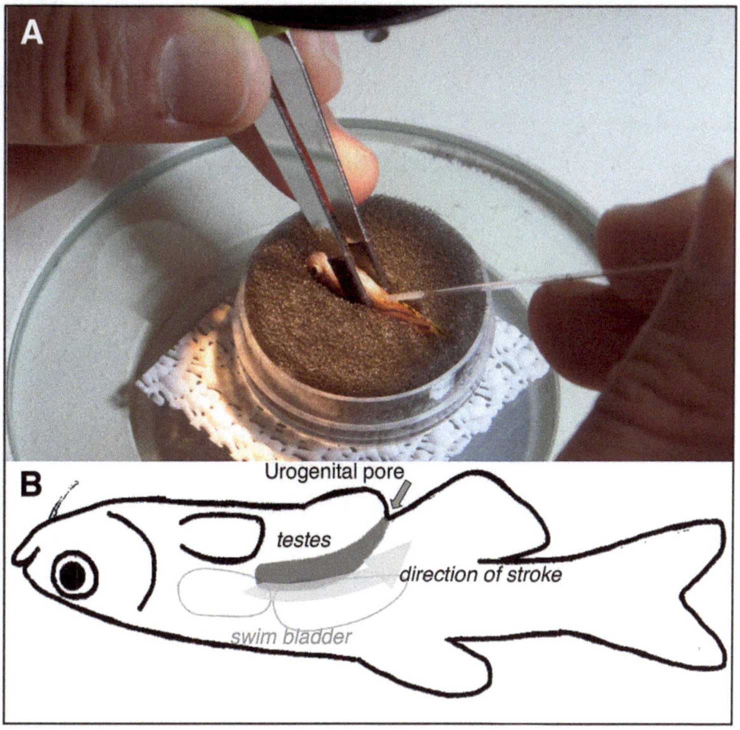

Fig. 1 Harvesting of zebrafish milt. (**a**) Male is placed belly-up in a PBS-dampened sponge slit in a 35 mm petri-dish on rubber support to prevent slipping. (**b**) The approximate location of testes (dark gray) between the swim bladder (grey line) and gut. The light gray arrow indicates the approximate placement and movement of the rubber-tipped forceps toward the urogenital pore

for a 1:5 dilution) E400G into a 0.6 mL microcentrifuge tube, and add 1 μL of the sperm suspension.

3. Mix by flicking the tube and hold at room temperature. Optional: Use separate (colored) microcentrifuge tubes for each line (*see* **Note** 7).
4. Calibrate A_{400} absorption of a blank E400G sample using the NanoDrop's *Cell Cultures* menu:
   (a) Set the absorbance cursor to 400 nm.
   (b) Add 1.5 μL E400G to the spectrophotometer.
   (c) Make sure the data is logged and displayed in a useful manner for later use.
   (d) Read the E400G blank sample to verify it has set the instrument's readings close to 0.
   (e) If a blank sample reads higher than 0.005, recalibrate with fresh 1.5 μL of E400G.

(f) When prompted, save the NanoDrop data, with the date of the freeze event and line name(s).

5. To measure the absorption of a sample, enter the sample ID on the measure cell cultures page.
6. Mix the sample well using a vortex mixer set at intermediate speed (~1300 RPM).
7. Immediately load 1.5 μL of the diluted sperm, and read (select the measure button in the left-hand corner) the cursor absorbance (A_{OD400}):

   (a) Any reading above 0.3 is acceptable and you can proceed with the dilution.

   (b) Any reading at or below 0.2 should be diluted 1:5 and repeated.

8. Measure at least 3, up to 5 times, and calculate the average A_{400} for all samples. Occasional errant readings can be disregarded; however a minimum of three successful readings is optimal.
9. Clean the NanoDrop using a clean Kimwipe and deionized water. Wipe the top arm (mirror) and the bottom lens with a moistened Kimwipe, and then dry completely with a Kimwipe before closing the arm.
10. Calculate the cell density for each sample with the averaged A_{400}, using the Excel cell density calculator.
11. Dilute the sperm with E400G according to the desired number of samples or sperm concentration. Sperm cell densities should be between 4.0×10^8 and 1.6×10^9 cells/mL. This will result in samples with 2.0×10^6–8.0×10^6 cells/sample (*see* **Notes 8–10**).

3.3 Pooled Sperm Freezing

1. Prepare a cooler filled with powdered dry ice made from liquid CO_2 [19] (*see* **Note 11**).
2. Arrange 15 mL Falcon tubes each with an empty Matrix cryovial tube functioning as a spacer and Falcon tube caps so that they are ready to hold samples immediately after the vials have been capped (Fig. 2a (1)).
3. Prepare labeled sample cryovials as needed prior to freezing (all samples should be labeled with the freeze date and allele number and/or line identification number, Fig. 2a (2)).
4. If you are using a multi-capper, prepare it with the appropriate caps already attached before adding cryoprotectant (RMMB) to sperm (Fig. 2a (3)).
5. For each sample, determine the volume of RMMB (RMMB volume = 3 × the sperm volume; *see* **Note 12**).
6. Add RMMB to sperm and mix by pipetting.
7. Aliquot 20 μL immediately into pre-labeled 0.5 mL cryovials.

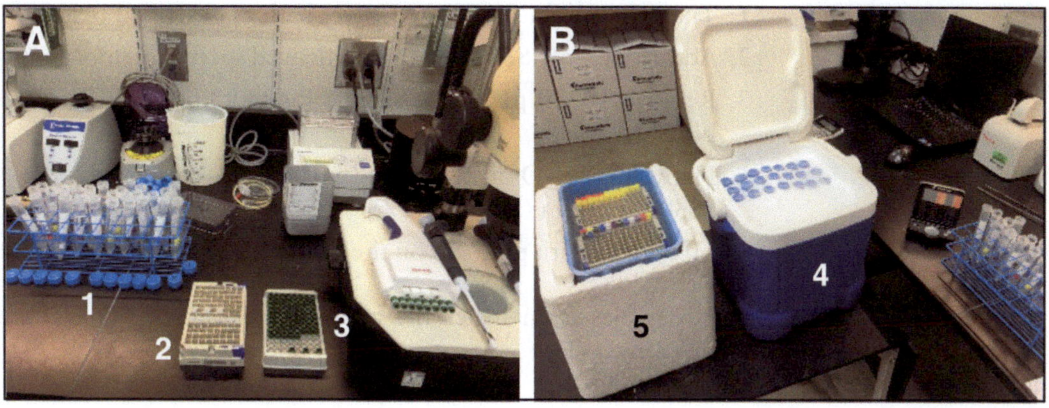

Fig. 2 Preparation of the cryopreservation workplace. (**a**) Falcon tubes with spacer inserts (1), empty Matrix cryopreservation vials in cryogenic boxes (2), and caps (or loaded multi-capper, if available, (3) are prepared. (**b**) After aliquoting 20 μL of sperm sample with RMMB into cryovials, cryovials are capped and placed on top of spacer vials in 15 mL Falcon tubes (1). Falcon tubes are pushed into powdered dry ice (4) for at least 20 min (panel **b**, right). After cryopreservation, samples are transferred to liquid-nitrogen pre-cooled cryoboxes (5, panel **b**, left) and then placed in racks into vapor- or liquid-phase cryogenic nitrogen freezers

8. Without delay, cap cryovials (using the automated capper for rows of eight tubes, if available Fig. 2a (3)), and place into the 15 mL conical tubes (containing a Matrix cryovial spacer, Fig. 2a (1)).
9. Cap the conical tubes, and drive tubes into the dry ice until caps are flush with the surface (*see* Fig. 2b (4) and **Notes 13** and **14**).
10. Freeze samples in dry ice for 20–50 min, and then quickly transfer samples to a cryogenic freezer box submerged in LN_2 (Fig. 2b (5)).

3.4 Sperm Motility Assessment

The best way to assess the quality of sperm is to observe its motility. Computer-assisted sperm analysis (CASA) software systems provide the most objective and comprehensive quantification of density and motility parameters, but a manual, subjective assessment is sufficient for most sperm freezing applications. A compound microscope is all that is needed. A 20× objective and DIC or dark field illumination is ideal. Osmolality affects both the speed and duration of sperm motility. Fresh (pre-freeze) sperm will be faster and have a higher percentage of cells motile than post-thaw samples.

3.4.1 Pre-freeze Motility

1. Place 6 μL dH_2O on a microscope slide.
2. Add 0.5–1 μL of the final sperm dilution to the drop, mix, and spread quickly with the pipet tip.
3. Observe immediately (*see* **Note 15**).

3.4.2 Post-thaw Sperm Motility

1. Thaw the sperm sample in a water bath as described below.
2. Add 150 μL of SS300 solution to thawed sperm and mix gently.
3. Remove 10–20 μL of the sample for motility assessment (*see* **Note 16**).
4. Place 5 μL dH_2O on a slide, and add 4.25 μL of thawed sperm/SS300 solution (*see* **Note 17**).
5. Mix briefly with pipet tip on the slide and observe immediately.

3.5 Obtaining Eggs, Thawing Sperm, and In Vitro Fertilization (IVF) Procedure

Isolate females from males the afternoon before egg collection, thawing, and IVF.

3.5.1 Egg Collection

1. Place females in a tank with pre-anesthesia solution at least 10 min before full anesthesia. However, fish can be held in pre-anesthesia solution for several hours (*see* **Note 18**).
2. Anesthetize females in tricaine/MS-222 solution.
3. Rinse fish in isotonic PBS and blot dry by gently rolling on paper towel (Fig. 3a).
4. Place fish on its side in a small petri dish (35 or 60 mm).
5. Dampen fingers in PBS fish rinse.
6. Strip eggs by applying gentle (light) finger pressure on the ventral abdomen, and move your finger from anterior to posterior. Eggs will be expelled readily if the female is ready (*see* **Note 19**; Fig. 3b).
7. Transfer the female to a recovery tank.
8. Combine several clutches of eggs if needed by gently moving eggs to another dish with a fine-tipped paint brush dampened in isotonic PBS (*see* **Note 20**, Fig. 3c, d).
9. Maintain pooled eggs in a closed dish in a moisture chamber at room temperature no longer than 5–10 min before in vitro fertilization.

3.5.2 Sperm Sample Thawing

1. Remove the cryovial from the LN_2 and quickly open cap to vent any LN_2 in the vial (*see* **Note 21**).
2. Thaw cryovial in a 38 °C water bath until the frozen pellet is less than 3 mm in diameter (takes approximately 10–15 s, Fig. 4a).
3. Add 150 μL room-temperature SS300 solution to the cryovial (Fig. 4b). If you are thawing sperm that was frozen without milk, add 2 mg/mL Difco Skim Milk (Difco #232100) to the SS300.

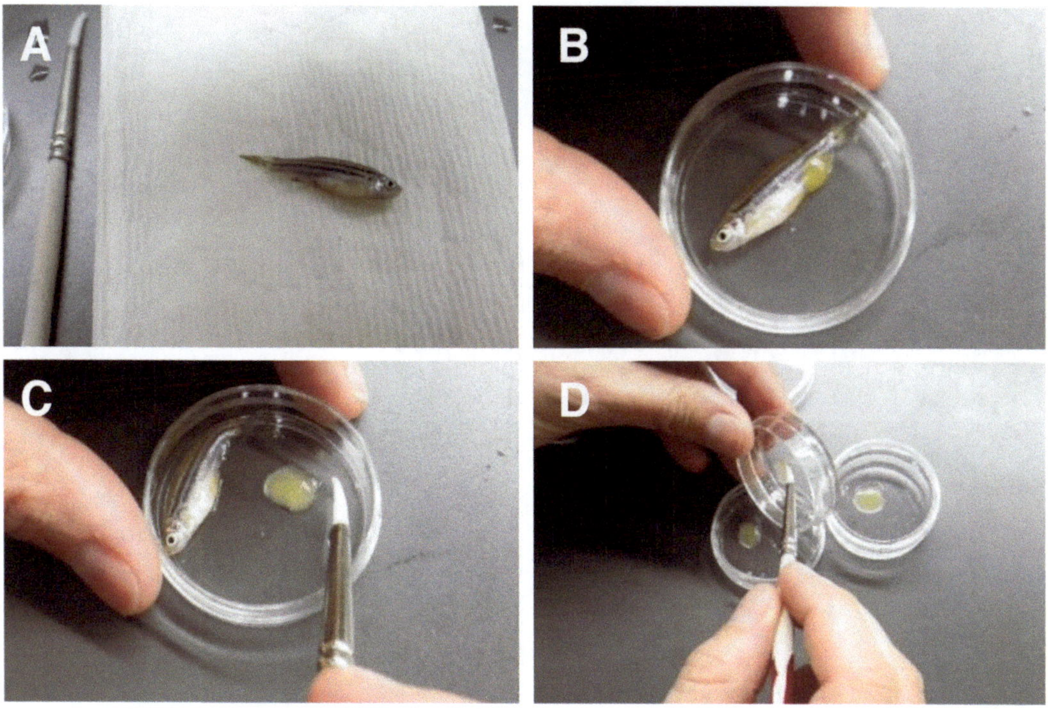

Fig. 3 Squeezing and manipulating eggs. (**a**) Anesthetized female is gently blotted on paper towel after immersion in isotonic PBS. (**b**) Good-quality eggs are uniformly yellowish and somewhat sticky. Bad-quality eggs are whitish and watery. Sometimes a few white eggs are mixed with good ones; the batch might still be usable. (**c**) To manipulate eggs without activating them, use a fine paint brush dipped into isotonic PBS. (**d**) Several clutches of eggs can be pooled with a PBS-moistened paint brush. Avoid metal objects such as spatulas to manipulate eggs!

4. *Optional note*: For post-thaw motility assessment, a small portion (10–20 μL) of the sperm/SS300 mix can be removed and held in a microcentrifuge tube at RT if assessed immediately, or on ice if stored for later (*see* Subheading 3.4).

3.5.3 In Vitro Fertilization

1. Add 200 μL dH_2O to the cryovial to activate the sperm. Gently mix sperm 1–2 times with a micropipetter, and *transfer into* eggs: Slide pipet tip sideways along the bottom of the petri dish into the pile of eggs and to the center. Expel the activated sperm into the mass of eggs (not on top of the eggs).
2. Start a 2-min countdown timer.
3. *Do not* move, mix, or swirl the dish—let it sit completely undisturbed for 2 min (Fig. 4c), and then flood the dish with embryo medium (Fig. 4d).
4. Determine the fertilization rate 2–4 h postfertilization, or as soon as cell divisions are clearly recognizable. Count embryos and remove unfertilized eggs (*see* **Note 22**).

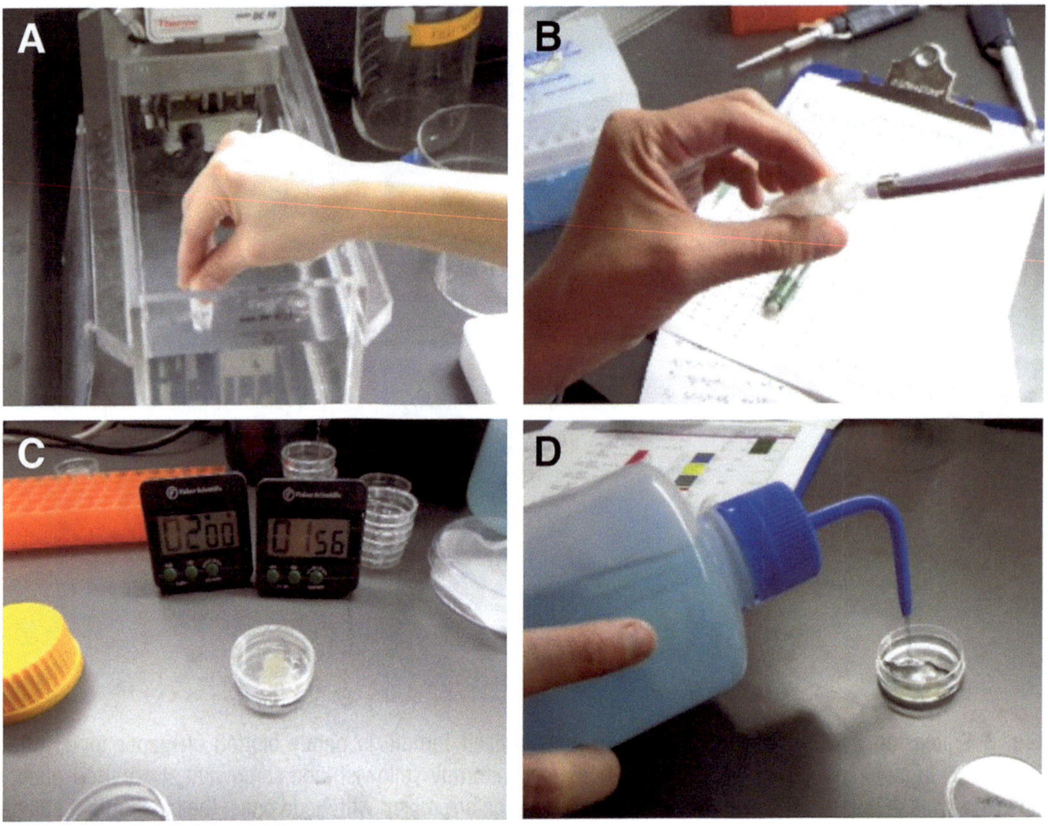

Fig. 4 Sample thawing and in vitro fertilization. (**a**) Thaw sperm samples at 38 °C for 10–15 s or until the pellet is ~3 mm diameter. (**b**) Addition of 150 μL SS300 to vial followed by 200 μL dH_2O and mixing of activated sperm by gently pipetting 1–2 times. (**c**) In a dish prepared with fresh-squeezed eggs, the tip of the pipette is placed next to eggs and is moved sideways into the center of the pile, along the bottom of the dish. Sperm is released into the pile of eggs. The dish is closed and left undisturbed (no mixing!) for 2 min. (**d**) After 2 min, the dish is flooded with embryo medium

4 Notes

1. SS300+ Milk is used for thawing sperm samples that were frozen with a cryopreservation medium not containing skim milk powder (other protocols). The milk helps prevent sperm tails from sticking and tangling.
2. A few general guidelines for the starting volume of E400G into which sperm is collected. This will ensure that the sperm concentration is as optimal as possible in the resulting samples without activating the cells. The starting volume of E400G, depends on:
   (a) The number of available males (short-fin fish typically give more sperm than long-fin fish).
   (b) The number of desired samples.

(c) To calculate a conservative starting volume of E400G:
   - Short-fin fish: (# males − 2) × 5 μL.
   - Long-fin fish: (# males − 4) × 5 μL.

3. Pre-anesthesia/sedation of males is recommended for lines sensitive to bleeding from the gills during anesthesia. Males should be placed in pre-anesthesia tricaine solution (48 mg/L MS-222 in fish water) for 10–30 min prior to anesthesia.
4. Eggs and sperm are activated by water and it is virtually impossible to dry fish completely. Therefore, it is important to rinse anesthetized males and females with isotonic PBS solution before gamete collection to avoid premature activation of sperm or unfertilized eggs.
5. The density of sperm cells *in extender and cryoprotective medium* plays a crucial role for optimal cryopreservation, thawing, and reactivation. While satisfying results can be achieved without cell density estimation and adjustment, we strongly recommend measuring concentrations prior to freezing to obtain the best possible results for recovery. Too few cells (or too many) present in relation to the available solution ingredients will not freeze as well as a well-adjusted cell density. There are two key benefits for adjusting cell densities: (a) Further dilution of sperm (if possible) generates more samples and generates a longer lasting resource. (b) Optimal equilibrium of cells and available salt and buffer molecules ensures that solution toxicity and osmotic shock are reduced, and as a result cells suffer less cryogenic-induced damage during freezing and thawing.
6. This calculator helps to estimate and adjust cell densities in most cases. However, because several variables can influence the readout, it is recommended to develop a calibration curve and formula like this with your own equipment. Download the complete ZIRC cryopreservation and IVF protocol (PDF) and Excel sperm density calculator files from the ZIRC website: https://zebrafish.org/wiki/protocols/start.

   This worksheet/calculator has been generated with ZIRC equipment, and it should generate results in the correct order of magnitude. It may or may not be necessary to calibrate a new A_{400} absorption curve for your own spectrophotometer for a more accurate conversion. To generate a curve specific for your equipment, use a hemocytometer or Makler chamber to determine cell densities microscopically. Determine the A_{400} of a corresponding dilution series with your spectrophotometer, and generate a curve from the dilution series. Calculate a best-fit curve (or line) of your absorption data, and use the formula in a worksheet to determine the cell density that corresponds to the A_{400} of unknown cell densities [20].

7. Using a designated colored microcentrifuge tube for the NanoDrop dilution makes it easily distinguishable from samples of other stocks.
8. If your goal is to achieve a particular number of samples, you can dilute the sample as far as 4.0×10^8 cells/mL (2.0×10^6 cells/sample). If your priority is to ensure that a line is reactivated with only a single thaw, you may want to leave the concentration higher, and add less E400G for a lower total volume. Cell densities should not be lower than 4.0×10^8 cells/mL for best results.
9. If more than 12 samples are going to be frozen, divide the final sperm volume into 2 aliquots. In addition, 0.5–1 μL of the final diluted sperm suspension can be removed for an optional pre-freeze motility assay (*see* Subheading 3.4).
10. Computer-assisted sperm analysis (CASA) or microscopic visualization can be used before and/or after freezing samples to assess cell sperm cell motility. When complete absence of motile sperm cells is observed, it indicates that cells may have been irreversibly damaged. In this case, in vitro fertilization can be used to determine whether a pool of samples fertilizes at all. Sometimes, low fertilization rates can be still obtained sometimes from low-motility/nonmotile samples.
11. A cooling rate between 10 and 15 °C/min was determined as the optimal range of cooling for the materials and solutions used in this protocol [18]. Using powdered dry ice with stacked cryovials in a 15 mL Falcon tube will achieve a cooling rate of 10–15 °C/min. If possible avoid changing specified materials in this protocol. If the specified materials are not available (e.g., the 15 mL vial or cryovials with the same wall material and thickness as specified here), you need to retest which freezing rate provides optimal post-thaw fertilization rates with the replacements. You may also have to adjust the thawing protocol to match the new freezing rate. The description for making powdered dry ice is included in the online protocol: zebrafish.org/zirc/documents/protocols/pdf/Cryopreservation_IVF/ZIRCCryo&%20IVF.pdf
12. Before use, thaw RMMB aliquots in a water bath or heating block at 45–50 °C. Raffinose precipitates if the RMMB solution is kept on ice. If this occurs, heat solution slightly to get it back into solution prior to use. Cool to room temp before mixing with sperm. Keep diluted sperm and RMMB at room temperature.
13. Our research suggests that cell viability is not affected for up to 5 min in cryoprotective medium before freezing. However, because the RMMB cryoprotective medium is toxic to sperm cells at longer holding times, **steps 6–9** should be carried out

swiftly, with as little delay as possible, once the RMMB is mixed with the sperm.

14. If more than 12 samples are to be frozen, divide the final sperm volume into 2 smaller aliquots, and freeze them in 2 rounds (**steps 6–9**) to ensure rapid transition of the mix into dry ice after mixing with the RMMB cryoprotective medium.
15. Examining the remainder of the NanoDrop dilution, if performed, is a good use of the sperm. Checking pre-freeze motility confirms the viability and concentration of the sperm being frozen.
16. At this point, a small portion (10–20 μL) of the sample can be removed and held on ice for motility assessment. The remainder of the sample can be used for IVF as described above. It is best to view motility as soon as possible after thawing, but samples are typically stable on ice for 10–20 min.
17. For post-thaw motility, activating the sperm on a slide in the same relative proportions as in the IVF procedure gives a consistent method and provides a good sense of sperm concentration and motility as it is applied to the eggs.
18. Egg collection and in vitro fertilization (IVF) of eggs should be performed early in the morning. Egg quality will be optimal the first few hours after the lights have been turned on. Resorption of eggs begins 2–3 h after daybreak.
19. Good eggs will be golden in color and have little fluid and there should be no opaque or white eggs intermixed (Fig. 3b–d). Eggs can be moved away from the fish using a fine paint brush (Fig. 3c, d).
20. If females do not provide large enough clutches or in order to maximize the number of embryos produced by IVF, it is helpful to combine several clutches of eggs. A fine paint brush is an effective and gentle tool for moving and mixing eggs. Dampen fingers and paint brush in the isotonic PBS solution before squeezing females or manipulating eggs.
21. Before squeezing females, sperm samples can be retrieved from the liquid nitrogen freezer and maintained in a tray of liquid nitrogen within a small Styrofoam cooler until eggs are available.
22. A fertilized egg will be visible by the swelling of the first embryonic cell (zygote) and the chorion, or by the first cell divisions. Because fertilization and the emergence of the first cell and its division are usually not perfectly synchronized, the number of fertilized eggs can be more conveniently determined during blastula stages, after several rounds of cell divisions. The fertilization rate can be obtained by dividing the number of embryos with proliferating cells (fertilized) by the number of total viable eggs (100×).

Acknowledgments

ZIRC is grateful for continued NIH support from the Office of Research Infrastructure Programs (ORIP) in collaboration with the Eunice Kennedy Shriver National Institute of Child Health and Human Development (NICHD). We thank the ZIRC staff for daily help and discussion, and we gratefully acknowledge Huiping Yang, Leticia Torres, Yue Liu, Mary Hagedorn, and Terry Tiersch for advice and collaboration during the research and development phases of this protocol.

The most recent version of the ZIRC RMMB Freezing Protocol, including recipes for the solutions mentioned in this chapter, can be found in the protocols section on the ZIRC website: https://zebrafish.org/documents/protocols.php.

References

1. Haffter P, Granato M, Brand M, Mullins MC, Hammerschmidt M, Kane DA, Odenthal J, van Eeden FJ, Jiang YJ, Heisenberg CP, Kelsh RN, Furutani-Seiki M, Vogelsang E, Beuchle D, Schach U, Fabian C, Nüsslein-Volhard C (1996) The identification of genes with unique and essential functions in the development of the zebrafish, Danio rerio. Development 123:1–36
2. Driever W, Solnica-Krezel L, Schier AF, Neuhauss SC, Malicki J, Stemple DL, Stainier DY, Zwartkruis F, Abdelilah S, Rangini Z, Belak J, Boggs C (1996) A genetic screen for mutations affecting embryogenesis in zebrafish. Development 123:37–46
3. Amsterdam A, Burgess S, Golling G, Chen W, Sun Z, Townsend K, Farrington S, Haldi M, Hopkins N (1999) A large-scale insertional mutagenesis screen in zebrafish. Genes Dev 13(20):2713–2724
4. Kettleborough RN, Busch-Nentwich EM, Harvey SA, Dooley CM, de Bruijn E, van Eeden F, Sealy I, White RJ, Herd C, Nijman IJ, Fenyes F, Mehroke S, Scahill C, Gibbons R, Wali N, Carruthers S, Hall A, Yen J, Cuppen E, Stemple DL (2013) A systematic genome-wide analysis of zebrafish protein-coding gene function. Nature 496(7446):494–497. https://doi.org/10.1038/nature11992
5. Varshney GK, Lu J, Gildea DE, Huang H, Pei W, Yang Z, Huang SC, Schoenfeld D, Pho NH, Casero D, Hirase T, Mosbrook-Davis D, Zhang S, Jao LE, Zhang B, Woods IG, Zimmerman S, Schier AF, Wolfsberg TG, Pellegrini M, Burgess SM, Lin S (2013) A large-scale zebrafish gene knockout resource for the genome-wide study of gene function. Genome Res 23(4):727–735. https://doi.org/10.1101/gr.151464.112
6. Varshney GK, Pei W, LaFave MC, Idol J, Xu L, Gallardo V, Carrington B, Bishop K, Jones M, Li M, Harper U, Huang SC, Prakash A, Chen W, Sood R, Ledin J, Burgess SM (2015) High-throughput gene targeting and phenotyping in zebrafish using CRISPR/Cas9. Genome Res 25(7):1030–1042. https://doi.org/10.1101/gr.186379.114
7. Varshney GK, Sood R, Burgess SM (2015) Understanding and editing the zebrafish genome. Adv Genet 92:1–52. https://doi.org/10.1016/bs.adgen.2015.09.002
8. Auer TO, Del Bene F (2014) CRISPR/Cas9 and TALEN-mediated knock-in approaches in zebrafish. Methods 69(2):142–150. https://doi.org/10.1016/j.ymeth.2014.03.027
9. Lin S, Yang S, Hopkins N (1994) lacZ expression in germline transgenic zebrafish can be detected in living embryos. Dev Biol 161(1):77–83
10. Stuart GS, Vielkind JR, McMurray JV, Westerfield M (1990) Stable lines of transgenic zebrafish exhibiting reproducible tissue-specific transgene expression patterns. Development 109:577–584
11. Long Q, Meng A, Wang H, Jessen J, Farrell M, Lin S (1997) GATA-1 expression pattern can be recapitulated in living transgenic zebrafish using GFP reporter gene. Development 124(20):4105–4111
12. Yin L, Maddison LA, Li M, Kara N, LaFave MC, Varshney GK, Burgess SM, Patton JG, Chen W (2015) Multiplex conditional mutagenesis using transgenic expression of Cas9 and

sgRNAs. Genetics 200(2):431–441. https://doi.org/10.1534/genetics.115.176917

13. Varga ZM (2016) Aquaculture, husbandry, and shipping at the Zebrafish International Resource Center. Methods Cell Biol 135:509–534
14. Alestrom P, D'Angelo L, Midtlyng PJ, Schorderet DF, Schulte-Merker S, Sohm F, Warner S (2019) Zebrafish: housing and husbandry recommendations. Lab Anim. https://doi.org/10.1177/0023677219869037
15. Varga ZM, Tiersch TR (2012) Workshop and panel discussion: high-throughput cryopreservation of germplasm as an exchange currency for genetic resources. Comp Biochem Physiol 155(1):167–168. https://doi.org/10.1016/j.cbpc.2011.06.008
16. Westerfield M (2007) The zebrafish book: a guide for the laboratory use of zebrafish Danio (Brachydanio) rerio. Institute of Neuroscience, University of Oregon, Oregon. http://zfin.org/zf_info/zfbook/zfbk.html
17. Freeman A, Holland R, Hwang-Shum J-J, Lains D, Matthews J, Murray KN, Nasiadka A, Quinn E, Varga ZM, Westerfield M (2019) The Zebrafish International Resource Center. In: Jarret RL, McCluskey K (eds) The biological resources of model organisms. CRC Press, New York, pp 113–143
18. Matthews JL, Murphy JM, Carmichael C, Yang H, Tiersch T, Westerfield M, Varga ZM (2018) Changes to extender, cryoprotective medium, and in vitro fertilization improve zebrafish sperm cryopreservation. Zebrafish 15 (3):279–290. https://doi.org/10.1089/zeb.2017.1521
19. Carmichael C, Westerfield M, Varga ZM (2009) Cryopreservation and in vitro fertilization at the zebrafish international resource center. Methods Mol Biol 546:45–65
20. Tan E, Yang H, Tiersch TR (2010) Determination of sperm concentration for small-bodied biomedical model fishes by use of microspectrophotometry. Zebrafish 7(2):233–240. https://doi.org/10.1089/zeb.2010.0655

Chapter 10

Quantifying Tissue Tension in the Granulosa Layer After Laser Surgery

Peng Xia and Carl-Philipp Heisenberg

Abstract

Tissue morphogenesis is driven by mechanical forces triggering cell movements and shape changes. Quantitatively measuring tension within tissues is of great importance for understanding the role of mechanical signals acting on the cell and tissue level during morphogenesis. Here we introduce laser ablation as a useful tool to probe tissue tension within the granulosa layer, an epithelial monolayer of somatic cells that surround the zebrafish female gamete during folliculogenesis. We describe in detail how to isolate follicles, mount samples, perform laser surgery, and analyze the data.

Key words Tissue tension, Morphogenesis, Laser ablation, Zebrafish folliculogenesis, Granulosa cells

1 Introduction

During tissue morphogenesis, cells both generate cell-intrinsic mechanical forces through cytoskeletal rearrangements and adhesion to other cells and the extracellular matrix and are exposed to extrinsic forces from the environment in which they are placed. These forces lead to cell movements and changes in cell shape, which in turn can feed back on cell fate, acquisition in a process termed "mechanosensation" [1, 2]. Therefore, measuring the forces within a tissue is key to understand the mechanical regulation of tissue morphogenesis. To achieve this, diverse methods were successfully developed by combining high-resolution imaging with biophysical tools, such as micropipette aspiration, traction force microscopy, atomic force microscopy, magnetic tweezers, optical tweezers, and laser ablation [3]. In this section, we will describe of how to use laser ablation to quantitatively measure tissue tension in the granulosa cell layer of zebrafish ovarian follicles.

During zebrafish folliculogenesis, the oocyte is surrounded by a single squamous epithelial layer of granulosa cells [4]. Communication between the oocyte and granulosa cells through signaling is

Roland Dosch (ed.), *Germline Development in the Zebrafish: Methods and Protocols*, Methods in Molecular Biology, vol. 2218, https://doi.org/10.1007/978-1-0716-0970-5_10, © Springer Science+Business Media, LLC, part of Springer Nature 2021

critical for the development of the follicle [5]. In stage II follicles, a single granulosa cell at the animal pole of the oocyte differentiates into a micropyle precursor cell (MPC), which later on generates the micropyle within the egg shell or chorion in zebrafish [4, 6]. The micropyle guides the sperm to pass through the chorion during fertilization [7]. MPC differentiation has recently been found to be dependent of the expression of the co-transcriptional regulator TAZ [8–10]. Specifically, during MPC differentiation the MPC grows bigger and thereby mechanically compresses its surrounding cells, which again inhibits the potential of those cells to differentiate into another MPC, by suppressing their TAZ activity [8]. To analyze the mechanical compression of the surrounding granulosa cells by the growing MPC, we performed ultraviolet (UV) laser ablation of cell-cell junctions between granulosa cells to investigate tension distribution between those cells [8].

Pulsed near-infrared (NIR) or ultraviolet (UV) laser beam can generate plasma in tissues, by which spatially confined ablations of cells or subcellular structures, such as cytoskeletal elements, can be achieved [11]. In tensed structures, the ablation breaks the force balance, resulting in transient recoil of the surrounding tissue. The initial recoil velocity is considered as a proxy for the tension within the probed structure [12, 13]. In this chapter, we will represent the setup for plasma-mediated UV laser ablation on zebrafish ovarian granulosa cells. We will provide detailed protocols for medium preparation, ovarian follicle isolation, sample mounting, laser-cutter adjustment, UV laser ablation, high-speed imaging, and data analysis.

2 Materials

2.1 Transgenic Fish Lines

1. Double transgenic line Tg(buc:buc-gfp)(krt18:krt18-gfp) expressing GFP-tagged Bucky ball (Buc) to label the vegetal pole of oocyte and GFP-tagged Krt18 to label the micropyle precursor cell (MPC) and granulosa cell boundary [8].

2.2 Media

1. Tricaine medium: 2.5 g of tricaine (3-amino benzoic acidethylester) was dissolved in 500 ml tank water, supplemented with 2.5 g of KH powder, and stored at room temperature (*see* **Notes 1** and **2**).
2. Follicle culture medium: 9.7 g of Leibovitz's L-15 medium powder (final concentration is 70%) was dissolved in H_2O, supplemented with 50 U/ml penicillin-streptomycin and 0.5% bovine serum albumin, calibrated to pH 8.0, sterile filtered, and stored at 4 °C.

2.3 Ovarian Follicle Isolation

1. Operating scissors for decapitation, 14 cm.
2. Iris scissors for dissection, 10 cm.
3. Scalpel handle with blade for dissection.
4. 92 × 16 mm plastic petri dish for dissection.
5. 40 × 12 mm glass petri dish for keeping the ovary.
6. 150 mm glass Pasteur pipette.
7. Pipette nipple.
8. Stereomicroscope equipped with transmitted light and eyepiece reticles.
9. Stage micrometer.
10. Two pairs of sharp forceps (Dumont #55).

2.4 Laser-Cutter Microscope Setup

1. Inverted microscope with a 63× water immersion objective.
2. Spinning disk microscope with EMCCD for fast acquisition.
3. Solid-state 355 nm UV-A pulsed laser unit with dual-axis galvanometric scanners.
4. 488 nm laser unit with emission filters for imaging.
5. FRAPPA photobleaching module that allows to bleach an arbitrary area and observe recovery of luminescence.
6. Motorized piezo stage.
7. Heating unit for stage and objectives.
8. LabVIEW software to control the cutter and imaging system.

2.5 Sample Mounting

1. 1% low melting point (LMP) agarose dissolved in 70% Leibovitz's L-15 medium, pH 8.0. Boil the agarose-medium mixture in a microwave until it is dissolved, and distribute 0.5 ml aliquots in 1.5 ml Eppendorf tubes, and store at 4 °C.
2. Heating block at 70 °C.
3. Heating block at 40 °C.
4. 35 mm cell culture dishes with #1.5H thickness (170 μm) glass bottom.
5. Stereomicroscope equipped with transmitted light and broad-spectrum LED illumination.

2.6 Data Analysis

1. ImageJ software for image processing.
2. Excel and GraphPad Prism software for quantitative analysis.

3 Methods

3.1 Ovary Dissection

1. Add 2 ml of follicle culture medium to the 40 × 12 mm glass petri dish for keeping the ovary after dissection (*see* **Notes 3** and **4**).
2. Euthanize the female fish in tricaine medium.
3. Net the fish out from the tricaine medium.
4. Use tissue paper to absorb the liquid on the surface of the fish body, and place the fish on the cover of 92 × 16 mm plastic petri dish.
5. Decapitate the fish with 14 cm operating scissors.
6. Cut the body with iris scissors along the ventral midline from the pectoral to anal fin (Fig. 1a) (*see* **Note 5**).
7. Use the forceps to cut the gastrointestinal tract from the anal side, and remove the gastrointestinal tract, liver, and also swim bladder if necessary.
8. Carefully remove the ovary from fish body with scalpel blade.
9. Gently move the ovary to the follicle culture medium in 40 × 12 mm glass petri dish.
10. Clean the dissection devices with pure water, and then decontaminate the devices with 70% ethanol.

3.2 Follicle Isolation

1. Pipette the ovary with 150 mm glass Pasteur pipette to dissociate the follicles.
2. Calibrate the eyepiece reticles in stereomicroscope with stage micrometer (*see* **Note 6**).

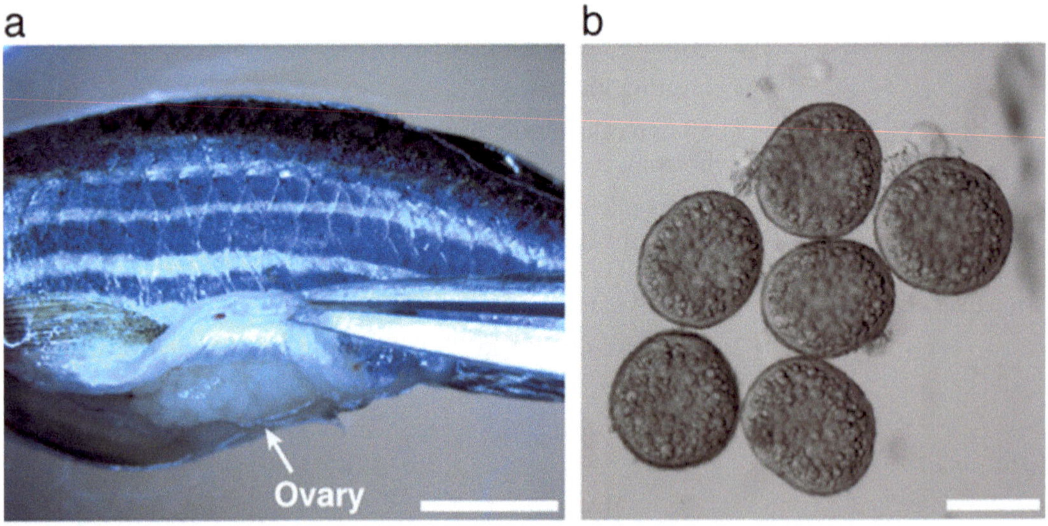

Fig. 1 Zebrafish ovary dissection and follicle isolation. (**a**) Ovary after cutting the body with iris scissors along the ventral midline. Scale bar, 0.5 cm. (**b**) Late stage II follicles were isolated individually after pipetting. Scale bar, 200 μm

3. Sort the follicles with desired diameter using forceps under the stereomicroscope (Fig. 1b) (*see* **Note 7**).
4. Prepare another 40 × 12 mm glass petri dish with fresh follicle culture medium, and transfer the sorted follicles to the dish with 150 mm glass Pasteur pipette.

3.3 Prepare the Laser-Cutter Microscope

1. Switch on red warning sign at the door of the microscope room to prohibit the entrance during operation of classes 3B and 4 lasers.
2. Wear laser safety goggles to avoid potential contact of eyes with classes 3B and 4 lasers.
3. Start the microscope, computer, and laser system (Fig. 2a) (*see* **Note 8**).
4. Switch on heating unit for stage and objectives, and set the temperature to 26 °C.
5. Start LabVIEW software to initialize the controlling system (*see* **Note 9**).
6. Prepare a dish for calibrating the laser beam by painting with a marker pen on the inner side of the bottom of a 35 mm cell culture dish (Fig. 2b) (*see* **Note 10**).
7. Put a drop of water on the 63× water immersion objective (*see* **Note 11**).
8. Mount the calibration dish on the stage, and focus on the marker pen traces on the glass bottom through eyepieces with bright field (*see* **Note 12**).
9. In the LabVIEW software, set up the exposure time and frame rates to 150 ms, the gain to 150; choose the correct laser and filters.
10. Set up the laser parameter to 1000 Hz pulse rate, 25 pulses per ablation points, 2 points/μm, and laser power to 1% (6.8 μW).
11. Open the camera shutter, start live preview, and precisely focus on the marker pen traces in bright field.
12. Open the UV laser shutter.
13. Draw a ROI in the LabVIEW software, apply laser ablation, and align the laser beam to the ROI (Fig. 2c) (*see* **Note 13**).
14. After laser beam alignment, stop live preview, and close the shutters for UV laser and the camera.

3.4 Mounting Follicles in Agarose

1. Turn on the 70 °C and 40 °C heating block (*see* **Note 14**).
2. Melt the 0.5 ml aliquot of 1% LMP agarose at 70 °C.
3. After melting the LMP agarose, quickly add 0.5 ml room temperature-stored follicle culture medium, and immediately

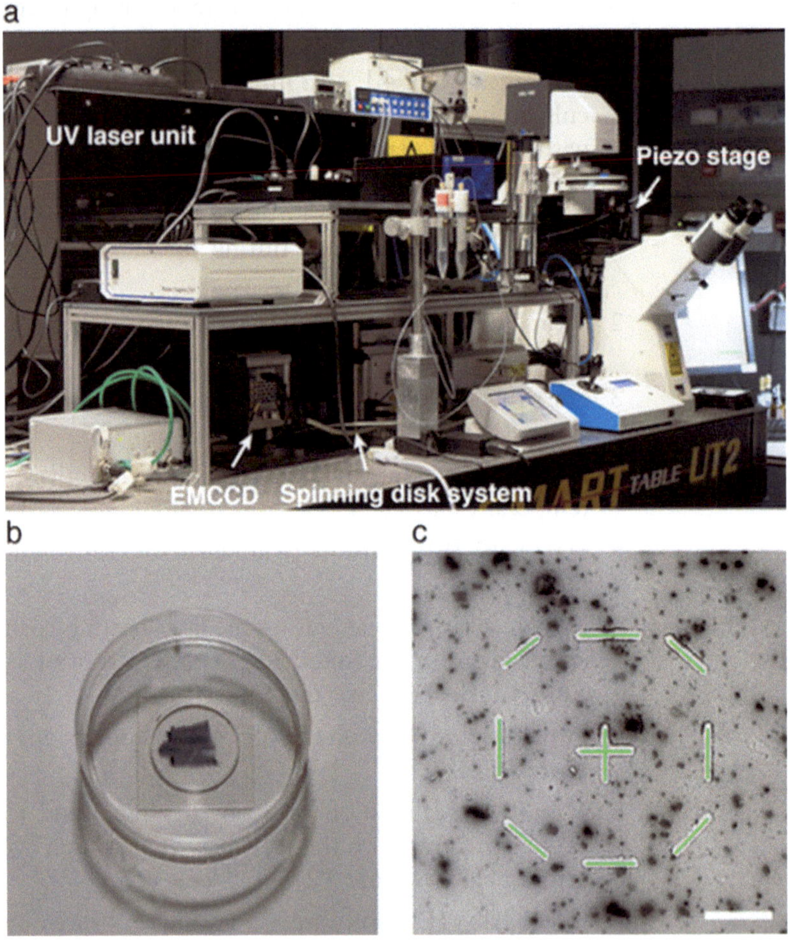

Fig. 2 Calibration of the UV laser-cutter microscope. (**a**) Setup of the laser-cutter system mounted on a spinning disk microscope. (**b**) Glass bottom cell culture dish with marker pen traces for calibrating the laser beam. (**c**) ROI (green lines) and laser ablation on the marker pen traces after beam calibration. Scale bar, 20 μm

put the Eppendorf tube onto 40 °C heating block. The final concentration of LMP should be 0.5% (*see* **Note 15**).

4. Put a 35 mm cell culture dish with glass bottom under the stereomicroscope, and focus on the bottom using bright-field imaging.
5. Turn on the LED illumination, and set up the filter for GFP signal.
6. Transfer three to five sorted follicles into the Eppendorf tube with 40 °C LMP (*see* **Note 16**).
7. Quickly transfer the follicles with agarose solution onto the glass bottom of the cell culture dish.

8. Under the stereomicroscope, the transgenic Buc-GFP signal is very bright and can be utilized as a vegetal pole marker [14]. Orient the follicles with forceps to make the animal pole facing the glass bottom.
9. Wait 1–2 min for the agarose to solidify.

3.5 UV Laser Ablation of Granulosa Cellular Junctions

1. Mount the sample dish on the stage of the laser-cutter microscope.
2. Find the follicles in bright field.
3. Open the UV laser and camera shutters, and start live preview.
4. Adjust intensity of the 488 nm laser, so that the Krt18-GFP signal can be clearly detected but is not oversaturated (*see* **Note 17**).
5. Acquire z stack reference images before performing the laser ablation. Count how many cell rows from the MPC the cutting site will be located (Fig. 3a) (*see* **Note 18**).
6. Draw a line ROI perpendicular to the desired junction.
7. In the LabVIEW software, reshape the ROI length to 3 μm, and confirm in the live preview that the final ROI pass through the desired junction.
8. Start acquiring the images (*see* **Note 19**).
9. Perform laser-cut, and record images for 2 min (Fig. 3b) (*see* **Note 20**).
10. Acquire z stack reference images after the laser ablation to confirm that no unspecific damage to the follicle occurred during the ablation (*see* **Note 21**).
11. Close the UV laser and camera shutters.
12. Mount next dish of follicles and repeat the laser ablation processes until completing the experiments.
13. Switch off the 40 and 70 °C heating blocks.
14. Close the software. Turn off the laser system, and transfer the data.
15. Switch off the computer, microscope, and heating unit for stages and objectives.
16. Clean the objectives, stage, heating blocks, and the bench.
17. Switch off the red warning sign for class 3B and 4 lasers.

3.6 Data Analysis

1. Load image sequence to ImageJ and generate a time-lapse stack.
2. Draw a line ROI along the pre-cut junction. It should be perpendicular to the laser ablation ROI (Fig. 3c).

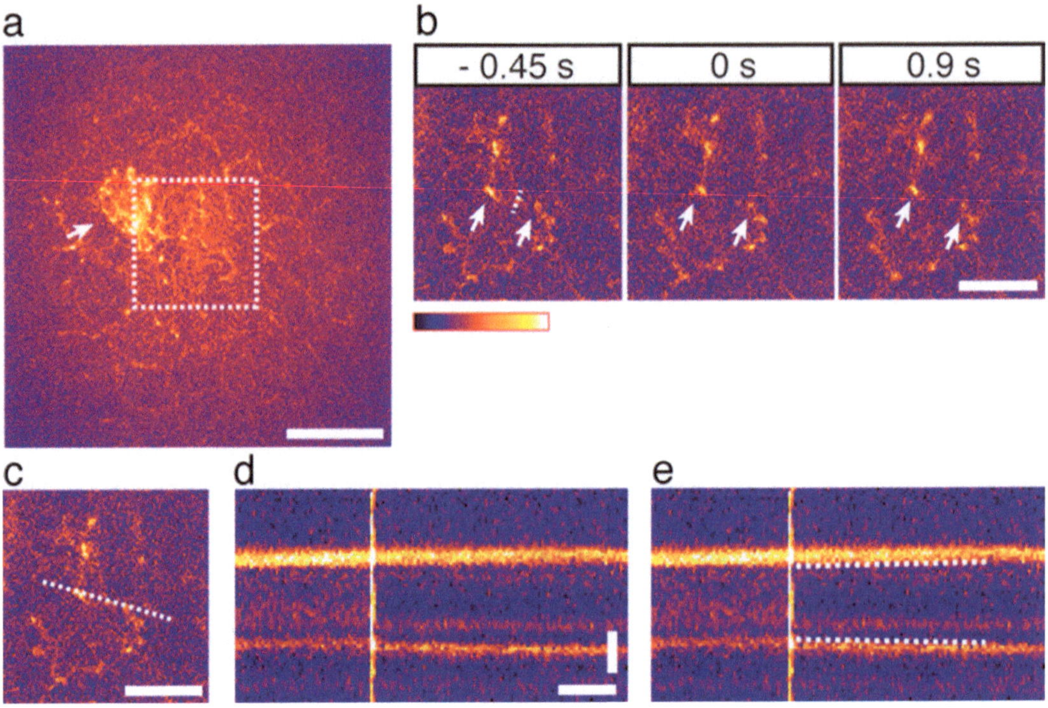

Fig. 3 Laser ablation of cell-cell junction in granulosa layer of late stage II follicles. (**a**) Maximum intensity projection of z-stack spinning disk confocal images shows Krt18-GFP in the animal pole of a late stage II follicle. Arrow indicates the MPC. The region in (**b, c**) is outlined by white dashed box. Scale bar, 20 μm. (**b**) Single-section images during UV laser-cutting. Junctions are shown before (−0.45 s) and after (0 s and 0.9 s) cutting. The cut was 3 μm long (dashed line in the left panel) and performed on junction that oriented perpendicular to the MPC. The junction was cut between −0.45 s and 0 s. Arrows indicate the positions of vertices. Scale bar, 10 μm. (**c**) Image of the junction before cut. The dashed line indicates the ROI for generating kymograph in (**d, e**). Scale bar, 10 μm. (**d**) Kymograph showing the evolution of junction opening following the cut. Horizontal scale bar, 3 s. Perpendicular scale bar, 3 μm. (**e**) Kymograph with dashed lines outlining the movements of vertices. The slopes of the two lines were utilized in calculating the recoil velocity

3. Generate a kymograph with "Reslice" function (Image → Stacks → Reslice; or use hotkey **Shift**+/). Check "Rotate 90 degrees" and "Avoid interpolation." The outcome kymograph will represent pre- and post-cut states of the junction. The *x*-axis of the kymograph represents the temporal property (1 frame/pixel) and the *y*-axis the spatial property (pixel size is the same as in the original time-lapse images).
4. Use false color (e.g., in the ImageJ Startup Tools, find LUT → Fire), and elevate contrast (Image → Adjust → Brightness/Contrast; or use hotkey **Command** + **Shift** + **C**) to increase the visibility of the junction (Fig. 3d).
5. Draw a line ROI along the direction of initial opening at one side of the junction (Fig. 3e), and measure the slope (angle value) of the line (Analyze → Measure; or use hotkey **M**). In

the "Results" panel, find the angle value of the line ROI. The initial opening velocity equals to tangent of the absolute value of the slope multiplied with spatial pixel resolution in the y-axis of the kymograph and then divided by temporal pixel resolution in the x-axis.

6. Apply the same measurement to the other side of the junction.
7. If the slopes in a kymograph are for one side of the junction positive and for the other negative, the initial recoil is the addition of the initial opening velocities of the two sides of the junction. If the slopes are both positive or if they are both negative, the initial recoil is the subtraction of the initial opening velocities.
8. Perform statistical analysis with Excel and GraphPad Prism.

4 Notes

1. Tricaine 5 g/L medium with a pH of 6.5–7.5 is lethal for the adult zebrafish. KH (sodium bicarbonate) is added to neutralize the pH value.
2. Tricaine medium is normally working for at least 1 month at room temperature after preparing. Fish should immediately stop opercular and body movements in efficient tricaine medium. Sometimes, fish will bleed from the gill after euthanasia, which can contaminate the tricaine buffer and decrease its efficiency. So it is recommended to get the fish out as soon as they are euthanized. If the fish do not stop movements quickly in the tricaine medium, make fresh medium to ensure they are euthanized properly before decapitation.
3. If the culture medium is stored at 4 °C, warm it up to room temperature before using.
4. Always cover the petri dish with culture medium to avoid contamination by bacterial or fungus from the air.
5. When cutting the fish body, be careful not to damage the heart so as to avoid excessive bleeding and not to damage the gastrointestinal tract system so as to minimize contamination.
6. Avoid changing focus plane after eyepiece reticles are calibrated; to achieve this, the stage micrometer should be placed in the same vertical level as the follicles in the dish when calibrating the eyepiece reticles.
7. Be careful not to damage the follicles with sharp forceps.
8. Make sure all the objective mounts are blocked by objectives or blocking plugs before switching on the system.

9. Make sure the UV laser shutter is closed during any action before laser ablation, such as cleaning the objectives and stage, changing objectives or filters, mounting dish, focusing on the specimen, and finding sample position.
10. The dish made for calibrating the laser beam should be the same as the dish for the specimen. The exact dish made for calibration can be reused for many times.
11. Be careful not to put too much water on the objective; otherwise the water will run off the lens. In that case, use a piece of dry tissue paper to clean the objective and mount immediately.
12. For objectives with high numerical apertures, the thickness of the cover glass significantly affects the aberrations. Most objectives are designed for the standard #1.5H thickness (170 μm) cover glass. Some objectives have a correction collar so that cover glass with other thickness can be used. Make sure the correction collar slider is adjusted to the correct value before mounting the calibration dish.
13. Laser beam calibration should start after temperature balancing of the objective. In our case, the preheating takes 20 min. The calibration should be performed in two dimensions.
14. It will take some time for the heating blocks to warm up. One can turn them on before preparing the laser-cutter microscope to save some time.
15. Before mixing with the follicle culture medium, shake the Eppendorf tube with LMP agarose to check if the agarose is completely melted. In our case, it usually takes less than 5 min for melting the agarose.
16. Transfer the follicles with minimum culture medium to avoid diluting the agarose. Discard the remaining follicle culture medium once all the follicles in the glass Pasteur pipette are delivered.
17. As the Buc-GFP preferentially locates to the vegetal pole of the oocyte, the Krt18-GFP signal in the granulosa cells can only be detected at the animal pole of the follicles.
18. The MPC can be distinguished from the surrounding granulosa cells by its higher expression of Krt18-GFP [8].
19. Initiate the time-lapse imaging 5–15 s before the laser-cut, to make sure the position of the tissue is stable during the experiment, without the sample drifting.
20. The parameters for laser ablation in the granulosa layer need to be optimized when cutting for the first time. Try to find the proper intensity of UV laser that can efficiently ablate the junction without permanently damaging the membrane and the oocyte. In our case, we use 1000 Hz pulse rate, 25 pulses

per ablation points, and 6 equidistant sites along a 3 μm long line. Laser power was 1.1% (~7.6 μW). The exposure time and frame rates are 150 ms. The gain is 150. Laser-cut lasts for around 0.45 s.

21. A successful laser ablation experiment on the granulosa layer should display immediate recoil without shifting of the focus plane, or leak of cellular components. Each follicle can be cut only once. Data with severe membrane and oocyte damages should be discarded from the measurement.

Acknowledgments

We thank Prof. Masazumi Tada and Roland Dosch for providing transgenic zebrafish lines, the Heisenberg lab for technical assistance and feedback on the manuscript, and the Bioimaging and Fish facilities of IST Austria for continuous support. This work was funded by an ERC advanced grant (MECSPEC to C.-P.H.).

References

1. Heisenberg C-P, Bellaiche Y (2013) Forces in tissue morphogenesis and patterning. Cell 153:948–962. https://doi.org/10.1016/j.cell.2013.05.008
2. Chan CJ, Heisenberg C-P, Hiiragi T (2017) Coordination of morphogenesis and cell-fate specification in development. Curr Biol 27: R1024–R1035. https://doi.org/10.1016/j.cub.2017.07.010
3. Pinheiro D, Bellaiche Y (2018) Mechanical force-driven adherens junction remodeling and epithelial dynamics. Dev Cell 47:3–19. https://doi.org/10.1016/j.devcel.2018.09.014
4. Selman K, Wallace RA, Sarka A, Qi X (1993) Stages of oocyte development in the zebrafish, Brachydanio rerio. J Morphol 218:1–22. https://doi.org/10.1002/jmor.1052180209
5. Dranow DB, Hu K, Bird AM et al (2016) Bmp15 is an oocyte-produced signal required for maintenance of the adult female sexual phenotype in zebrafish. PLoS Genet 12: e1006323. https://doi.org/10.1371/journal.pgen.1006323
6. Marlow FL, Mullins MC (2008) Bucky ball functions in Balbiani body assembly and animal-vegetal polarity in the oocyte and follicle cell layer in zebrafish. Dev Biol 321:40–50. https://doi.org/10.1016/j.ydbio.2008.05.557
7. Yanagimachi R, Cherr G, Matsubara T et al (2013) Sperm attractant in the micropyle region of fish and insect eggs. Biol Reprod 88:47–47. https://doi.org/10.1095/biolreprod.112.105072
8. Xia P, Gütl D, Zheden V, Heisenberg C-P (2019) Lateral inhibition in cell specification mediated by mechanical signals modulating TAZ activity. Cell 176:1379–1392.e14. https://doi.org/10.1016/j.cell.2019.01.019
9. Yi X, Yu J, Ma C et al (2019) The effector of Hippo signaling, Taz, is required for formation of the micropyle and fertilization in zebrafish. PLoS Genet 15:e1007408. https://doi.org/10.1371/journal.pgen.1007408
10. Dingare C, Niedzwetzki A, Klemmt PA et al (2018) The Hippo pathway effector Taz is required for cell morphogenesis and fertilization in zebrafish. Development 145: dev167023. https://doi.org/10.1242/dev.167023
11. Vogel A, Venugopalan V (2003) Mechanisms of pulsed laser ablation of biological tissues. Chem Rev 103:577–644. https://doi.org/10.1021/cr010379n
12. Smutny M, Behrndt M, Campinho P et al (2015) UV laser ablation to measure cell and tissue-generated forces in the zebrafish embryo in vivo and ex vivo. Methods Mol Biol 1189:219–235. https://doi.org/10.1007/978-1-4939-1164-6_15

13. Campinho P, Behrndt M, Ranft J et al (2013) Tension-oriented cell divisions limit anisotropic tissue tension in epithelial spreading during zebrafish epiboly. Nat Cell Biol 15:1405–1414. https://doi.org/10.1038/ncb2869

14. Bontems F, Stein A, Marlow F et al (2009) Bucky ball organizes germ plasm assembly in zebrafish. Curr Biol 19:414–422. https://doi.org/10.1016/j.cub.2009.01.038

Chapter 11

In Vivo and Ex Vivo CT Imaging of Zebrafish Gonads

Christian Dullin and Louisa Habich

Abstract

3D imaging of the gonads in adult zebrafish in vivo is of great interest, as it allows to follow up on their development and/or the egg development in the same individual over time. Optical-based imaging methods can hardly be applied on the adult zebrafish, due to their limited transparency. In this chapter, we will demonstrate the application of micro computer tomography (CT) imaging for in vivo 3D imaging of the gonads in adult zebrafish. We explain how the limited soft-tissue contrast in CT can be overcome and which X-ray dose levels can be expected using this technique. Moreover, we will use high-resolution microCT to perform ex vivo 3D virtual histology of the adult zebrafish, which allows a simple quantitative analysis of the gonad regions, malformation or alterations in the development of the follicles.

Key words In vivo computed tomography, Anesthesia, Phase-contrast CT

1 Introduction

Whereas optical imaging techniques can be used on zebrafish embryos due to their transparency [1], in the adult zebrafish, other methods must be applied. Moreover, studying the gonad regions in adult zebrafish especially in a longitudinal manner is of great interest to follow up egg development, for instance. Thus, we used in vivo microCT, a technique commonly applied in small animal models including mice [2]. Since the contrast of CT is based on difference in the absorption of X-rays, energy is transferred to the body (the so-called dose) which leads to the known harmful side effects like an increased risk of cancer development [3]. Especially in small animal imaging, the required high spatial resolution can only be realized at elevated dose levels, and thus an impact on the well-being of the animal or an influence on the performed study needs to be carefully observed [4]. In addition, CT imaging requires a static subject/specimen during data acquisition which makes an optimized anesthesia protocol crucial for successful imaging [5]. Moreover, the intrinsic contrast of the

Roland Dosch (ed.), *Germline Development in the Zebrafish: Methods and Protocols*, Methods in Molecular Biology, vol. 2218, https://doi.org/10.1007/978-1-0716-0970-5_11,

© Springer Science+Business Media, LLC, part of Springer Nature 2021

gonad region—like other soft tissues—is rather poor in CT. Therefore, contrast agents need to be applied.

Here, we present our in vivo CT imaging protocol of anesthetized adult zebrafish injected with clinical (and therefore cheap) CT contrast agents in combination with an ex vivo staining and microCT imaging approach for validation of the results. We utilized a state-of-the-art in vivo small animal CT; thus the presented approach can also be used in combination with other devices.

2 Materials

2.1 Devices

1. Stereomicroscope.
2. In vivo small animal microCT Quantum FX (PerkinElmer, Massachusetts, USA).
3. White beam microCT setup at the SYRMEP beamline Italian synchrotron "Elettra," Trieste, Italy.

2.2 Consumables

1. 5 mL reaction tubes.
2. Insulin syringe.
3. Sponge.

2.3 Chemicals

1. Tricaine methanesulfonate.
2. Isoflurane.
3. Propofol.
4. Ultravist® 150 Iodine (Bayer Vital GmbH, Leverkusen, Germany).
5. 4% iodine in 100% ethanol.

3 Methods

3.1 Anesthesia for In Vivo CT Scans of Zebrafish

As mentioned above, CT requires a subject which is static during the acquisition. Therefore, anesthesia needs to be applied to enable a reliable immobilization of the individual.

1. Place zebrafish in tank with appropriate amount of anesthetic solution consisting of 2.5 ng/μL Propofol diluted with water (1 μL Propofol stock solution in 100 ml $H_2O = 9.62$ μg/mL).
2. Wait until complete immobilization of the fish occurred (approx. 2 min).
3. Take fish out of water and perform injection with 3 μL contrast agent in the gonads or the heart.
4. Place fish back in freshwater to revive them.
5. Observe the recovery following the injection.

6. Shortly before the scan: place fish back in the anesthetic solution.
7. Place fish after immobilization in 5 mL reaction tube filled with anesthetic to prevent the fish from drying out.
8. Perform scan for circa 5 min.

This anesthesia protocol successfully resulted in complete immobilization of the fish without causing cardiac arrest (*see* **Note 1**).

3.2 In Vivo CT Scanning

1. After anesthesia, place the zebrafish in the Quantum FX in vivo microCT.
2. Perform 360° scan with a field of view (FOV) of $20 \times 20 \times 20$ mm^3 for 4.5 min operating the X-ray tube at 90 kV and 200 μA.
3. Obtained 3D data sets will have a pixel size of $40 \times 40 \times 40$ μm^3 (*see* **Note 2**).

3.3 X-Ray Contrast Agents for In Vivo Imaging of Gonads

For visualization of the gonads, perform:

1. Injection of 3 μL Ultravist 150, a clinically used CT contrast agent that contained 150 mg/mL iodine, directly into the gonads under a stereomicroscope.
2. Perform CT scan.

Figure 1 shows exemplary cut sections through 3D CT data sets of (a) a female and (b) a male zebrafish, both after injection of Ultravist 150 and scanned in vivo in the microCT within 4.5 min. The reconstructed data sets had an isotropic pixel size of 40 μm. Clearly, the ovaries in the female fish together with the single eggs can be visualized (ov). Even the testis of the male zebrafish (Fig. 1b, te) can be depicted as validated by histology (Fig. 1c). Figure 1d shows a volume rendering of a CT scan depicting a male zebrafish. The data is virtually cut open to display the level of detail that could be achieved for the inner organs (*see* **Notes 3–5**).

3.4 Ex Vivo Staining of the Adult Zebrafish for High-Resolution MicroCT Acquisition

1. After performing the in vivo microCT scan, euthanize the zebrafish using the ice killing method [6].
2. Transfer fish to a 4% iodine in 100% ethanol solution.
3. Keep fish inside the solution for 9 days under continuous shaking.
4. Remove fish from the solution afterward and remove excessive staining solution.
5. Rinse fish with water and embed them in 1% agarose in cryo-vials to provide stability for high-resolution CT.

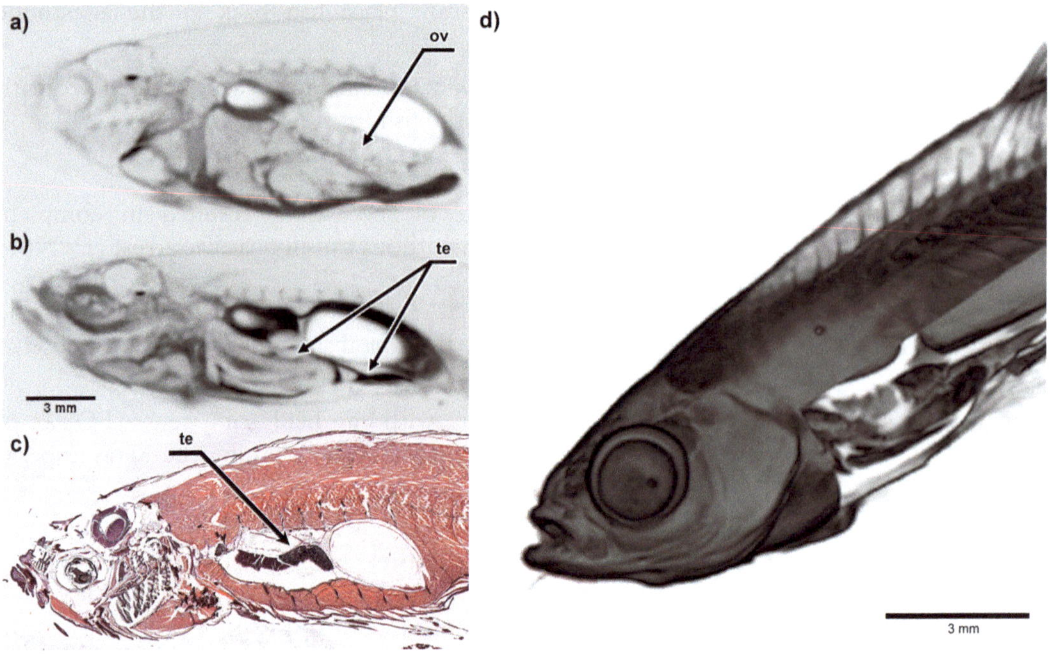

Fig. 1 In vivo CT imaging results of adult anesthetized zebrafish in water injected with 3 μL Ultravist 150 intracardially and intraperitoneally. (**a**) Virtual cross section through a female. Ovaries and single eggs are visible (ov). (**b**) Virtual cross section through a male. Due to the presence of the contrast agent, the testis region is clearly delineated (te). (**c**) H&E-stained cross section through a male zebrafish to validate the location of the testis (te). (**d**) Volume rendering of an in vivo CT scan of a male zebrafish to demonstrate the level of detail that can be achieved

3.5 High-Resolution Ex Vivo Synchrotron CT Imaging of Adult Zebrafish

1. For the high-resolution CT, apply the "white beam" setup of the SYRMEP beamline at the Italian synchrotron "Elettra" with the following settings: sample-to-detector 250 mm, pixel size 1.98 μm, white beam setup in combination with 0.5 mm silica and 1 mm aluminum filter resulting in an average beam energy of 16.7 keV, 2400 angular projection in an off-center scan over 360° leading to the generation of a FOV of $7 \times 7 \times 3.6$ mm^3.
2. Mount fish in upright position.
3. Perform seven consecutive scans for each specimen.
4. Apply TIE-Hom single-distance phase retrieval algorithm implemented in the free software STP (https://github.com/ElettraSciComp/STP-Gui.git) using a delta-to-beta ratio of 100 to exploit the phase effects caused by the coherence of the X-ray at the SYRMEP.

Examples are shown for a male and female fish following data reconstruction after image acquisition at the Italian synchrotron: In Figs. 2 and 3, imaging results for the gonad regions of a female and male adult zebrafish are shown, respectively. Figure 2a

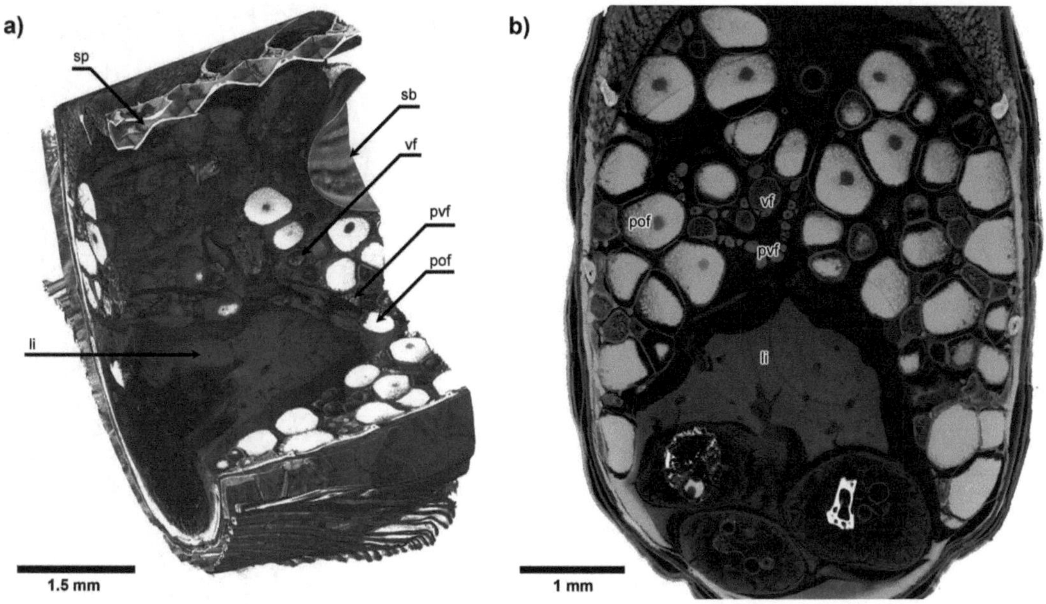

Fig. 2 Ex vivo CT imaging results of a female zebrafish stained with 4% iodine in 100% ethanol for 9 days, embedded in 1% agarose gel and imaged at the SYRMEP beamline of the Italian synchrotron "Elettra" with approx. 2 μm resolution. (**a**) Volume rendering representation of the ovary region (virtually cut open to better display the inner organs). (**b**) 2D cut section through the same data set. Clearly inner organs like the liver (li), spine (sp), and swim bladder (sb) can be seen. Moreover, the ovary tract is displayed in detail showing pre-vitellogenic follicle (pvf), vitellogenic follicle (vf), and preovulatory follicle (pof). The dark spot in the preovulatory follicle shows the germinal vesicle, the nucleus of the oocyte, highlighting the high resolution achieved by microCT imaging

demonstrates that not only the ovary regions of a female zebrafish can be displayed in detail, but also single follicles in different development state can be observed. In Fig. 3 the testis of a male zebrafish is shown, also depicting individual cells (*see* **Note 6**).

4 Notes

1. In order to immobilize the zebrafish during the scan procedure, we tried out different methods including a sponge, 5 mL reaction tube, and a falcon tube cut in a half and covered with foil from the inside. In our opinion, the best method for fixation was the 5 mL tube as the falcon tube resulted in too much pressure from the outside leading to the death of the fish and the sponge tend to soak up the contrast agent.
2. The applied X-ray imaging protocols deliver a dose of about 3 Gy, which was found low enough to allow for in vivo imaging but should be carefully considered for experiments that will require longitudinal imaging. Moreover, a potential influence

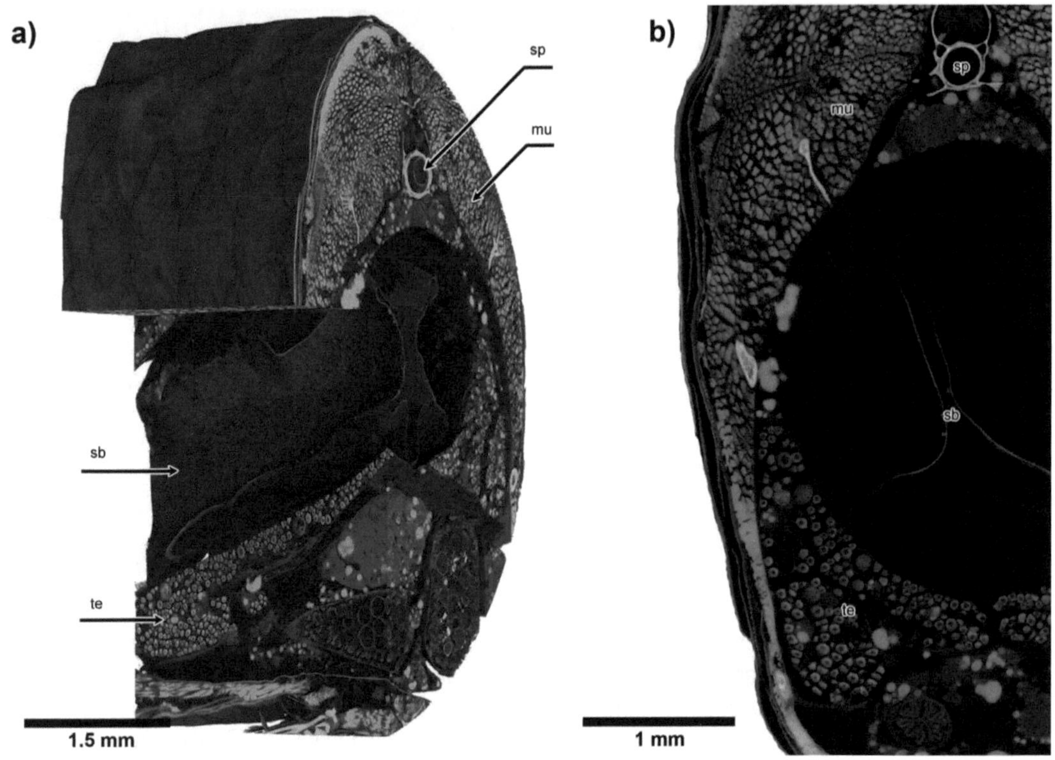

Fig. 3 Ex vivo CT imaging results of a male zebrafish stained with 4% iodine in 100% ethanol for 9 days, embedded in 1% agarose gel and imaged at the SYRMEP beamline of the Italian synchrotron "Elettra" with approx. 2 μm resolution. (**a**) Volume rendering representation of the testis region (virtually cut open to better display the inner organs). (**b**) 2D cut section through the same data set. Clearly inner organs like the spine (sp), the swim bladder (sb), and the skeletal muscles (mu) can be seen. Moreover, the testis is displayed in detail (te)

onto the development of the eggs or onto the fertilization state of the males needs still to be evaluated.

3. In order to optimize the concentration of contrast agent, we tried different concentrations of the contrast agent. In our opinion, the optimal contrast was reached for 3 μL of the contrast agent containing 150 mg/mL iodine.
4. Since the injection of the contrast agent into the heart of the zebrafish is a delicate procedure, we often found different patterns of contrast agent distribution as a result of an improper injection. Thus, a larger cohort of fish should be used to account for that.
5. We advise against injection of the contrast agent into the belly of highly gravid females. The high pressure causes the eggs to spill out at the injection site.
6. Due to the staining with iodine, the X-ray absorption of the specimen is greatly increased, which in combination with the

extremely high flux of a white beam synchrotron imaging setup will generate a high-dose deposition within the specimen which can lead to air bubble formation in the agarose gel used for embedding. Thus either embedding in resin should be considered or the total acquisition time should be kept at a minimum.

Acknowledgments

The authors thank the staff of the SYRMEP beamline especially Dr. Giuliana Tromba for the help with the synchrotron CT experiments.

References

1. Omar M, Rebling J, Wicker K et al (2017) Optical imaging of post-embryonic zebrafish using multi orientation raster scan optoacoustic mesoscopy. Light Sci Appl 6:e16186–e16186. https://doi.org/10.1038/lsa.2016.186
2. Holdsworth DW, Thornton MM (2002) Micro-CT in small animal and specimen imaging. Trends Biotechnol 20:S34–S39. https://doi.org/10.1016/S0167-7799(02)02004-8
3. Brenner DJ, Hall EJ (2007) Computed tomography – an increasing source of radiation exposure. N Engl J Med 357:2277–2284. https://doi.org/10.1056/NEJMra072149
4. Carlson SK, Classic KL, Bender CE, Russell SJ (2007) Small animal absorbed radiation dose from serial micro-computed tomography imaging. Mol Imaging Biol 9:78–82
5. Vasquez SX, Shah N, Hoberman AM (2013) Small animal imaging and examination by micro-CT. In: Barrow PC (ed) Teratogenicity testing: methods and protocols. Humana, Totowa, NJ, pp 223–231
6. Valentim AM, Eeden FJ, Strähle U, Olsson IAS (2016) Euthanizing zebrafish legally in Europe: are the approved methods of euthanizing zebrafish appropriate to research reality and animal welfare? EMBO Rep 17:1688–1689. https://doi.org/10.15252/embr.201643153

Chapter 12

Live and Time-Lapse Imaging of Early Oogenesis and Meiotic Chromosomal Dynamics in Cultured Juvenile Zebrafish Ovaries

Avishag Mytlis and Yaniv M. Elkouby

Abstract

Oocyte production is crucial for sexual reproduction. Recent findings in zebrafish and other established model organisms emphasize that the early steps of oogenesis involve the coordination of simultaneous and tightly sequential processes across cellular compartments and between sister cells. To fully understand the mechanistic framework of these coordinated processes, cellular and morphological analysis in high temporal resolution is required. Here, we provide a protocol for four-dimensional live time-lapse analysis of cultured juvenile zebrafish ovaries. We describe how multiple-stage oocytes can be simultaneously analyzed in single ovaries, and several ovaries can be processed in single experiments. In addition, we detail adequate conditions for quantitative image acquisition. Finally, we demonstrate that using this protocol, we successfully capture rapid meiotic chromosomal movements in early prophase for the first time in zebrafish oocytes, in four dimensions and in vivo. Our protocol expands the use of the zebrafish as a model system to understand germ cell and ovarian development in postembryonic stages.

Key words Zebrafish oogenesis, Ovary culture, Live time-lapse microscopy, Quantitative imaging, Zygotene chromosomal bouquet, Meiosis, Chromosomal pairing, Balbiani body, Oocyte polarity, Symmetry breaking, Postembryonic development

1 Introduction

Oogenesis, the production of oocytes, is key for sexual reproduction. The early steps of oocyte production are compiled of intricate and highly dynamic processes (rev. in [1]) that simultaneously span major cellular compartments, involve special intercellular interactions and organizations, and cooperate in the proper development of oocyte morphology and function. Oogenesis begins during postembryonic development in the juvenile zebrafish [1], when germline stem cells in the gonads give rise to mitotic precursor

Electronic supplementary material: The online version of this chapter (https://doi.org/10.1007/978-1-0716-0970-5_12) contains supplementary material, which is available to authorized users.

Roland Dosch (ed.), *Germline Development in the Zebrafish: Methods and Protocols*, Methods in Molecular Biology, vol. 2218, https://doi.org/10.1007/978-1-0716-0970-5_12, © Springer Science+Business Media, LLC, part of Springer Nature 2021

cells called oogonia [2], which divide several times before producing differentiating oocytes [3, 4]. Oogonial divisions are incomplete, maintaining cytoplasmic bridges between sister cells [4–6], likely by inhibition of the abscission step of cytokinesis as was shown in the mouse [7, 8]. These specialized oogonial divisions form a cellular organization that is conserved from insects to humans, called the germline cyst [4]. In the germline cyst, oogonia are clustered, connected by cytoplasmic bridges, and enwrapped by somatic pre-granulosa cells [4].

In the germline cyst, the induction of meiosis next transforms oogonia to early differentiating oocytes. A hallmark of meiosis is recombination between homologous chromosomes. Homologous recombination (HR) repairs induced DNA double-strand breaks and increases genetic diversity of gametes (rev. in [9–11]), shaping the genome for the next generation. HR depends on chromosomes first finding and pairing with their homolog via synaptonemal complexes (SC) [9–12]. Chromosomal pairing takes place in the first stages of prophase I, the leptotene and zygotene stages, while HR is executed at the subsequent pachytene stage [12–16]. In pairing, telomeres execute a unique and universally conserved meiotic function. Telomeres tether to Sun/KASH complexes on the nuclear envelope (NE) that associate with perinuclear microtubules (MTs) via dynein [12, 17–29]. Telomere association with MTs facilitates their rotations around the NE, which in turn shuffle chromosomes and drive their search for homologs. Telomeres ultimately cluster at one pole on the NE, while the free looping ends of the chromosomes face the other side—a configuration called the zygotene chromosomal bouquet [30]. The bouquet configuration is conserved from yeast to mammals with a common mechanism of telomere-Sun-NE attachment [12, 17–29]. Telomere clustering is thought to stabilize proper pairing between homologous chromosomes and is essential for synapsis, HR, and fertility [12, 17–20, 26, 31–34].

An additional key feature of oocytes is formation of a large mRNA-protein (mRNP) granule, called the Balbiani body (Bb), that is conserved from insects to humans [6, 35–51]. In most vertebrates, including zebrafish, the oocyte is polarized along an animal-vegetal axis [52]. The oocyte animal-vegetal axis is established by the Bb, which first polarizes the oocyte by specifying its vegetal pole [5, 6, 52–54]. The oocyte vegetal pole is key for embryonic development, harboring patterning factors that establish the global embryonic body axes and the germline lineage [52, 55–64]. Earlier in oogenesis, the Bb forms adjacent to the nucleus and translocates to the oocyte cortex, where it unloads its mRNPs to specify the vegetal pole of the oocyte [6, 53, 55, 62, 65, 66]. Loss of the Bb in zebrafish results in radially symmetrical eggs and early embryonic lethality [6]. In mice, where oocyte polarization is not known to regulate embryonic patterning, the Bb is associated with proper primordial follicle formation [38].

We have previously identified the oocyte symmetry-breaking event that leads to Bb formation during the zygotene chromosomal bouquet configuration [5]. In bouquet formation, MTs emanate from the centrosome, organize perinuclearly, and mediate telomere rotation, which ultimately cluster toward the centrosome [5, 27, 67]. We found that while Bb granules are distributed radially in oogonial cells, they first become polarized in subsequent zygotene stages, when they localize around the bouquet centrosome [5]. Moreover, Bb granule localization to the centrosome at this stage requires bouquet MTs [5], showing that bouquet and Bb formation are mechanistically coupled [5]. Similar polarized aggregation of early forming Bb components was detected apposing the presumptive telomere cluster in oocytes of the insect *T. domestica* [68] and around the centrosome in mice oocytes [38], suggesting that the coupling of bouquet and Bb formation is widely conserved [4].

Furthermore, we found that earlier in the germline cyst, the centrosome localizes to the cytoplasm adjacent to the cytoplasmic bridge [5]. This suggests that the centrosome is aligned to an earlier oogonial division plane and indicates a mechanistic link between the cyst organization and oocyte polarization [1, 5]. These findings emphasize that early oocyte differentiation requires the coordination of simultaneous and/or tightly sequential processes across cellular compartments and between sister cells in the cyst. They further stress that in order to fully understand early oogenesis, we need to analyze these processes in higher temporal resolution. For example, characterizing the events that govern bouquet formation by visualizing key players in real time will resolve its mechanistic framework and interplay with linked processes.

Powerful methodologies in established model organisms significantly advanced our mechanistic understanding of bouquet formation and chromosomal pairing. Genetic analysis in the mouse pioneered the identification of the long-sought after meiotic complex that tethers telomeres to the Sun protein in vertebrates [19, 26]. In *Drosophila*, genetic studies recently discovered key novel meiotic cohesions [69, 70]. In *C. elegans*, the role of dynein was first implicated in regulating telomere rotation [20], and live time-lapse analysis resolved chromosomal dynamics and pairing in high resolution [71–73]. Live time-lapse studies traced chromosomal rotation in mice spermatocytes, revealing similar bouquet formation dynamics [27, 74]. Collectively, these findings established common principles for meiotic chromosomal homology searches and pairing by chromosomal rotation in bouquet formation.

However, variations from these principles exist between species and sexes. For example, in *C. elegans,* chromosomes are tethered to the NE through chromosomal regions called pairing centers and not telomeres [20], and the oocyte symmetry is only broken upon fertilization [75]. In *Drosophila*, chromosomal pairing utilizes centromeres instead of telomeres and intriguingly occurs during the mitotic division preceding meiosis [15, 76, 77]. Finally, in mice spermatocytes a Bb does not form, and other processes that are not directly involved in chromosomal paring have not been associated with the bouquet configuration. Therefore, to address the coordination of bouquet formation with other processes like cellular polarization in early oogenesis, live time-lapse analysis of ovaries is required. However, to our knowledge, this has yet to be reported in vertebrates.

Here we report a protocol for live time-lapse image analysis of cultured juvenile zebrafish ovaries. We demonstrate that this protocol successfully records bouquet chromosomal dynamics for the first time in vertebrate oocytes in sufficient spatial and temporal resolution, in real time, and in vivo. We show that multiple-stage oocytes can be simultaneously analyzed within the same ovary and that multiple ovaries can be processed in single experiments. We detail our protocol and provide adequately consistent experimental conditions suitable for quantitative image acquisition and analysis. Combining live time-lapse analysis with transgenic reporter and mutant lines will provide a powerful approach to investigate early oogenesis. Recent efforts in zebrafish characterized and dissected key events in early oogenesis [4, 53], established the sequential events of chromosomal pairing [12, 67], identified functional mutants [6, 54, 78–81], generated key transgenic reporter lines for microscopy and biochemical experiments [63, 64, 82], and developed methodologies for analysis in vivo [83]. Our protocol expands further the use of the zebrafish model system to understand germ cell and ovarian development in postembryonic stages.

2 Materials

All solutions are made with autoclaved sterile water and analytical grade reagents. Make all solutions fresh (unless otherwise indicated) and prewarmed to 28 °C (unless otherwise indicated).

2.1 Ovary Dissection and Culture

1. Dissecting dish. Can be replaced with in-house made options (*see* **Note 1**).
2. Forceps #5.
2. Micro-scissors.
3. Tricaine. Make 0.4% stock in sterile water.

4. Cooling incubator set to 28 °C.
5. Hank's solution: 0.137 M NaCl, 5.4 mM KCl, 0.25 mM Na_2H PO_4, 0.44 mM KH_2 PO_4, 1.3 mM $CaCl_2$, 1.0 mM Mg SO_4, 4.2 mM NaH CO_3. Follow the *Zebrafish Book* for making and storage [84] (*see* **Note 2**).
6. 2× L-15, without L-glutamine and phenol red. Store at 4 °C (*see* **Note 3**).
7. GlutaMax 100×. Store at RT.
8. HL-15 solution: 50% Hank's, 50% L-15, 1:100 GlutaMax.
9. Hoechst H33342. Final solution 6.66 μM in HL-15.
10. Mitotracker Red CMXRos. Final solution 500 nM in HL-15 (*see* **Note 4**).
11. $DiOC_6$. Final solution 0.001 mg/ml.
12. Glass Petri dish, 60 × 15 mm.
13. Glass 9-well plate.

2.2 Mounting

1. Low-melt agarose (gelling temperature, 27.4 °C).
2. Mounting Solution A: 1% low-melt agarose in Hank's solution. Store at 4 °C (*see* **Note 5**).
3. Mounting Solution B: make 500 μl of 490 μl of 2× L-15 (no L-glutamine, no phenol red), 10 μl GlutaMax. This is equivalent to a 2× HL-15 solution. Make fresh and keep at 28 °C.
4. Final Mounting Medium: mix 500 μl of solution A with 500 μl of solution B to make a final solution of 0.5% low-melt agarose in 1× HL-15.
5. Glass bottom dish, 35 mm, or CELLview cell culture dish with glass bottom, 35 mm, four compartments (*see* **Notes 6** and 7).

2.3 Microscopy

1. Zeiss LSM 880 confocal microscope or equivalent.
2. Environmental cage for controlled conditions of the stage and objectives (*see* **Note 8**).
3. Environmental stage chamber for controlled incubation of samples.
4. Z-piezo component for faster z-stack scanning—optional.
5. Immersing media for 40× water objective—Immersol W *ne* = 1334 (Zeiss)—store at 28 °C in the microscope cage (*see* **Note 9**).
6. Immersing media for 63× oil objective 518 F *ne* = 1518 (Zeiss); store at 28 °C in the microscope cage.

3 Methods

3.1 Ovarian Isolation, Staining, and Culturing

1. Collect fish at the desired age between 4 and 8 weeks post-fertilization (wpf), and anesthetize single fish in 0.02% tricaine (*see* **Note 10**).
2. Transfer the fish to a dissecting dish under a dissecting microscope.
3. Measure the fish standard length (SL; distance from snout to base of the tail [85]), and confirm SL = 10–20 mm (Fig. 1Aa; *see* **Notes 10** and **11**).
4. To prepare for dissecting out the ovaries, pin the fish to the plate through the eye and the base of the tail.
5. To expose the visceral cavity and access the ovaries, remove the pectoral fin, and cut the fish body wall as follows using micro-scissors:
   (a) From the cloaca to the gills along the ventral midline.
   (b) From the cloaca dorsally.
   (c) From the anterior end of cut "a" dorsally.
   (d) Hold the cut body wall and remove it by cutting along the dorsal ends of cuts "b" and "c."
6. Cut off and discard the head and the tail.
7. Transfer the remaining trunk piece that contains the visceral organs and ovaries into prewarmed (28 °C) Hank's solution in a glass dish.
8. Hold the trunk piece at the bottom of the dish with one hand using forceps. Keep holding and remove the digestive system using forceps with your other hand. Figure 1Ab shows the trunk piece after removal of the digestive system.
9. Locate the ovaries: the two ovaries are bilaterally found dorso-laterally to the swim bladder and under a layer of lipid cells (Fig. 1Ab). The ovary that is closer to the microscope objective (Fig. 1Ab, black open arrowhead) may appear "floating." This ovary was connected to the body wall that was removed in **step 5**. The ovary that is closer to the microscope stage (Fig. 1Ab, magenta open arrowhead) will be still connected to the remaining body wall on the other side of the fish. The two ovaries will be connected to each other at their posterior ends that merged into the oviduct.
10. Punctuate the swim bladder to deflate it and remove (*see* **Note 12**).
11. Find the anterior and posterior connection points of the ovaries to the body.

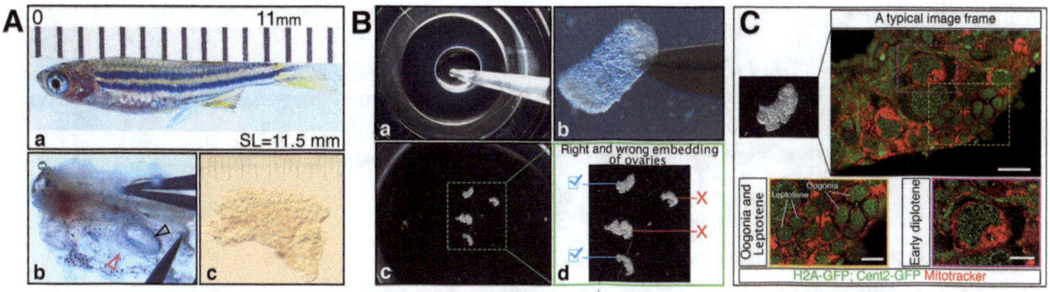

Fig. 1 Protocol for dissection, culture, and mounting ovaries for live time-lapse microscopy. (**A**) Ovary dissection from a juvenile zebrafish. (a) A fish with SL = 11.5 mm is shown. Ruler grades are 1 mm. (b) **Steps 5–9** in Subheading 3.1 result in a trunk piece (anterior to the left), where the digestive system is removed exposing the two ovaries laterally to the swim bladder. The "floating" ovary (black open arrowhead) is closer to the microscope objective plane, and the other ovary (magenta open arrowhead) is closer to the microscope stage plane. (c) **Step 15** in Subheading 3.1 results in a clean ovary that is ready for culturing and embedding. Scale long grades are 10 μm. Panels a and b show images of the same fish, and panel c is a representative ovary from a different fish of the same SL, sacrificed for photography purposes. (**B**) Mounting and embedding ovaries for culture and imaging. (**a**) Loading the mounting medium on a glass bottom dish (**step 3** in Subheading 3.2). (**b**) Embedding the ovary in the medium using the outer surface of the forceps (**step 5** in Subheading 3.2). (**c**) Embedded ovaries (**steps 5** and **6** in Subheading 3.2). (**d**) Examples of correct embedding of ovaries (enlargement of the green dashed box in c). Correct embedding of ovaries (blue lines and check marks) results in flattened ovaries that are adjacent to the glass bottom of the dish. Incorrect embedding (red lines and X marks) results in ovaries that are not flattened and/or are curled up or folded. (**C**) Typical confocal imaging frames of live ovaries. Top panel shows a typical ×0.8 zoom frame imaged by a × 40 objective as described in Subheading 3.3 **step 6**. Note multiple stage oocytes are detected, including oogonia and leptotene (orange dashed box, enlarged in bottom left panel) and early diplotene (magenta dashed box, enlarged in bottom right panel). Ovaries were taken from a double transgenic reporter line *Tg(h2a:h2a-gfp);Tg(bAct:cetn2-gfp)*, labeling histone and centrosomes, respectively (both in green), and stained with Mitotracker (red). Scale bars are top panel 20 μm and bottom panels 10 μm. The ovary in the top left bright field panel is not the ovary imaged in the remaining confocal panels and is shown as a representative for illustration purpose only

12. Gently lift the ovaries and cut the connection points to remove the ovaries. Cut the connective tissues at these connection points and not directly the ovaries (*see* **Note 13**).
13. Gently hold the ovaries at the bottom of the glass dish (*see* **Note 14**).
14. The two ovaries are connected in their posterior ends where they merge into the oviduct. Cut the posterior tip to separate the two ovaries (*see* **Note 15**).
15. Working on one ovary at a time, remove the lipid cells and connective tissues from around the ovary as much as possible. Avoid touching the ovaries directly to prevent damage, and only pull out the lipid and connective tissues (*see* **Note 16**). Figure 1Ac shows a clean ovary.
16. Transfer the clean ovaries into a glass well plate containing prewarmed (28 °C) HL-15, and store in a 28 °C incubator (*see* **Note 17**).

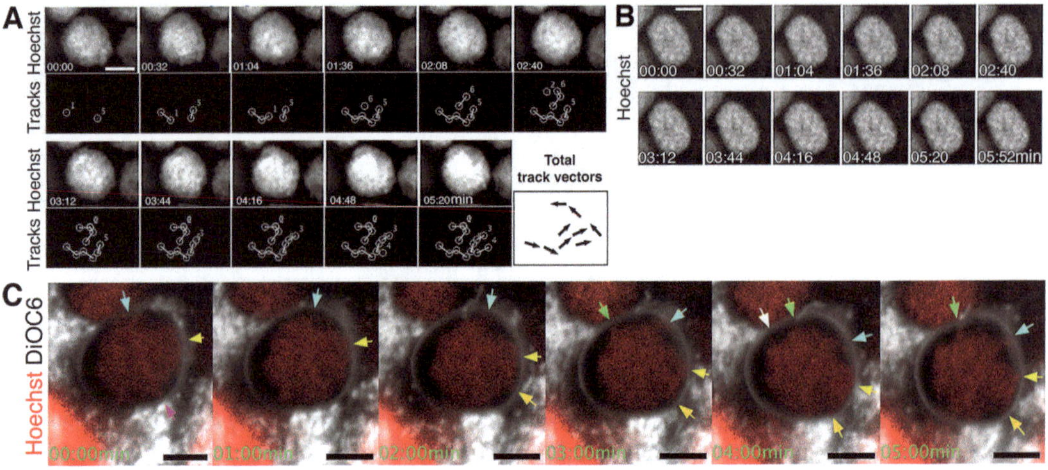

Fig. 2 Live time-lapse imaging of cultured ovaries reveals rapid meiotic chromosomal rotations during bouquet formation. (**a**) Live time-lapse images of chromosomal (Hoechst, gray) rotations during bouquet formation in zygotene oocytes. Lower panels show tracks of representative chromosomal regions, whose vector trajectories are summarized in the right panel. Images are snapshots from Supplemental Video 1. Scale bar is 5 μm. Time points are indicated in each panel. (**b**) Chromosomes do not rotate in premeiotic oogonial cells from the same ovary in (**a**). Images are snapshots from Supplemental Video 1. Scale bar is 5 μm. Time points are indicated in each panel. (**c**) Live time-lapse images of chromosomal (Hoechst, red) movements and the nuclear envelope (NE; labeled by the membrane detecting lipid dye DiOC6, gray) reveal presumptive telomere rotations in zygotene oocytes. Arrows indicate chromosomal arms that project to and interface the NE and likely represent telomere-NE attachment points that are characteristic to bouquet formation at this stage. Arrows marking specific chromosomal arms and presumptive telomere-NE attachment points throughout time are color-coded and demonstrate a clockwise rotation in this example. Images are snapshots from Supplemental Video 2. Scale bar is 5 μm. Time points are indicated in each panel. An additional representative example of zygotene oocyte is shown in Supplemental Video 3. In addition to the NE, DiOC6 (gray) labels membrane organelles in the cytoplasm. All images in **a–c** and Supplemental Videos 1, 2, and 3 are recorded as Z-stacks and are partial Z projections around the equatorial plane of the nuclei shown. Representatives of zygotene oocytes and oogonia from seven and six ovaries, respectively, from five independent experiments are shown. The detected zygotene chromosomal rotations and their rates are consistent with these features as reported in mice spermatocytes [27, 74]

17. Repeat **steps 1–15** with the desired number of fish. The cleaned ovaries from multiple fish of the same genotype/condition can be pooled in the glass well plate (*see* **Note 18**).
18. After all ovaries are collected, incubate them at 28 °C for 1 h.
19. Cultured ovaries can be stained with vital dyes, such as Mitotracker (Fig. 1C), Hoechst (Fig. 2), and DiOC6 (Fig. 2). To stain ovaries with vital dyes, replace the HL-15 with HL-15 containing the vital dyes for the incubation in **step 18**, and protect from light. The following steps will then need to be performed in HL-15 that contains the dyes used.
20. Under these conditions, ovaries can be cultured for several hours.

3.2 Mounting Ovaries for Live Imaging

1. Prepare Mounting Solutions A and B in advance:
   (a) Boil Mounting Solution A in a microwave, cool down on the bench for 10 min, and store 1 ml in 42 °C to keep it depolymerized.
   (b) Make 500 μl of Mounting Solution B and keep at 28 °C.
2. When you are ready to mount, rinse a glass bottom dish with Hank's solution.
3. Mix 500 μl of solution A with 500 μl of solution B, resulting in the Final Mounting Medium. Mix well and load 100 μl on the glass bottom of the dish under a dissecting scope (Fig. 1Ba). Spread evenly.
4. Allow the loaded Final Mounting Medium to cool down for 1 min. The solution will become mildly viscous and when embedding the ovaries will allow their easier manipulation and orientation.
5. Embed one ovary from Subheading 3.1 into the Final Mounting Medium: gently push the ovary down to the glass bottom, and use the outside surface of the forceps to flatten the ovary (Fig. 1Bb). Ideally, an ovary should lay straight with its flat side tightly against the glass (Fig. 1Bc–d, blue lines and check marks; *see* **Note 19**).
6. Repeat **step 5** to embed more ovaries. If using a four-compartment glass bottom dish (e.g., CELLview), up to four ovaries per compartment can be embedded. If using a single-compartment (regular) glass bottom dish, up to six ovaries can be embedded per dish (Fig. 1Bc; *see* **Note 20**).
7. Let the Final Mounting Medium solidify, and add HL-15 gently against the dish walls to prevent disturbing the solidified medium. If using a four-compartment glass bottom dish (CELLview), add 1 ml HL-15. If using a single-compartment glass bottom dish, add 4 ml HL-15 (*see* **Note 21**).
8. Store the dish at 28 °C and prepare for imaging in the next section. Protect from light (for optimal signal of vital dyes and/or transgenic fluorescent reporters).

3.3 Microscopy and Live Time-Lapse Image Acquisition

The following describes imaging on a Zeiss LSM880 confocal microscope and should be suitable in principle for similar confocal and spinning disc confocal microscopes.

1. While the ovaries are stored during **step 8** of Subheading 3.2, turn on the microscope, and set the desired objective (we use a 40× water or 63× oil objective). Images shown in Figs. 1C and

Fig. 2, and Supplemental Videos 1, 2, and 3 were acquired using a 40× water objective.

2. Set the incubation temperatures—microscope cage, 28 °C; stage environmental chamber bottom part, 29 °C; top lid part, 30 °C. Let all parts reach their set temperatures (*see* **Note 22**).
3. Load your sample, and close the environmental chamber and cage. Locate an ovary and adjust your focus.
4. Once focused, allow ~30–45 min for the sample and all microscope parts to equilibrate. This equilibration is required to prevent focal drifts during image acquisition.
5. At the end of **step 4**, the samples are ready for imaging.
6. Set the following image acquisition parameters (*see* **Note 23**):
   - *XY* = 1104 × 1104 pixels.
   - Zoom = ×0.8–×1 (Fig. 1C top panel shows a typical frame view).
   - 12 bit.
   - Pixel dwell time = 0.59 s.
   - Sampling averaging = ×2.
   - Laser power = 2.5–6% depending on the reporter fluorophore and vital dyes that are being imaged (*see* **Note 24**).
   - Gain = 380–650 depending on the reporter fluorophore and vital dyes that are being imaged.
   - Use the range indicator function to confirm that the power and gain set values do not reach signal saturation conditions.
   - Z thickness of each channel = 1.1 μm (*see* **Note 25**).
   - Z increments for stack acquisition = 0.53 μm (should be set to approximately half the Z thickness).
7. Set time-lapse parameters to:
   - Time intervals = 15–100 s depending on image acquisition parameters, size of Z-stack, and number of channels used.
   - Total time = 5–60 min.
   - (*see* **Note 26**)

     Images in Figs. 1C and 2, and Supplemental Videos 1, 2, and 3 were acquired using these parameters.
8. Using the "Live" scan, locate a region of interest (Fig. 1C), adjust the setting in **step** 7, and acquire images (*see* **Note 27**).
9. After image acquisition, time-lapse recordings are analyzed and formatted into TIFF and video files using the Fiji software. Figure 2 shows snapshots from Supplemental Videos 1, 2, and 3, which were recorded in the settings described here.

These four-dimensional recordings reveal rapid chromosomal movements during zygotene stages of early meiotic prophase in intact cultured ovaries (*see* **Note 28**).

4 Notes

1. Dissecting dishes can be made in-house by casting plastic Petri dishes with animal-proof nontoxic silicone for reusable dishes or with 2–3% agarose in Hank's solution for single-use dishes.
2. Hank's final solution is stored at 4 °C, and prewarm sufficient amount to 28 °C.
3. L-glutamine is not stable and is added fresh from a stock (GlutaMax). Avoid the use of phenol red since it is autofluorescent and will increase background during image acquisition.
4. This Mitotracker version selectively stains active mitochondria and can serve as indication for cell viability.
5. This solution can be made in larger volumes, stored at 4 °C, and boiled to make fresh aliquots for every experiment.
6. For any glass bottom dish, it is critical to confirm that the thickness of the glass is 0.17 mm (#1.5 glass). Confocal microscopes usually correct for spherical aberrations along the *Z* axis for a 0.17 mm thick glass. Different thicknesses might result in spherical aberrations during image acquisition.
7. The CELLview four-compartment dish is more cost-effective since it requires smaller amounts of medium and reagents but is also advantageous for experiments where control and experiment samples can be cultured side by side in different compartments of the same dish for maximal consistency of conditions within experiments.
8. If an environmental cage is not available, use an objective heater in addition to the stage-insert environmental chamber. Microscope rooms are often chilled, and therefore without a cage, the objective is cold and when touching the sample will act as a heat sink and cool down your sample.
9. During longer imaging sessions, immersing water might evaporate. Changes in immersion volume or its complete evaporation might affect light refraction and image acquisition. Immersol has a refraction index that is equivalent to that of water, but does not evaporate.
10. The age of the fish used can be determined according to the oogenesis stage of interest. Since juvenile growth varies between fish facilities and also between individual fish that are raised in the same tank, the standard length (SL, distance from snout to base of the tail) value is an accurate and consistent

measure of fish postembryonic development [85]. In our facility, at 4 wpf (SL ≅ 10 mm) ovaries include mostly germline stem cells, oogonia, and early meiotic onset oocytes, and subsequent stages of meiotic oocytes of ≤60 μm in diameter (St. I-diplotene) progressively appear through 5–6 wpf (SL = 10–14 mm) and 7–8 wpf (SL = 15–21 mm) [5, 83, 86]. At 7–8 wpf and beyond, ovaries might convert to testes and become thinner and more elongated and appear milky. These gonads contain apoptotic oocytes and should not be used to analyze oocyte development. This protocol is suitable for fish SL = 10–20 mm (Fig. 1 A).

11. For reproducibility, it is recommended to document the SL of fish used in experiments.
12. The "floating" ovary might be removed or injured during the removal of the swim bladder; be careful to gently remove only the swim bladder. If the ovary is removed with it, you can still detach it from the bladder and clean as in **steps 13** and **15**. While the ovary is still connected to lipid cells, it tends to float to the surface of the Hank's solution in the dish; try to hold it to avoid losing it.
13. Live ovaries are soft; be very careful not to injure them. In most cases, one can dissect two intact ovaries from a single fish, but this could be challenging. Ovaries might be lost during the removal of the body wall (**step 5**) or the digestive system (**step 8**) or the swim bladder (**step 10**). In each step, look carefully to check if you can still locate the ovaries. Check the removed parts as well. Often, even if ovaries are removed with them, they can still be intact and good for analysis.
14. As long as lipid cells surround the ovaries, they will tend to float and might be lost. Since in **step 15** you will work on one ovary at a time, remove some of the lipid cells now before separating the ovaries in **step 14**, to avoid losing the ovary you are not working on. Only remove lipid cells until the ovaries do not float too much; you will clean them more thoroughly in **step 15**. Live ovaries tend to curl up once detached from the body; you will straighten them later during mounting (Fig. 1B).
15. Sometimes it is easier to separate the two ovaries before removing them completely in **step 12**. In such case, separate and remove the "floating" ovary, perform **steps 15** and **16**, and go back to the same trunk piece to remove the other ovary for **steps 15** and **16**.
16. Live ovaries tend to curl up once detached from the body; gently straighten them to remove lipid and connective tissues. Remove all lipid cells to avoid light scattering through them during imaging. Remove as much connective tissue as possible to improve vital dye penetration and make the tissue less thick,

which is also better for imaging. However, it is better to leave a little bit of connective tissue than to injure the ovary. Figure 1Ac, Bb–d shows properly clean ovaries.

17. Transferring the ovaries can be done by gently lifting them up using forceps, or by aspiring them using a glass Pasteur pipette (do not use plastic as the ovaries will adhere to it and might get stuck in the pipette).
18. Keep the plate containing dissected ovaries at 28 °C, and only take it out when the next ovary is ready. Immediately place the plate with the newly dissected ovary back to 28 °C.
19. Flattening the ovary to be straight adjacently to the glass bottom of the dish is critical. The more the ovary is flattened and the closer it is to the glass, the better it is for image acquisition. In confocal microscopy, the quality of the image decreases with the distance of the tissue from the objective. Sometimes ovaries will not be perfectly flat. If the ovary is still close to the glass, it can still be imaged; however, only the ovary portions that are adjacent to the glass can be imaged at sufficient quality. Figure 1Ad shows properly flattened ovaries (blue lines and check marks) and ovaries that are not properly flattened (red lines and X marks).
20. When embedding additional ovaries, embed them in a sufficient distance from previously embedded ones, and be careful not to move them to avoid disrupting their orientation.
21. If using vital dyes, add HL-15 that contains them.
22. Keeping the sample temperature at 28 °C during imaging is critical for live experiments in order to keep conditions as close to physiological as possible. This is especially important when examining temperature-sensitive processes like ones that depend on microtubules, which depolymerize under chilled conditions.
23. For quantitative imaging and for equivalent and reproducible image acquisition within and between samples and experiments, these parameters need to be set to consistent values in all images throughout all experiments. We recommend using the Zeiss "Reuse" button that allows to import all image acquisition parameters from saved acquired images and apply them to new images.
24. Keep laser power as low as possible to prevent tissue damage and photobleaching.
25. Determine the size of the pinhole to provide a thickness of 1.1 μm to all channels for a consistent Z resolution. An Airy unit is associated with the pinhole size but is wavelength dependent. Therefore, the same Airy unit value will provide different *Z* thicknesses for different channels and disrupt

consistent data acquisition. Do not adjust the Airy unit to be consistent.

26. In confocal microscopy there is a tradeoff between the described parameters. To record in higher temporal resolution if needed, image acquisition needs to be faster. A faster image acquisition can be achieved by reducing parameters like the bit of the image, *XY* pixel number, number of *Z* slices per stack, number of channels recorded, and *Z* increments in stacks when acceptable. However, these will reduce the spatial resolution in large frames. It is recommended to use these faster parameters when imaging smaller frames of specific cells of interest.
27. A great advantage of the juvenile zebrafish ovary is that it is morphologically flat and translucent and that it contains all the early stages of oogenesis, from the germline stem cells, through oogonia and early meiotic oocytes, to St. I-diplotene oocytes [5, 83]. In the settings described here, frames covering sufficiently large areas of the ovary can be recorded that contain many or all of these stages in high spatial resolution (Fig. 1C, top panel). This allows you to later zoom in to specific regions while preserving spatial resolution (Fig. 1C, bottom panels). This approach enables a robust analysis of multiple stages in high numbers.
28. Our protocol detects meiotic chromosomal movements that are characteristic to early prophase I chromosomal homology searches (Fig. 2 and Supplemental Videos 1, 2, and 3). We detected rapid chromosomal rotational movements specifically during zygotene stages (Fig. 2a, Supplemental Video 1; oocytes are staged according to [83]). These movements were not observed in other oocyte stages, as shown here in oogonial cells that were imaged from the same ovary as Fig. 2a as control (Fig. 2b, Supplemental Video 1). Since chromosomal homology search movements are led by NE-tethered telomeres during bouquet formation in zygotene stages, we recorded ovaries co-labeled for chromosomes (Hoechst) and the NE (DiOC6). We observed that during rotations, chromosomal arms project to and interface the NE (Fig. 2c, arrowheads). The sites of chromosome-NE interface at this stage likely represent telomere-NE attachment points, as previously shown in fixed samples and by ultrastructure analysis [5]. The detected presumptive telomere-NE attachment points (distinct points are color-coded through time in Fig. 2c) rotated in the same direction and rate as the bulk chromosomal rotation movements, as expected in bouquet dynamics [27, 74]. An additional example of zygotene oocytes exhibiting chromosomal movements along the inner side of the NE is shown in Supplemental Video 3. The chromosomal rotational movements we detect are consistent with bouquet dynamics as

recorded in mice spermatocytes [27, 74]. To our knowledge, these are the first live recordings of zygotene chromosomal movements in oocytes within intact cultured ovaries in a vertebrate species.

References

1. Elkouby YM (2017) All in one - integrating cell polarity, meiosis, mitosis and mechanical forces in early oocyte differentiation in vertebrates. Int J Dev Biol 61(3–5):179–193. https://doi.org/10.1387/ijdb.170030ye
2. Selman K, Wallace RA, Sarka A, Qi XP (1993) Stages of oocyte development in the zebrafish, brachydanio-rerio. J Morphol 218 (2):203–224. https://doi.org/10.1002/Jmor.1052180209
3. Leu DH, Draper BW (2010) The ziwi promoter drives germline-specific gene expression in zebrafish. Dev Dyn 239(10):2714–2721. https://doi.org/10.1002/dvdy.22404
4. Elkouby YM, Mullins MC (2017) Coordination of cellular differentiation, polarity, mitosis and meiosis – new findings from early vertebrate oogenesis. Dev Biol 430(2):275–287. https://doi.org/10.1016/j.ydbio.2017.06.029
5. Elkouby YM, Jamieson-Lucy A, Mullins MC (2016) Oocyte polarization is coupled to the chromosomal bouquet, a conserved polarized nuclear configuration in meiosis. PLoS Biol 14 (1):e1002335. https://doi.org/10.1371/journal.pbio.1002335
6. Marlow FL, Mullins MC (2008) Bucky ball functions in Balbiani body assembly and animal-vegetal polarity in the oocyte and follicle cell layer in zebrafish. Dev Biol 321 (1):40–50. https://doi.org/10.1016/j.ydbio.2008.05.557
7. Greenbaum MP, Ma L, Matzuk MM (2007) Conversion of midbodies into germ cell intercellular bridges. Dev Biol 305(2):389–396. https://doi.org/10.1016/j.ydbio.2007.02.025
8. Greenbaum MP, Yan W, Wu MH, Lin YN, Agno JE, Sharma M, Braun RE, Rajkovic A, Matzuk MM (2006) TEX14 is essential for intercellular bridges and fertility in male mice. Proc Natl Acad Sci U S A 103(13):4982–4987. https://doi.org/10.1073/pnas.0505123103
9. Gray S, Cohen PE (2016) Control of meiotic crossovers: from double-strand break formation to designation. Annu Rev Genet 50 (1):175–210. https://doi.org/10.1146/annurev-genet-120215-035111
10. Hughes SE, Miller DE, Miller AL, Hawley RS (2018) Female meiosis: synapsis, recombination, and segregation in *Drosophila melanogaster*. Genetics 208(3):875–908. https://doi.org/10.1534/genetics.117.300081
11. Zickler D, Kleckner N (2015) Recombination, pairing, and synapsis of homologs during meiosis. Cold Spring Harb Perspect Biol 7(6). https://doi.org/10.1101/cshperspect.a016626
12. Blokhina YP, Nguyen AD, Draper BW, Burgess SM (2019) The telomere bouquet is a hub where meiotic double-strand breaks, synapsis, and stable homolog juxtaposition are coordinated in the zebrafish, *Danio rerio*. PLoS Genet 15(1):e1007730. https://doi.org/10.1371/journal.pgen.1007730
13. Bolcun-Filas E, Rinaldi VD, White ME, Schimenti JC (2014) Reversal of female infertility by Chk2 ablation reveals the oocyte DNA damage checkpoint pathway. Science 343 (6170):533–536. https://doi.org/10.1126/science.1247671
14. Burgoyne PS, Mahadevaiah SK, Turner JM (2009) The consequences of asynapsis for mammalian meiosis. Nat Rev Genet 10 (3):207–216. https://doi.org/10.1038/nrg2505
15. Lake CM, Hawley RS (2012) The molecular control of meiotic chromosomal behavior: events in early meiotic prophase in Drosophila oocytes. Annu Rev Physiol 74:425–451. https://doi.org/10.1146/annurev-physiol-020911-153342
16. MacQueen AJ, Hochwagen A (2011) Checkpoint mechanisms: the puppet masters of meiotic prophase. Trends Cell Biol 21 (7):393–400. https://doi.org/10.1016/j.tcb.2011.03.004
17. Ding X, Xu R, Yu J, Xu T, Zhuang Y, Han M (2007) SUN1 is required for telomere attachment to nuclear envelope and gametogenesis in mice. Dev Cell 12(6):863–872. https://doi.org/10.1016/j.devcel.2007.03.018
18. Penkner AM, Fridkin A, Gloggnitzer J, Baudrimont A, Machacek T, Woglar A, Csaszar E, Pasierbek P, Ammerer G, Gruenbaum Y, Jantsch V (2009) Meiotic chromosome homology search involves

modifications of the nuclear envelope protein matefin/SUN-1. Cell 139(5):920–933. https://doi.org/10.1016/j.cell.2009.10.045

19. Shibuya H, Ishiguro K, Watanabe Y (2014) The TRF1-binding protein TERB1 promotes chromosome movement and telomere rigidity in meiosis. Nat Cell Biol 16(2):145–156. https://doi.org/10.1038/ncb2896
20. Sato A, Isaac B, Phillips CM, Rillo R, Carlton PM, Wynne DJ, Kasad RA, Dernburg AF (2009) Cytoskeletal forces span the nuclear envelope to coordinate meiotic chromosome pairing and synapsis. Cell 139(5):907–919. https://doi.org/10.1016/j.cell.2009.10.039
21. Chikashige Y, Tsutsumi C, Yamane M, Okamasa K, Haraguchi T, Hiraoka Y (2006) Meiotic proteins bqt1 and bqt2 tether telomeres to form the bouquet arrangement of chromosomes. Cell 125(1):59–69. https://doi.org/10.1016/j.cell.2006.01.048
22. Conrad MN, Dominguez AM, Dresser ME (1997) Ndj1p, a meiotic telomere protein required for normal chromosome synapsis and segregation in yeast. Science 276 (5316):1252–1255
23. Trelles-Sticken E, Dresser ME, Scherthan H (2000) Meiotic telomere protein Ndj1p is required for meiosis-specific telomere distribution, bouquet formation and efficient homologue pairing. J Cell Biol 151(1):95–106
24. Horn HF, Kim DI, Wright GD, Wong ES, Stewart CL, Burke B, Roux KJ (2013) A mammalian KASH domain protein coupling meiotic chromosomes to the cytoskeleton. J Cell Biol 202(7):1023–1039. https://doi.org/10.1083/jcb.201304004
25. Morimoto A, Shibuya H, Zhu X, Kim J, Ishiguro K, Han M, Watanabe Y (2012) A conserved KASH domain protein associates with telomeres, SUN1, and dynactin during mammalian meiosis. J Cell Biol 198 (2):165–172. https://doi.org/10.1083/jcb.201204085
26. Shibuya H, Hernandez-Hernandez A, Morimoto A, Negishi L, Hoog C, Watanabe Y (2015) MAJIN links telomeric DNA to the nuclear membrane by exchanging telomere cap. Cell 163(5):1252–1266. https://doi.org/10.1016/j.cell.2015.10.030
27. Shibuya H, Morimoto A, Watanabe Y (2014) The dissection of meiotic chromosome movement in mice using an in vivo electroporation technique. PLoS Genet 10(12):e1004821. https://doi.org/10.1371/journal.pgen.1004821
28. Long J, Huang C, Chen Y, Zhang Y, Shi S, Wu L, Liu Y, Liu C, Wu J, Lei M (2017) Telomeric TERB1-TRF1 interaction is crucial for male meiosis. Nat Struct Mol Biol. https://doi.org/10.1038/nsmb.3496
29. Pendlebury DF, Fujiwara Y, Tesmer VM, Smith EM, Shibuya H, Watanabe Y, Nandakumar J (2017) Dissecting the telomere-inner nuclear membrane interface formed in meiosis. Nat Struct Mol Biol. https://doi.org/10.1038/nsmb.3493
30. Scherthan H (2001) A bouquet makes ends meet. Nat Rev Mol Cell Biol 2(8):621–627. https://doi.org/10.1038/35085086
31. Ortega S, Prieto I, Odajima J, Martin A, Dubus P, Sotillo R, Barbero JL, Malumbres M, Barbacid M (2003) Cyclin-dependent kinase 2 is essential for meiosis but not for mitotic cell division in mice. Nat Genet 35(1):25–31. https://doi.org/10.1038/ng1232
32. Viera A, Alsheimer M, Gomez R, Berenguer I, Ortega S, Symonds CE, Santamaria D, Benavente R, Suja JA (2015) CDK2 regulates nuclear envelope protein dynamics and telomere attachment in mouse meiotic prophase. J Cell Sci 128(1):88–99. https://doi.org/10.1242/jcs.154922
33. Viera A, Rufas JS, Martinez I, Barbero JL, Ortega S, Suja JA (2009) CDK2 is required for proper homologous pairing, recombination and sex-body formation during male mouse meiosis. J Cell Sci 122(Pt 12):2149–2159. https://doi.org/10.1242/jcs.046706
34. Blokhina YP, Nguyen AD, Draper BW, Burgess SM (2018) The telomere bouquet is a hub where meiotic double-strand breaks, synapsis, and stable homolog juxtaposition are coordinated in the zebrafish, *Danio rerio*. bioRxiv 2018:428086. https://doi.org/10.1101/428086
35. Cox RT, Spradling AC (2003) A Balbiani body and the fusome mediate mitochondrial inheritance during Drosophila oogenesis. Development 130(8):1579–1590
36. Kloc M, Jedrzejowska I, Tworzydlo W, Bilinski SM (2014) Balbiani body, nuage and sponge bodies--term plasm pathway players. Arthropod Struct Dev 43(4):341–348. https://doi.org/10.1016/j.asd.2013.12.003
37. Pepling ME, Wilhelm JE, O'Hara AL, Gephardt GW, Spradling AC (2007) Mouse oocytes within germ cell cysts and primordial follicles contain a Balbiani body. Proc Natl Acad Sci U S A 104(1):187–192. https://doi.org/10.1073/pnas.0609923104
38. Lei L, Spradling AC (2016) Mouse oocytes differentiate through organelle enrichment from sister cyst germ cells. Science 352

(6281):95–99. https://doi.org/10.1126/science.aad2156

39. Jaglarz MK, Nowak Z, Bilinski SM (2003) The Balbiani body and generation of early asymmetry in the oocyte of a tiger beetle. Differentiation 71(2):142–151. https://doi.org/10.1046/j.1432-0436.2003.710205.x
40. Jedrzejowska I, Kubrakiewicz J (2007) The Balbiani body in the oocytes of a common cellar spider, *Pholcus phalangioides* (Araneae: Pholcidae). Arthropod Struct Dev 36 (3):317–326. https://doi.org/10.1016/j.asd.2007.01.004
41. VonWittich W (1845) Dissertatio sistens observations quaedam de sistens aranearum ex ovo evolutione. Halis Saxonum
42. Marlow FL, Mullins MC (2008) Bucky ball functions in Balbiani body assembly and animal-vegetal polarity in the oocyte and follicle cell layer in zebrafish. Dev Biol 321 (1):40–50. https://doi.org/10.1016/j.ydbio.2008.05.557
43. Dumont JN (1978) Oogenesis in *Xenopus laevis* (Daudin). VI The route of injected tracer transport in the follicle and developing oocyte. J Exp Zool 204(2):193–217. https://doi.org/10.1002/jez.1402040208
44. Carlson JL, Bakst MR, Ottinger MA (1996) Developmental stages of primary oocytes in turkeys. Poult Sci 75(12):1569–1578
45. Rodler D, Sinowatz F (2013) Expression of intermediate filaments in the Balbiani body and ovarian follicular wall of the Japanese quail (*Coturnix japonica*). Cells Tissues Organs 197(4):298–311. https://doi.org/10.1159/000346048
46. Ukeshima A, Fujimoto T (1991) A fine morphological study of germ cells in asymmetrically developing right and left ovaries of the chick. Anat Rec 230(3):378–386. https://doi.org/10.1002/ar.1092300311
47. Weakley BS (1967) "Balbiani's body" in the oocyte of the golden hamster. Zeitschr Zellforsch Mikroskop Anat 83(4):583–588
48. Barton BR, Hertig AT (1972) Ultrastructure of annulate lamellae in primary oocytes of chimpanzees (*Pan troglodytes*). Biol Reprod 6 (1):98–108
49. Albamonte MI, Albamonte MS, Stella I, Zuccardi L, Vitullo AD (2013) The infant and pubertal human ovary: Balbiani's body-associated VASA expression, immunohistochemical detection of apoptosis-related BCL2 and BAX proteins, and DNA fragmentation. Hum Reprod 28(3):698–706. https://doi.org/10.1093/humrep/des453
50. Hertig AT (1968) The primary human oocyte: some observations on the fine structure of Balbiani's vitelline body and the origin of the annulate lamellae. Am J Anat 122 (1):107–137. https://doi.org/10.1002/aja.1001220107
51. Tworzydlo W, Marek M, Kisiel E, Bilinski SM (2016) Meiosis, Balbiani body and early asymmetry of Thermobia oocyte. Protoplasma. https://doi.org/10.1007/s00709-016-0978-7
52. Escobar-Aguirre M, Elkouby YM, Mullins MC (2017) Localization in oogenesis of maternal regulators of embryonic development. Adv Exp Med Biol 953:173–207
53. Escobar-Aguirre M, Zhang H, Jamieson-Lucy A, Mullins MC (2017) Microtubule-actin crosslinking factor 1 (Macf1) domain function in Balbiani body dissociation and nuclear positioning. PLoS Genet 13(9): e1006983. https://doi.org/10.1371/journal.pgen.1006983
54. Gupta T, Marlow FL, Ferriola D, Mackiewicz K, Dapprich J, Monos D, Mullins MC (2010) Microtubule actin crosslinking factor 1 regulates the Balbiani body and animal-vegetal polarity of the zebrafish oocyte. PLoS Genet 6(8):e1001073. https://doi.org/10.1371/journal.pgen.1001073
55. Langdon YG, Mullins MC (2011) Maternal and zygotic control of zebrafish dorsoventral axial patterning. Annu Rev Genet 45:357–377. https://doi.org/10.1146/annurev-genet-110410-132517
56. Lu FI, Thisse C, Thisse B (2011) Identification and mechanism of regulation of the zebrafish dorsal determinant. Proc Natl Acad Sci U S A 108(38):15876–15880. https://doi.org/10.1073/pnas.1106801108
57. Tran LD, Hino H, Quach H, Lim S, Shindo A, Mimori-Kiyosue Y, Mione M, Ueno N, Winkler C, Hibi M, Sampath K (2012) Dynamic microtubules at the vegetal cortex predict the embryonic axis in zebrafish. Development 139(19):3644–3652. https://doi.org/10.1242/dev.082362
58. Nojima H, Rothhamel S, Shimizu T, Kim CH, Yonemura S, Marlow FL, Hibi M (2010) Syntabulin, a motor protein linker, controls dorsal determination. Development 137 (6):923–933. https://doi.org/10.1242/dev.046425
59. Colozza G, De Robertis EM (2014) Maternal syntabulin is required for dorsal axis formation and is a germ plasm component in Xenopus. Differentiation 88(1):17–26. https://doi.org/10.1016/j.diff.2014.03.002

60. Extavour CG, Akam M (2003) Mechanisms of germ cell specification across the metazoans: epigenesis and preformation. Development 130(24):5869–5884. https://doi.org/10.1242/dev.00804
61. Kobayashi S, Amikura R, Okada M (1994) Localization of mitochondrial large rRNA in germinal granules and the consequent segregation of germ line. Int J Dev Biol 38 (2):193–199
62. Kosaka K, Kawakami K, Sakamoto H, Inoue K (2007) Spatiotemporal localization of germ plasm RNAs during zebrafish oogenesis. Mech Dev 124(4):279–289. https://doi.org/10.1016/j.mod.2007.01.003
63. Krishnakumar P, Riemer S, Perera R, Lingner T, Goloborodko A, Khalifa H, Bontems F, Kaufholz F, El-Brolosy MA, Dosch R (2018) Functional equivalence of germ plasm organizers. PLoS Genet 14(11): e1007696. https://doi.org/10.1371/journal.pgen.1007696
64. Riemer S, Bontems F, Krishnakumar P, Gomann J, Dosch R (2015) A functional Bucky ball-GFP transgene visualizes germ plasm in living zebrafish. Gene Expr Patterns 18(1–2):44–52. https://doi.org/10.1016/j.gep.2015.05.003
65. Kloc M, Bilinski S, Etkin LD (2004) The Balbiani body and germ cell determinants: 150 years later. Curr Top Dev Biol 59:1–36. https://doi.org/10.1016/S0070-2153(04)59001-4
66. Houston DW (2013) Regulation of cell polarity and RNA localization in vertebrate oocytes. Int Rev Cell Mol Biol 306:127–185. https://doi.org/10.1016/B978-0-12-407694-5.00004-3
67. Saito K, Sakai C, Kawasaki T, Sakai N (2014) Telomere distribution pattern and synapsis initiation during spermatogenesis in zebrafish. Dev Dyn 243(11):1448–1456. https://doi.org/10.1002/dvdy.24166
68. Tworzydlo W, Kisiel E, Jankowska W, Witwicka A, Bilinski SM (2016) Exclusion of dysfunctional mitochondria from Balbiani body during early oogenesis of Thermobia. Cell Tissue Res. https://doi.org/10.1007/s00441-016-2414-x
69. Collins KA, Unruh JR, Slaughter BD, Yu Z, Lake CM, Nielsen RJ, Box KS, Miller DE, Blumenstiel JP, Perera AG, Malanowski KE, Hawley RS (2014) Corolla is a novel protein that contributes to the architecture of the synaptonemal complex of Drosophila. Genetics 198(1):219–228. https://doi.org/10.1534/genetics.114.165290
70. Lake CM, Nielsen RJ, Guo F, Unruh JR, Slaughter BD, Hawley RS (2015) Vilya, a component of the recombination nodule, is required for meiotic double-strand break formation in Drosophila. eLife 4:e08287. https://doi.org/10.7554/eLife.08287
71. Kohler S, Wojcik M, Xu K, Dernburg AF (2017) Superresolution microscopy reveals the three-dimensional organization of meiotic chromosome axes in intact *Caenorhabditis elegans* tissue. Proc Natl Acad Sci U S A 114(24): E4734–E4743. https://doi.org/10.1073/pnas.1702312114
72. Rog O, Dernburg AF (2015) Direct visualization reveals kinetics of meiotic chromosome synapsis. Cell Rep. https://doi.org/10.1016/j.celrep.2015.02.032
73. Rog O, Kohler S, Dernburg AF (2017) The synaptonemal complex has liquid crystalline properties and spatially regulates meiotic recombination factors. eLife 6. https://doi.org/10.7554/eLife.21455
74. Lee CY, Horn HF, Stewart CL, Burke B, Bolcun-Filas E, Schimenti JC, Dresser ME, Pezza RJ (2015) Mechanism and regulation of rapid telomere prophase movements in mouse meiotic chromosomes. Cell Rep. https://doi.org/10.1016/j.celrep.2015.03.045
75. Mikl M, Cowan CR (2015) Cell polarity in one-cell *C. elegans* embryos: ensuring an accurate and precise spatial axis during development. In: Ebnet K (ed) Cell polarity 2: role in development and disease. Springer, Cham, pp 3–32. https://doi.org/10.1007/978-3-319-14466-5_1
76. Christophorou N, Rubin T, Bonnet I, Piolot T, Arnaud M, Huynh JR (2015) Microtubule-driven nuclear rotations promote meiotic chromosome dynamics. Nat Cell Biol 17 (11):1388–1400. https://doi.org/10.1038/ncb3249
77. Christophorou N, Rubin T, Huynh JR (2013) Synaptonemal complex components promote centromere pairing in pre-meiotic germ cells. PLoS Genet 9(12):e1004012. https://doi.org/10.1371/journal.pgen.1004012
78. Saito K, Siegfried KR, Nusslein-Volhard C, Sakai N (2011) Isolation and cytogenetic characterization of zebrafish meiotic prophase I mutants. Dev Dyn 240(7):1779–1792. https://doi.org/10.1002/dvdy.22661
79. Webster KA, Henke K, Ingalls DM, Nahrin A, Harris MP, Siegfried KR (2018) Cyclin-dependent kinase 21 is a novel regulator of proliferation and meiosis in the male germline of zebrafish. Reproduction 157(4):383–398. https://doi.org/10.1530/REP-18-0386

80. Dingare C, Niedzwetzki A, Klemmt PA, Godbersen S, Fuentes R, Mullins MC, Lecaudey V (2018) The Hippo pathway effector Taz is required for cell morphogenesis and fertilization in zebrafish. Development 145(22): dev167023. https://doi.org/10.1242/dev.167023
81. Yi X, Yu J, Ma C, Dong G, Shi W, Li H, Li L, Luo L, Sampath K, Ruan H, Huang H (2019) The effector of Hippo signaling, Taz, is required for formation of the micropyle and fertilization in zebrafish. PLoS Genet 15(1): e1007408. https://doi.org/10.1371/journal.pgen.1007408
82. Roovers EF, Kaaij LJT, Redl S, Bronkhorst AW, Wiebrands K, de Jesus Domingues AM, Huang HY, Han CT, Riemer S, Dosch R, Salvenmoser W, Grun D, Butter F, van Oudenaarden A, Ketting RF (2018) Tdrd6a regulates the aggregation of buc into functional subcellular compartments that drive germ cell specification. Dev Cell 46 (3):285–301. e289. https://doi.org/10.1016/j.devcel.2018.07.009
83. Elkouby Y, Mullins M (2017) Methods for the analysis of early oogenesis in zebrafish. Dev Biol 430(2):310–324
84. Westerfield M (1993) The zebrafish book: a guide for the laboratory use of zebrafish (*Brachydanio rerio*). M. Westerfield, Eugene, OR
85. Parichy DM, Elizondo MR, Mills MG, Gordon TN, Engeszer RE (2009) Normal table of postembryonic zebrafish development: staging by externally visible anatomy of the living fish. Dev Dyn 238(12):2975–3015. https://doi.org/10.1002/dvdy.22113
86. Maack G, Segner H (2003) Morphological development of the gonads in zebrafish. J Fish Biol 62(4):895–906. https://doi.org/10.1046/J.1095-8649.2003.00074.X

Chapter 13

Detection of the Polar Body After Fertilization

Hua Ruan, Xiaogui Yi, and Honghui Huang

Abstract

The polar body, with haploid DNA, is a small cell produced during the meiosis of an oocyte. Here, we describe the detailed procedures for the detection of the second polar body in zebrafish (*Danio rerio*) embryos after 10 min post fertilization. A polar body can be easily distinguished as a small dot with a DAPI-stained nucleus surrounded by Phalloidin-labeled F-actin in each fertilized zebrafish embryo.

Key words Polar body, Meiosis, Zebrafish, Fertilization, Phalloidin, DAPI

1 Introduction

The polar body, also known as polar cell or polocyte, is a byproduct of an egg, the female gamete, and is produced during meiosis. In animals, polar bodies are sterile and short-lived. However, polar bodies in angiosperms (flowering plants) can be fertilized by sperm, and develop into endosperms to nourish embryos.

Eggs develop from oocytes by meiosis where two division cycles, meiosis I and II, happen [1–4]. As a teleost, zebrafish (*Danio rerio*) oocytes are divided into five developmental stages according to their sizes [5]. During oogenesis, zebrafish oocyte stays in the prophase of meiosis I for about 7 days from stage I to III [5–7], during which the oocyte grows greatly in size due to the accumulation of substances, such as proteins and mRNAs [8–10]. As meiosis proceeds, the first cell division is completed within 2 days when the oocyte at stage IV produces and expulses the first polar body [5]. Upon entering stage V, the oocyte becomes mature and then arrests at the metaphase of meiosis II [5, 11]. Once the mature oocyte is ovulated and activated, the second division of meiosis follows quickly, and completes by the extrusion of the second polar body [12, 13]. At the same time, the chorion, an acellular protective coat of the egg, expands promptly.

Since the first polar body produced by the oocyte at stage IV is inside the zebrafish ovary, it is difficult to detect it. However, the

Roland Dosch (ed.), *Germline Development in the Zebrafish: Methods and Protocols*, Methods in Molecular Biology, vol. 2218,
https://doi.org/10.1007/978-1-0716-0970-5_13, © Springer Science+Business Media, LLC, part of Springer Nature 2021

second polar body is extruded when the mature oocyte (at stage V) is ovulated and activated, allowing capturing it in a quite short time window from egg-laying until its degeneration. Although the polar body and egg contain the same number of chromosomes, the former is of much smaller size due to its remarkably less cytoplasm. Thus, this cell type exhibits a unique character: a lump of DNA with closely surrounded F-actin [13–15]. Since the pronuclei of the egg or sperm lack similar F-actin abundance around them, they are easily distinguishable from the polar body. Here, we present a protocol to detect the second polar body in the zebrafish embryo. We collected synchronous fertilized embryos at 10 min post fertilization, when the second polar body usually appears over the cytomembrane of embryo near the site of sperm penetration [12]. To mark the second polar body, we used Phalloidin and DAPI to label F-actin and DNA, respectively. In each fertilized wild-type embryo, we could detect one DAPI-stained polar body surrounded by Phalloidin-positive F-actin and two pronuclei, deriving respectively from an egg and a sperm, only labeled by DAPI (Fig. 1a, a′). Eggs produced by *taz* homozygous mutant females fail to be fertilized due to the lack of the micropyle, the sperm entry channel in oocyte chorion, despite that oogenesis in this mutant is largely normal [15, 16]. Here, we used *taz* mutants as a reference to demonstrate eggs without fertilization, yielding only one pronucleus and one polar body (Fig. 1b, b′).

2 Materials

Instruments, labware, and reagents provided here are for references, alternatives with similar functions should work.

2.1 Zebrafish and Fish Facility

Wild-type zebrafish (*Danio rerio*) AB strain and *taz* mutant [15] were raised in the fish facility according to standard procedures [17].

2.2 Zebrafish Embryo Collection

1. Two L crossing tanks with built-in meshes, separators, and lids.
2. Plastic Pasteur pipettes (capacity 3 mL).
3. Egg water: Water with 0.03% sea salt. Dissolve 0.3 g sea salt in 1 L autoclaved distilled water. Store at room temperature (*see* **Note 1**).
4. Plastic Petri dishes (diameter 90 mm × high 15 mm) with an agarose lining. Plastic Petri dishes laid with a thin layer of 1.5% agarose to prevent dechorionated embryos sticking to the plastic bottom of dishes and get damaged. Add 100 mL egg water to a 500 mL glass conical flask, weigh 1.5 g agarose and transfer it to the flask, melt agarose in a microwave oven and cool agarose to 50 °C. Lay each dish with about 10 mL agarose

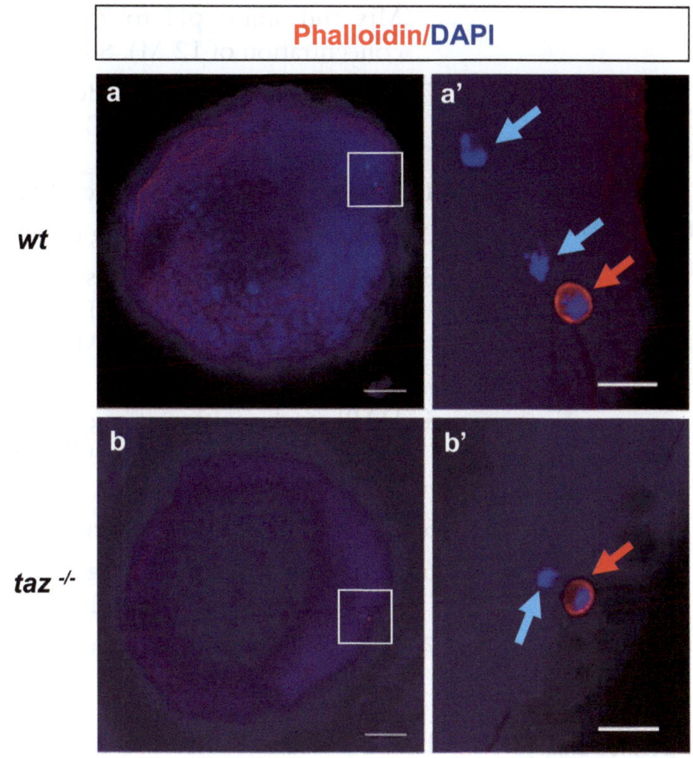

Fig. 1 Polar bodies in embryos produced by wild-type and *taz*$^{-/-}$ females at 10 min post fertilization. (**a**, **a**′) In a wild-type fertilized embryo (*wt*), there is one polar body, with DAPI-stained nucleus surrounded by Phalloidin-TRITC-labeled F-actins (red arrow), and two DAPI-positive pronuclei (blue arrow), which are from sperm and egg. (**b**, **b**′) In an egg from *taz*$^{-/-}$ female which fails to be fertilized, there is one polar body (red arrow) and one pronucleus (blue arrow) from the egg. Scale bar: 100 μm in a and b, and 20 μm in **a**′ and **b**′

solution and allow agarose solidification at room temperature. Store the dishes at 4 °C (*see* **Note 2**).

2.3 Dechorionated Embryo Fixation

1. A dissection microscope.
2. Two pairs of forceps (Dumont#5).
3. Glass Pasteur pipettes (length 150 mm) (*see* **Note 3**).
4. Screw glass bottles/tubes with caps (as the container for dechorionated embryo fixation, volume 2 mL).
5. Phosphate buffered saline stock: 10× PBS buffer. Dissolve 80 g NaCl, 2 g KCl, 14.4 g Na_2HPO_4 and 2.4 g NaH_2PO_4 in 800 mL distilled water, and then make up to 1000 mL with distilled water. Autoclave and store at 4 °C.
6. PBS (pH = 7.4). Measure 100 mL 10× PBS buffer and 900 mL autoclaved distilled water to a 1 L autoclaved bottle.

Mix and adjust pH to 7.4 with about 30 μL HCl (Stock concentration of 12 M). Store at 4 °C.

7. 4% PFA (paraformaldehyde) solution: Dissolve 4 g PFA in 100 mL PBS. Store at 4 °C (*see* **Note 4**).

2.4 Phalloidin and DAPI Staining

1. Embryo glass dish (diameter 30 mm × deep 12 mm).
2. Lightproof box for putting embryo glass dish.
3. Phalloidin-TRITC stock solution: 0.1 mg/mL Phalloidin-TRITC. Dissolve 0.1 mg Phalloidin-tetramethylrhodamine B isothiocyanate in 1 mL DMSO. Aliquot and store at −20 °C (*see* **Note 5**).
4. DAPI stock solution: 1 mg/mL DAPI. Dissolve 10 mg DAPI in 10 mL autoclaved distilled water. Aliquot and store at −80 °C (*see* **Note 6**).
5. PBST: PBS with 0.1% Tween 20.
6. Phalloidin-TRITC staining buffer: Dilute Phalloidin-TRITC stock solution in PBST at 1:400 (*see* **Note 7**).
7. DAPI staining buffer: Dilute DAPI stock solution in PBST at 1:1000 (*see* **Note 8**).

2.5 Stained Embryo Imaging

1. A fluorescence dissection microscope.
2. A confocal microscope.
3. Microscope slides (dimension 75 mm × 25 mm) and coverslips (dimension 22 mm × 22 mm).
4. A homemade bridged microscope slide: To keep spheroidal embryos from collapsing during confocal imaging, a bridge was constructed on a microscope slide with two bridge feet made of coverslips (Fig. 2a, b). Clean microscope slides and coverslips with 100% ethanol and lens paper. Paint a thin layer of colorless nail polish at two contact points on the microscope slide surface. Put a piece of coverslip on each point and gently press. Leave the set-up to dry on the bench for 30 min. Paint another thin layer of nail polish on the surfaces of new coverslips and stick them to the top of the previous ones and press gently. Put the above slides in a 37 °C oven for at least 60 min for complete drying (*see* **Note 9**).
5. A homemade soft hairbrush: Trim a fishing line (diameter 0.15 mm) into 1.5-cm segments. Melt paraffin wax in a 68 °C oven. Immerse the tip of a glass Pasteur pipette into melted paraffin wax until the tip fills to 1 cm length. Immediately insert one segment of the fishing line into the molten wax at the tip with tweezers before wax solidification (Fig. 2c) (*see* **Note 10**).

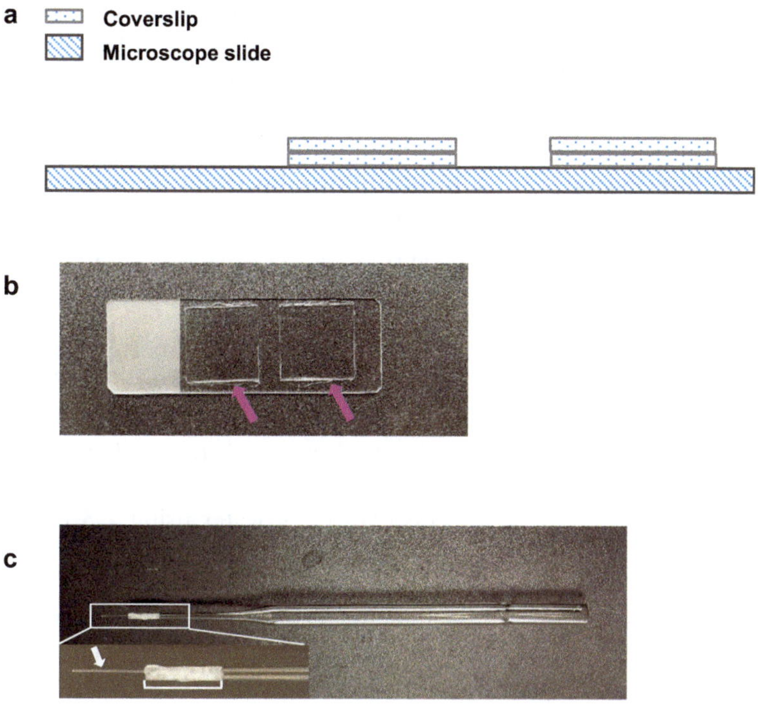

Fig. 2 Homemade devices for mounting embryos. (**a**) A lateral diagrammatic sketch of a bridged microscope glass. Four pieces of coverslip are adhered to a glass microscope slide by colorless nail polish to build 2 ft of the bridge. (**b**) A bridged microscope glass. Pink arrows point the coverslips on the glass microscope. (**c**) A soft hairbrush. A magnification of the tip area (white box) is showed in the inset. The fine fish line is inserted into the tip of a glass Pasteur pipette and fixed by paraffin wax (white line) with about 0.75 cm left outside (white arrow)

6. Mounting medium with anti-fluorescence quenching reagent: 80% glycerol, 3% 1,4-diazabicyclo[2.2.2]octane (DABCO), and 1% methyl cellulose in PBS. Weigh 0.3 g DABCO and 0.1 g methyl cellulose to a 15 mL tube wrapped by aluminum foil, add 1 mL 10× PBS buffer and dissolve them. Top up with 8 mL autoclaved 100% glycerol and autoclaved distilled water to 10 mL (*see* **Note 11**). Mix well and store at 4 °C.

3 Methods

3.1 Zebrafish Natural Spawning

In the late afternoon of the day of setting up crossing, the following steps are carried out in the fish facility.

1. Put the mesh into the crossing tank and partition the crossing tank into two compartments with a separator.
2. Fill crossing tank with system water until the mesh is covered with 3–5 cm high of water.

3. For each cross, one male and one female adult zebrafish are transferred individually into each compartment, kept apart by a separator (*see* **Note 12**).
4. Set up about five crosses (*see* **Note 13**).

3.2 Synchronous Fertilized Embryo Collection

Unless otherwise specified, all procedures are carried out in the fish facility or at room temperature.

1. Thirty minutes after the light in the fish facility comes on, transfer the mesh with fish and separator into a new crossing tank with system water (*see* **Note 14**).
2. Remove the separator to allow zebrafish mating (*see* **Note 15**).
3. Upon female fish starting to lay eggs, count 60 s and remove the mesh with fish to another new crossing tank with system water, and collect embryos immediately to an agarose lined dish with egg water using a 3 mL plastic Pasteur pipette (*see* **Note 16**).

3.3 Dechorionated Embryo Fixation

1. Put the agarose lined dish under a dissection microscope and remove the chorion of each embryo carefully with two pointed forceps (Dumont #5) (*see* **Note 17**).
2. At 10 min post fertilization, transfer dechorionated embryos to 2 mL glass bottles/tubes with fresh 4% PFA solution with a glass Pasteur pipette, change 4% PFA solution once. Fix embryos at room temperature for 2 h or at 4 °C for overnight (*see* **Note 18**).

3.4 Phalloidin and DAPI Staining

Unless specified otherwise, all solutions should be added gently using a 1 mL blue tip.

1. Remove 4% PFA solution as much as possible and add 1.5 mL PBST to rinse the fixed embryos once.
2. Discard PBST and add fresh 1.5 mL PBST into the glass tube. Allow to stand on the bench for 10 min at room temperature.
3. Repeat **step 2** twice.
4. Transfer embryos to an embryo glass dish with a glass Pasteur pipette.
5. Discard PBST as much as possible and add 0.5 mL Phalloidin-TRITC staining buffer to the embryo glass dish. Put the embryo glass dish into a lightproof box and stand on the bench for 10 min at room temperature. From this step onwards, embryos should be kept away from light.
6. Pipette off Phalloidin-TRITC staining buffer and rinse the embryos with 1.5 mL PBST.
7. Wash the embryos in 1.5 mL PBST three times.

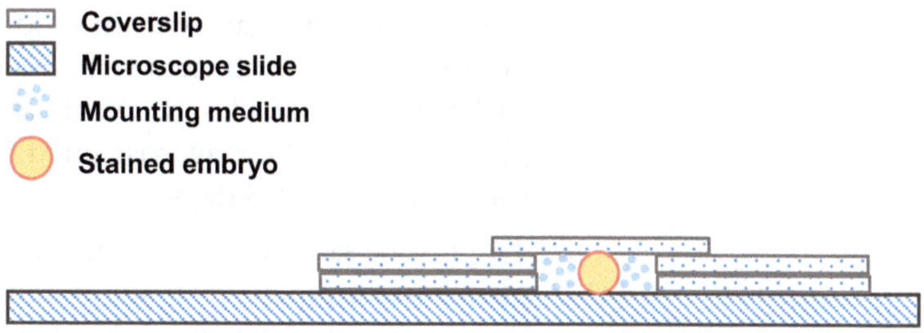

Fig. 3 A lateral diagrammatic sketch of a mounted embryo for imaging. Under a fluorescence dissection microscope, a stained embryo could be adjusted freely in the mounting medium between the two bridge feet

8. Remove PBST as much as possible and add 0.5 mL DAPI staining buffer. Keep the dish on the bench in the lightproof box for 10 min at room temperature.
9. Wash the embryos in 1.5 mL PBST three times.
10. Images of the stained embryos can be taken immediately, or samples can be kept at 4 °C in dark for a few days with steady fluorescence (*see* **Note 19**).

3.5 Stained Embryo Imaging

1. Drop the mounting medium with a glass Pasteur pipette in the indentation between bridge feet on a homemade bridged microscope slide (*see* **Note 20**).
2. Dip one stained embryo into the surface of the mounting medium with one glass Pasteur pipette while removing as much PBST floated on the mounting medium as possible (*see* **Note 21**).
3. Adjust the position of the embryo to make Phalloidin signals on the dorsal view with the homemade soft hairbrush under a fluorescence dissection microscope, and put a coverslip gently on the two cover glass piers (Fig. 3). Take images of embryos under a confocal microscope (*see* **Note 22**).

4 Notes

1. Sea salt is easily dissolved in distilled water by shaking. However, it is normal that a few fine impurities gradually sink to the bottom of the container. These impurities do not affect the follow-up experiments.
2. Heating is helpful to dissolve agarose in egg water in a microwave oven. It is necessary to prevent liquid overboiling from the flask and shake occasionally to prevent agarose from sitting at the bottom. Repeat heating and shaking till the solution becomes clear. While cooling the hot solution on the bench at room temperature, it is required to shake the flask for several

times to prevent agarose from solidifying at the bottom. About 10 mL agarose is sufficient to line one petri dish. After pouring, shake the dish quickly to evenly spread the solution over the plate bottom. These agarose lined dishes can be stored at 4 °C for no more than 3 days.

3. To prevent samples from being damaged by the sharp glass in glass Pasteur pipette, it is suggested to burn the tip of the glass Pasteur pipette over an alcohol lamp for a few seconds.
4. Freshly prepared 4% PFA solution is highly recommended to use for the fixation of embryos. Heating helps to dissolve PFA. To prepare 4% PFA solution, the bottle containing proportional PFA powder and PBS buffer is immersed in a water bath at 68 °C. Shake the bottle several times until the liquid is clear and transparent which usually takes 4–8 h.
5. According to the manufacturer's instructions, 1 mL DMSO is added into the tube containing 0.1 mg Phalloidin-TRITC powder. Pipette the solution up and down several times to mix and dissolve the powder. The solution is aliquoted into 100 μL in tubes wrapped by aluminum foil and stored at −20 °C or below.
6. According to the manufacturer's instructions, DAPI is suggested to dissolve in autoclaved distilled water. To prepare DAPI stock solution, proportional DAPI powder and water are put into a 15 mL tube wrapped by aluminum foil and shaken on a nutator for 10 min. The solution is aliquoted into 1 mL in tubes wrapped by aluminum foil and stored at −80 °C.
7. Phalloidin-TRITC staining buffer must be prepared freshly. During the last wash of fixed embryos in PBST, Phalloidin-TRITC stock solution can be diluted at 1:400 with PBST. Keep the buffer from light.
8. DAPI staining buffer requires fresh preparation. During the last wash of Phalloidin-TRITC-stained embryos in PBST, DAPI stock solution can be diluted at 1:1000 with PBST. Keep the buffer from light.
9. To build the two feet of the bridge on a microscope slide, the distance between them is suggested to keep at about 5 mm, which is sufficient for mounting embryos. To make coverslips fixed, it is important to totally dry nail polish, putting the bridged slides in a 37 °C oven is helpful.
10. To make a soft hairbrush, it is essential to operate as quickly as possible, especially during inserting the fishing line into paraffin wax. It is suggested to leave about 0.75 cm fishing line outside the paraffin wax for adjusting embryo position in the sticky mounting medium.

11. Methyl cellulose is easy to dissolve at a low temperature. Proportional DABCO and methyl cellulose powders with 10× PBS buffer are put into a 15 mL tube wrapped by aluminum foil and dissolved at 4 °C for overnight. Top up with autoclaved 100% glycerol and distilled water and mix well on a shaker.
12. Young and healthy zebrafish are highly suggested for the experiment. To achieve that, female and male have to be evaluated first according to the quality of eggs and fertilization rate by sperms. Healthy females should lay light yellow transparent eggs quickly once males pursue, and a qualified male fish should have a high fertilization rate which is more than 90%. These qualified male and female fish are suitable for experiments after raising separately in the fish facility for more than 1 week after evaluation.
13. Although healthy fish are pre-picked for experiments, about five crossings are suggested to ensure sufficient embryo supply.
14. It is important to transfer fish to a new crossing tank with fresh system water in which the fertilized embryos are clean without feces. To allow easy mating, the mesh is better covered by 1.5–2 cm water.
15. To detect the polar body just after fertilization, it is necessary to collect fertilized embryos as soon as possible. It is suggested to allow one pair of fish mating by removing the separator in a crossing tank at a time. If the female does not lay eggs, remove another separator until one female lays.
16. Once female fish begins to lay eggs, transfer fish quickly to another tank after 1 min. In that case, all embryos are synchronously fertilized. When all embryos sink to the bottom of the tank, carefully pour out most of water from the tank, and remove them from water using a 3 mL plastic Pasteur pipette to an agarose lined dish with egg water. It is not recommended to collect embryos with a net to prevent fragile embryos from being destroyed during water flushing to clean them.
17. The extraembryonic chorion of the egg will expand quickly after ovulation. To remove the chorion, one pair of sharp forceps is used to fix the chorion, while the other to make a small hole in the chorion. Gently tear this hole large enough for the embryo to fall out through the broken chorion when turned upside down.
18. At 10 min after fertilization, dechorionated embryos are gently removed by a glass Pasteur pipette. Quickly immerse into 1 mL fresh 4% PFA in a 2 mL glass bottle/tube to allow eggs to slowly fall into the solution. When they sink to the glass bottom, the 4% PFA solution should be removed quickly as much as possible by 1 mL pipette. Then 1.5 mL fresh 4% PFA solution is added along the glass wall slowly.

19. It is recommended to take images of the stained embryos as soon as possible.
20. Appropriate amount of the mounting medium should be added into the indentation of the bridge. Too much of the mounting medium in the indentation will cause the sample move after putting on the coverslip.
21. When the stained embryo is transferred to the mounting medium on the bridged slide, PBST transferred together with the embryo will float on the mounting medium due to its lower density, it is suggested to remove PBST carefully before mounting embryo.
22. Since the signal of Phalloidin is a very small dot against the large surface of an embryo, it is necessary to adjust the position of embryos with Phalloidin signal on the top of the embryo under a fluorescence dissection microscope using a homemade soft hairbrush.

Acknowledgments

We thank Li Jan Lo (Zhejiang University, China) for suggestions on the manuscript. This work is supported by the National Key R&D Program of China 2018YFA0800502, 2018YFA0801000, and Natural Science Foundation of China 31970762.

References

1. Page AW, Orr-Weaver TL (1997) Stopping and starting the meiotic cell cycle. Curr Opin Genet Dev 7(1):23–31
2. Nebreda AR, Ferby I (2000) Regulation of the meiotic cell cycle in oocytes. Curr Opin Cell Biol 12(6):666–675
3. Lubzens E, Young G, Bobe J et al (2010) Oogenesis in teleosts: how eggs are formed. Gen Comp Endocrinol 165(3):367–389
4. Suwa K, Yamashita M (2007) Regulatory mechanisms of oocyte maturation and ovulation. In: Babin PJ et al (eds) The fish oocyte: from basic studies to biotechnological applications. Springer, Dordrecht, pp 323–347
5. Selman K, Wallace RA, Sarka A et al (1993) Stages of oocyte development in the zebrafish, *Brachydanio rerio*. J Morphol 218(2):203–224
6. Nair S, Lindeman RE, Pelegri F (2013) In vitro oocyte culture-based manipulation of zebrafish maternal genes. Dev Dyn 242(1):44–52
7. Wang Y, Ge W (2004) Developmental profiles of activin βA, βB, and follistatin expression in the zebrafish ovary: evidence for their differential roles during sexual maturation and ovulatory cycle. Biol Reprod 71 (6):2056–2064
8. Elkouby YM, Jamieson-Lucy A, Mullins MC (2016) Oocyte polarization is coupled to the chromosomal bouquet, a conserved polarized nuclear configuration in meiosis. PLoS Biol 14 (1):e1002335
9. Elkouby YM, Mullins MC (2017) Coordination of cellular differentiation, polarity, mitosis and meiosis – new findings from early vertebrate oogenesis. Dev Biol 430(2):275–287
10. Elkouby YM, Mullins MC (2016) Methods for the analysis of early oogenesis in zebrafish. Dev Biol 430(2):310–324
11. Streisinger G, Walker C, Dower N et al (1981) Production of clones of homozygous diploid zebrafish (*Brachydanio rerio*). Nature 291 (5813):293–296
12. Wolenski JS, Hart NH (1987) Scanning electron microscope studies of sperm incorporation into the zebrafish (*Brachydanio*) egg. J Exp Zool 243(2):259–273

13. Dekens MP, Pelegri FJ, Maischein HM et al (2003) The maternal-effect gene futile cycle is essential for pronuclear congression and mitotic spindle assembly in the zebrafish zygote. Development 130(17):3907–3916
14. Marlow FL, Mullins MC (2008) Bucky ball functions in Balbiani body assembly and animal-vegetal polarity in the oocyte and follicle cell layer in zebrafish. Dev Biol 321 (1):40–50
15. Yi X, Yu J, Ma C et al (2019) The effector of Hippo signaling, Taz, is required for formation of the micropyle and fertilization in zebrafish. PLoS Genet 15(1):e1007408
16. Dingare C, Niedzwetzki A, Klemmt PA et al (2018) The Hippo pathway effector Taz is required for cell morphogenesis and fertilization in zebrafish. Development 145(22): dev167023
17. Westerfield M (2000) The zebrafish book. A guide for the laboratory use of zebrafish (Danio rerio), 4th edn. University of Oregon Press, Eugene, OR

Chapter 14

Labeling the Micropylar Cell in Zebrafish Whole-Mount and Cryo-sectioned Follicles

Chaitanya Dingare, Petra A. Klemmt, Alina Niedzwetzki, and Virginie Lecaudey

Abstract

In some animal species, fertilization occurs through a funnel-like canal called the "micropyle." In teleost fishes, the micropyle is formed by a very specialized follicle cell, called the micropylar cell (MC). Very little is known about the mechanisms underlying the specification and differentiation of the MC, a unique cell among hundreds that compose the follicle cell layer. The Hippo pathway effector Taz is essential for this process and is the first reported MC marker. Here, we describe a method to identify and mark the micropylar cell following the immunostaining procedure on cryosections or combining it with the RNA in situ hybridization on whole-mount follicles.

Key words Zebrafish, Micropyle, Micropylar cell, Follicle cell epithelium, Taz, In situ hybridization, Cryosection

1 Introduction

In the last decade, the zebrafish has been a choice of model organism for developmental biologists to study vertebrate development due to its amenability to forward and reverse genetics approaches as well as transparency of its embryos, making it an ideal system to manipulate and observe various developmental processes in vivo, even at the single cell level. In contrast, processes occurring in the juvenile or the adult zebrafish have been much less studied, possibly due to the lack of appropriate tools [1]. Oogenesis is one such process occurring in the juvenile and later in the adult zebrafish female. It is a complex process ranging from the differentiation of germ cells into oocytes to make them competent to be fertilized. However, how an oocyte acquires this competency is much less explored. Previous studies show that germline-soma interaction prepares an oocyte for fertilization [2–4]. How this interaction is mediated in the zebrafish ovary is completely unknown, one reason being the lack of specific molecular markers and techniques to label

Roland Dosch (ed.), *Germline Development in the Zebrafish: Methods and Protocols*, Methods in Molecular Biology, vol. 2218, https://doi.org/10.1007/978-1-0716-0970-5_14, © Springer Science+Business Media, LLC, part of Springer Nature 2021

the somatic, i.e., follicle cells and dissect their interaction with the developing oocyte. In this article, we have provided details on key molecular markers and protocols to use them to label the follicle cells and their interaction with the oocyte.

Recently, we and two other research groups have described, in detail, the differentiation of the micropylar cell (MC), a unique follicle cell, which forms the future sperm entry point or simply the "micropyle," making the oocyte capable of getting fertilized [5–7]. As oogenesis proceeds, only one cell among hundreds of follicle cells covering the oocyte grows much bigger in volume, acquires a typical mushroom shape, and remains in contact with the oocyte surface, thus creating a hole in the forming vitelline membrane. The Hippo pathway effector Taz was identified as a key regulator of the MC differentiation and as the earliest marker of this unique follicle cell. Furthermore, we have identified Keratin 18 and Desmoplakin as two additional, MC-specific markers.

Here, we describe novel methods to locate the MC either on cryosections by performing immuno-staining to label the molecular markers mentioned above or on whole-mount follicles by combining immuno-staining and RNA in situ hybridization to detect *cyclinb1* (*ccnb1*) mRNA, a marker of the animal pole of the oocyte, localized beneath the MC.

2 Materials

2.1 Ovary Dissection and Fixation

1. Tricaine stock solution: 2.1 g of Tricaine (ethyl-m-aminobenzoate methanesulfonate) in 450 ml of MilliQ water. Adjust pH to 7 with Tris–Cl (1 M, pH 9.0) and then adjust the final volume to 500 ml with MilliQ water (final concentration 4.2 g/l). Aliquot in 50 ml conical tubes and store frozen at −20 °C.
2. Tricaine solution to euthanize: Dilute 72 ml of stock solution in 1 l of fish system water (final concentration 300 mg/l).
3. 1 M Tris–Cl (pH = 7.4 and 8.0): dissolve 6.05 g of Tris base in 20 ml of autoclaved MilliQ water. Adjust the pH to the desired value using HCl and finally fill up the volume to 50 ml with MilliQ water.
4. 1× L-15 Medium with GlutaMAX and phenol red.
5. 10 mg/ml Collagenase: Dissolve lyophilized collagenase in autoclaved MilliQ water. Make 50 μl aliquots and store at −20 °C.
6. 10× Phosphate buffered saline (PBS): Dissolve sequentially 80 g NaCl, 2 g KCl, 17.3 g $Na_2HPO_4{\cdot}2H_2O$, and 2.4 g KH_2PO_4 in 500 ml of autoclaved MiliQ water. Adjust the pH to 7.4 and make up the volume to 1 l. Autoclave and store at

room temperature. Prepare a 1× PBS working solution by diluting tenfold with sterile water.

7. 5% formaldehyde: Dilute 0.5 ml of 37% formaldehyde in 9.5 ml of 1× PBS. Caution: the 37% formaldehyde stock solution (a) is considered as 100% stock solution and (b) should not contain any precipitate of para-formaldehyde.
8. 4% Paraformaldehyde (PFA) in PBS: 20 g PFA in 400 ml of autoclaved MilliQ water and 50 ml of 10× PBS. Heat up to 60 °C while stirring and add 1 N NaOH dropwise until para-formaldehyde is dissolved. After complete dissolution, the pH is adjusted to 7.4 if required and the final volume is adjusted to 500 ml with autoclaved MilliQ water. Aliquot in 50 ml conical tubes and store frozen at −20 °C for up to 1 year. Once thawed, use within 1 week.
9. Glacial acetic acid.

2.2 Cryosections

1. 25% Sucrose: Dissolve 2.5 g of sucrose in 4 ml of water and 1 ml of 10× PBS. Adjust the volume to 10 ml. Filter through a 0.22 μm filter and store at 4 °C.
2. 35% Sucrose: Dissolve 3.5 g of sucrose in 4 ml of water and 1 ml of 10× PBS. Adjust the volume to 10 ml. Filter through a 0.22 μm filter and store at 4 °C.
3. TissueTek.

2.3 Immunostaining

1. 10% Tween-20 (v/v): Dilute 10 ml of 100% Tween-20 in 90 ml of autoclaved MilliQ water and stir. Filter through the 0.22 μm filter and store at 4 °C.
2. 10% Triton X-100 (v/v): Dilute 10 ml of 100% Triton X-100 in 90 ml of autoclaved MilliQ water and stir. Store at 4 °C.
3. PBSTw (PBS with 0.1% Tween-20): Prepare 100 ml of PBSTw by diluting 10 ml of 10× PBS and 1 ml of 10% Stock of Tween-20 in 89 ml of autoclaved MilliQ water.
4. PBDT-I: 1 g of bovine serum albumin (BSA) in 10 ml of 10× PBS, 1 ml of dimethyl sulfoxide (DMSO), 5 ml of 10% Triton X-100 and MilliQ water up to 100 ml. Final concentration: 1% BSA (w/v), 1% DMSO (v/v), and 0.5% Triton X-100 (v/v) in 1× PBS.
5. PBDT-II: 1 g of BSA in 10 ml of 10× PBS, 1 ml of DMSO, 0.25 ml of 10% Triton X-100 and MilliQ water up to 100 ml. Final concentration: 1% BSA, 1% DMSO, and 0.025% Triton X-100 in 1× PBS.
6. Immunostaining blocking reagent:
   (a) 2% (v/v) Normal goat serum (NGS) in PBDT-I or
   (b) 5% (v/v) Normal goat serum (NGS) in PBDT-II.

7. Primary antibodies: Anti-Taz (1:200, Cell Signaling Technology, D24E4), Anti-Krt18 (1:200, Thermo Fischer Scientific, MA1-06326), and Anti-Dsp (Ready to use, Progen, 651155).
8. Secondary Antibodies: Anti-rabbit IgG and anti-mouse IgG coupled with Alexa Fluors (488, 561, and 660) (1:500, Thermo Fischer Scientific). Rhodamine-Phalloidin (1:100, Molecular Probes, R415) and DAPI (1 μg/ml) were used to label the F-actin cytoskeleton and the nuclei, respectively.

2.4 Whole-Mount In Situ Hybridization (WISH)

1. Prehybridization buffer (Prehyb): 250 ml of deionized, RNase and DNase-free Formamide, 125 ml of 20× SSC (final concentration:5×), 2.5 ml of 10 mg/ml stock of Heparin (final concentration:50 μg/ml) and 2.5 g of Torula RNA (final concentration:5 mg/ml).
2. RNA ISH probe: *cyclinb1* [5] diluted 1:200 in Prehyb.
3. 20× Saline sodium citrate (SSC): Dissolve sequentially 175.3 g of NaCl and 88.2 g of Na-Citrate in 500 ml of autoclaved MiliQ and adjust the volume to 1 l. Autoclave and store at RT.
4. ISH wash solutions:
   (a) Wash I: 50% formamide in 2× SSCTw (5 ml of deionized, RNase and DNase-free formamide, 1 ml of 20× SSC, 0.1 ml of 10% Tween-20 in MilliQ water up to 10 ml).
   (b) Wash II: 2× SSCTw (1 ml of 20× SSC, 0.1 ml of 10% Tween-20 in MilliQ water up to 10 ml).
   (c) Wash III: 0.2× SSCTw (0.1 ml of both 20× SSC and 10% Tween-20 and MilliQ water up to 10 ml).
5. TNT buffer (Tris–NaCl-Tween-20): 10 ml of 1 M Tris–Cl (pH = 7.4), 3 ml of 5 M NaCl, 5 ml of 10% Tween-20 and MilliQ water up to 100 ml.
6. 250 mM Maleic acid: Dissolve 29 g of maleic acid in 800 ml of MilliQ water. Slowly adjust the pH to 7.5 with 4 N NaOH solution. Stir the solution and measure the pH constantly while adding the NaOH to avoid a sudden surge in the value. Adjust the final volume to 1 l.
7. 100 mM maleic acid buffer (pH 7.5): Prepare 100 ml by mixing 40 ml maleic acid (250 mM) (pH 7.5), 3 ml NaCl (5 M), 1 ml 10% Tween-20 and 56 ml MilliQ water.
8. 10% BBR blocking solution (w/v) was prepared in maleic acid buffer by slowly heating at 60 °C in a water bath until the blocking reagent was completely dissolved.
9. 2% BBR (Boehringer blocking reagent) in TNT buffer: Dilute 1 ml of 10% BBR stock solution in 4 ml of freshly prepared TNT.

10. Anti-DIG-AP (alkaline phosphatase-conjugated anti-digoxigenin) antibody: Dilute 1:4000 in 2% BBR.
11. Equilibration buffer: 1 ml of Tris–Cl (1 M, pH 8.0), 0.1 ml of 10% Tween-20 and MiliQ water up to 10 ml.
12. Fast Red AP substrate was prepared as the manufacturer's instructions.

2.5 Mounting and Imaging

1. 1% Low melting agarose in 1× PBS (w/v):0.5 g of low melting agarose was dissolved in 50 ml of 1× PBS in the microwave. Melt in the microwave and keep in a 2 ml Eppendorf tube at 42 °C on a heating block before use.
2. Mowiol 4-88.

3 Methods

3.1 Zebrafish Husbandry

Procedures involving animals were conducted according to institutional guidelines and approved by the German authorities (veterinary department of the Regional Board of Darmstadt). Adult zebrafish were maintained under standard husbandry conditions with a 10 h/14 h dark-light cycle and a density of 5–7 adult fish/l [8].

3.2 Euthanizing a Zebrafish Female and Dissecting the Ovary

1. Euthanize a young, sexually mature female by an overdose of tricaine (*see* **Note 1**). Ten minutes after cessation of opercular movements, take the euthanized female out of the solution with a spoon and wipe off excess water with a soft tissue paper (Fig. 1a).
2. Transfer the euthanized female zebrafish in a 60 mm glass Petri dish or pin it down on a Styrofoam plate covered with a parafilm and a clean tissue paper (Fig. 1a–c), and place it under an epi-illuminated dissecting microscope.
3. Remove the scales near the cloaca using a sharp pair of forceps and expose the ovary by gently lifting up and tearing off the covering skin (Fig. 1b).
4. Scoop out the bi-lobed ovary with the help of a sharp forceps or a small spoon (Fig. 1c).

3.3 Preparing Ovarian Tissue for Cryosections

1. Immediately fix the dissected ovary in 2 ml cold 4% PFA at 4 °C on a rocker with gentle shaking for at least 12–14 h (overnight) (Fig. 1d, e).
2. Remove the PFA and wash the fixed ovarian tissue twice with 1× PBS for 5 min at room temperature under gentle shaking.
3. Transfer the washed tissue to the following solutions and let it sink to the bottom at 4 °C for about a day:

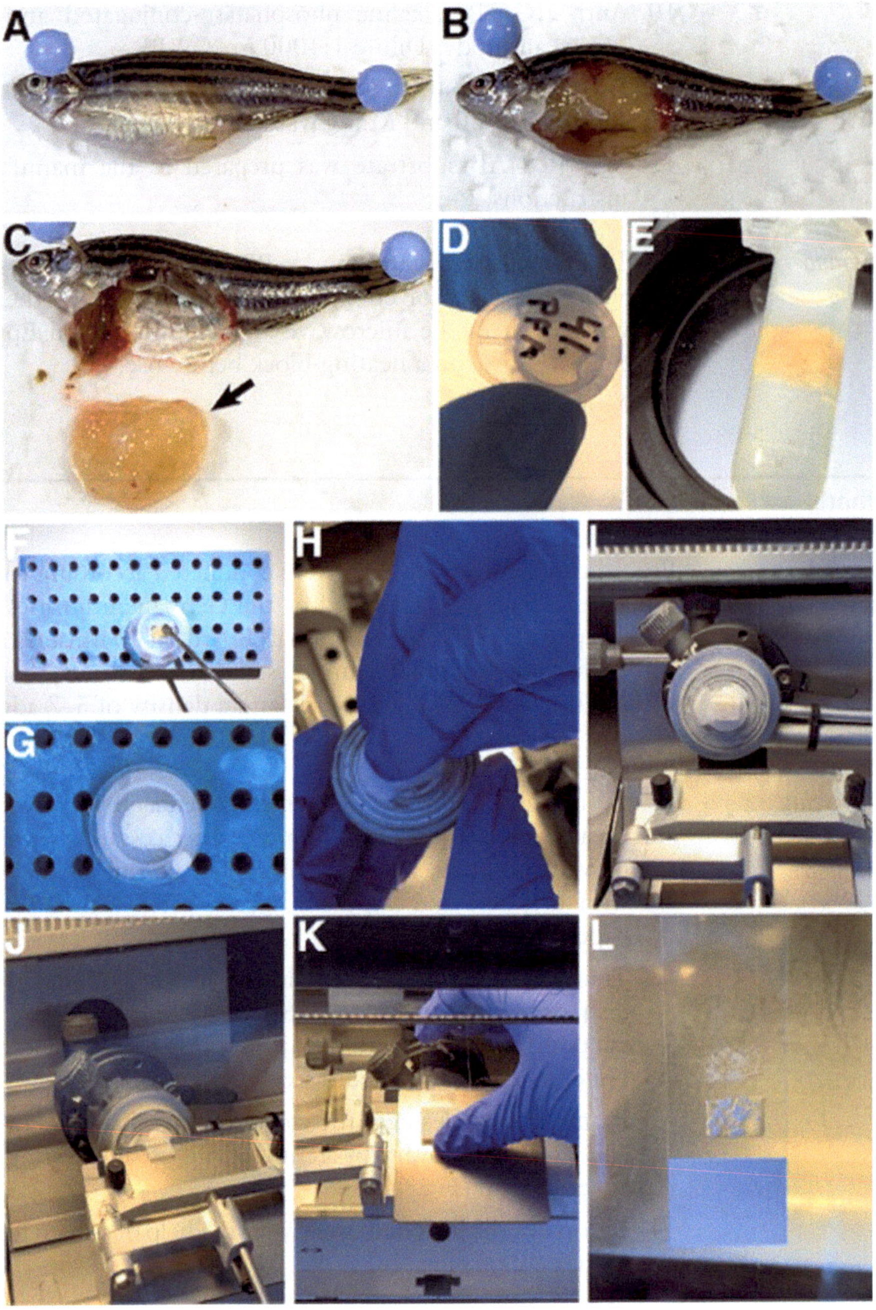

Fig. 1 Dissection, fixation, and cryo-sectioning of the ovarian tissue. (**a–c**) A sexually mature female was euthanized (**a**) and the ovary was exposed by removing the overlying skin (**b**). The ovary was dissected out intact as shown in (**c**, black arrow). (**d–e**). Dissected ovary was fixed with 4% PFA. (**f–l**) Following the fixation and the sucrose saturation, the ovarian tissue was transferred in a cryomold (**f**) and the cryomold containing the ovary immersed in TissueTek was kept on a pre-cooled metal stand to solidify (**g**). Upon solidification, the embedded tissue was separated from the mold in the cryochamber and mounted on the holder (**h**). The holder was then fixed onto the cryotome (**i**) and 10 μm sections were cut (**j**). Sections were collected individually on a slide (**k**) before being air-dried in a hood (**l**)

(a) 25% buffered sucrose solution.

(b) 35% buffered sucrose solution.

(c) Mix of 35% sucrose solution and TissueTek in 1:1 proportion (*see* **Note 2**).

(d) 100% TissueTek solution.

4. Replace with fresh TissueTek before embedding the tissue to get rid of any sucrose traces.
5. Transfer the tissue in a cryomold kept on dry ice or alternatively on a metal block pre-cooled at – 80 °C and cover with fresh TissueTek (Fig. 1f).
6. Keep the cryomold on dry ice or on the metal block until TissueTek is completely frozen (Fig. 1g).
7. Store the embedded tissue at −20 or −80 °C until cryo-sectioning.

3.4 Cryo-sectioning

1. Set the cryotome chamber at −20 °C.
2. Remove and trim the embedded tissue with a cold razor blade in the cryotome chamber.
3. Fix the trimmed tissue on a holder with excess TissueTek (Fig. 1h).
4. Fix the holder onto the block maintained at −15 °C (Fig. 1i).
5. Adjust the angle between the cryo-blade and the block and cut 10 μm-thick sections (Fig. 1j).
6. Place individual sections on ultra superfrost slides (*see* **Note 3**) (Fig. 1k).
7. Dry at room temperature for 2 h in the hood (*see* **Note 4**) (Fig. 1l).

3.5 Immunostaining on Cryosections

1. Before rehydration, surround all the sections together on a slide with a hydrophobic pen to ensure that the sections remain covered with the solutions.
2. Rehydrate the air-dried slides with 1 ml of PBSTw solution for 15 min and block with 5% NGS in PBDT- II solution for 30 min.
3. Incubate with 200 μl of primary antibody solution overnight at 4 °C without shaking in a humidified chamber.
4. Remove the antibody solution and replace by approximately 500 μl of PBDT-II solution.
5. Wash the slides three times for 10 min each with PBDT-II solution on a shaker at room temperature.
6. Immediately replace the last wash by 200 μl of secondary antibody solution at the recommended dilution and incubate

for 1 h at room temperature without shaking in a humidified chamber (*see* **Note 5**).

7. Wash the slides three times for 10 min each with PBDT-II solution on a shaker at room temperature.
8. Replace the PBDT-II solution with 1× PBS solution and wash each slide twice for 5 min each.
9. Remove excess PBS and add 50–70 μl Mowiol per slide.
10. Lower a coverslip onto the slide and gently press to spread the Mowiol and to remove bubbles.
11. Let the slides dry overnight under the hood.
12. Store slides at 4 °C until they are imaged (possibly within a week).

3.6 Imaging of Immunostained Cryosections

1. Image stained cryosections using a spinning disc or point-scanning confocal microscope.
2. First scan each section under a 20× objective using the oculars and wide-field fluorescence settings to screen for the micropylar cell using Taz (Fig. 2a, e), Keratin 18 or Desmoplakin (Fig. 2g) as markers (*see* **Note 6**).
3. Image the micropylar cell region with a 60× objective to visualize the subcellular details (Fig. 2a–h).
4. Process images using Fiji.

3.7 Preparation of Follicles for Whole-Mount Immuno-Fluorescence and In Situ Hybridization (WISH)

Since most of the antibodies do not work after in situ hybridization, the immuno-staining procedure is performed prior to the WISH procedure.

1. Dissect the ovary of a young, sexually mature zebrafish female as described in Subheading 3.1 (Fig. 1a–c).
2. Detangle the nests of follicles forming the ovaries with the help of sharp forceps (Fig. 3a, b) (*see* **Note 7**).
3. Wash quickly in L-15 medium.
4. Transfer all the follicles to a 2 ml Eppendorf tube and replace the medium with 1 ml of fresh L-15 medium.
5. Add 50 μl of 10 mg/ml collagenase to the follicles and incubate at room temperature for 10 min (Fig. 3c) (*see* **Note 8**).
6. Pipette up and down 10–15 times the follicles with a cut 1 mL pipette tip (*see* **Note 9**).
7. Discard the medium and add freshly prepared 5% formaldehyde (*see* **Note 10**) followed immediately by 8–10 drops of glacial acetic acid (Fig. 3d) (*see* **Note 11**, Fig. 3e, f).
8. Incubate in fixative at room temperature for 2 h on a shaker (*see* **Note 12**).

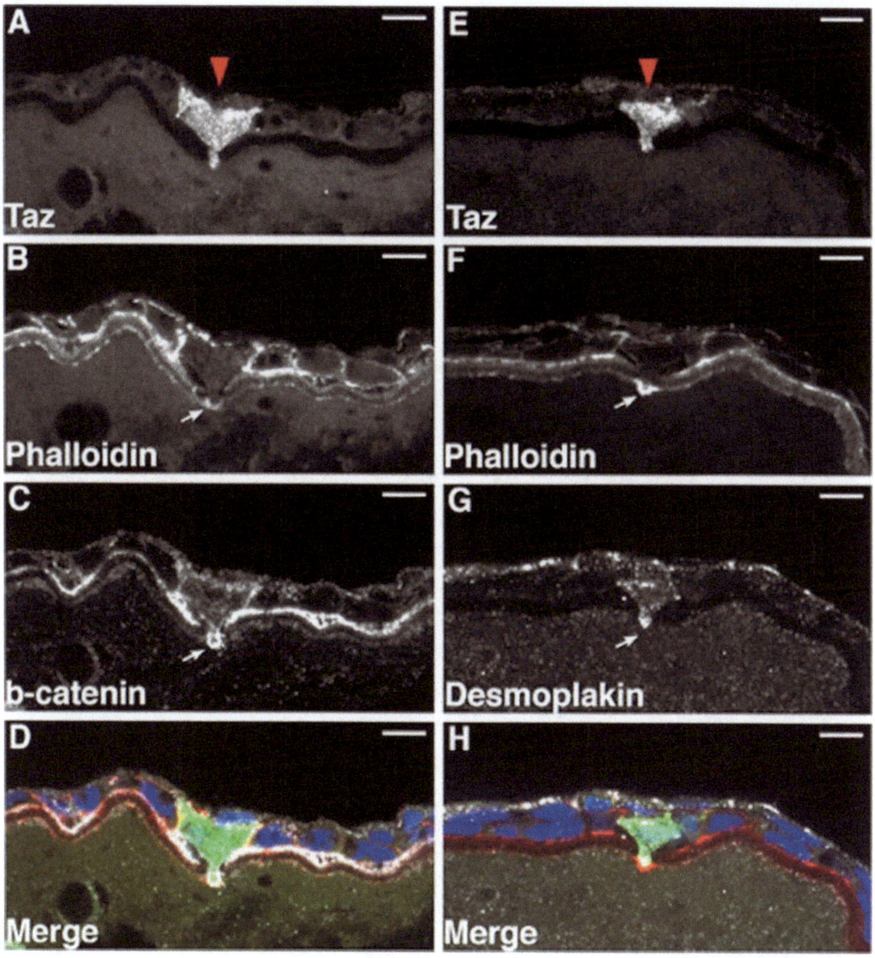

Fig. 2 Immunostaining on cryosections to label the micropylar cell. Cryosections were stained with Taz (**a**, **e**) and β-catenin (**c**) or Desmoplakin (**g**) antibodies to mark the micropylar cell (red arrowheads in **a** and **e**). Phalloidin staining revealed the F-actin enrichment at the contact point (white arrow in **b** and **f**) between the micropylar cell and the oocyte. This contact point also showed enriched beta-catenin and Desmoplakin (white arrow in **c** and **g**). The cryosections were also counterstained with DAPI to label the nuclei as shown on the merged panels (**d**, **h**)

9. Wash the follicles five times for 5 min each with PBSTw.
10. Sort the follicles according to their sizes (follicles of different sizes are shown in Fig. 3g for reference). Take only the individual follicles and not the one which remained in the nest.
11. Transfer follicles ranging from mid-stage II to early-stage IV of oogenesis to a fresh tube (*see* **Notes 13** and **14**).

3.8 Whole-Mount Immuno-Fluorescence

1. Replace the previous PBSTw solution with 200 µl of 2% NGS in PBDT-I solution to block the follicles for 30 min. At this stage, Ribolock RNase inhibitor (1:200) is added to the PBDT-I to prevent RNA degradation.

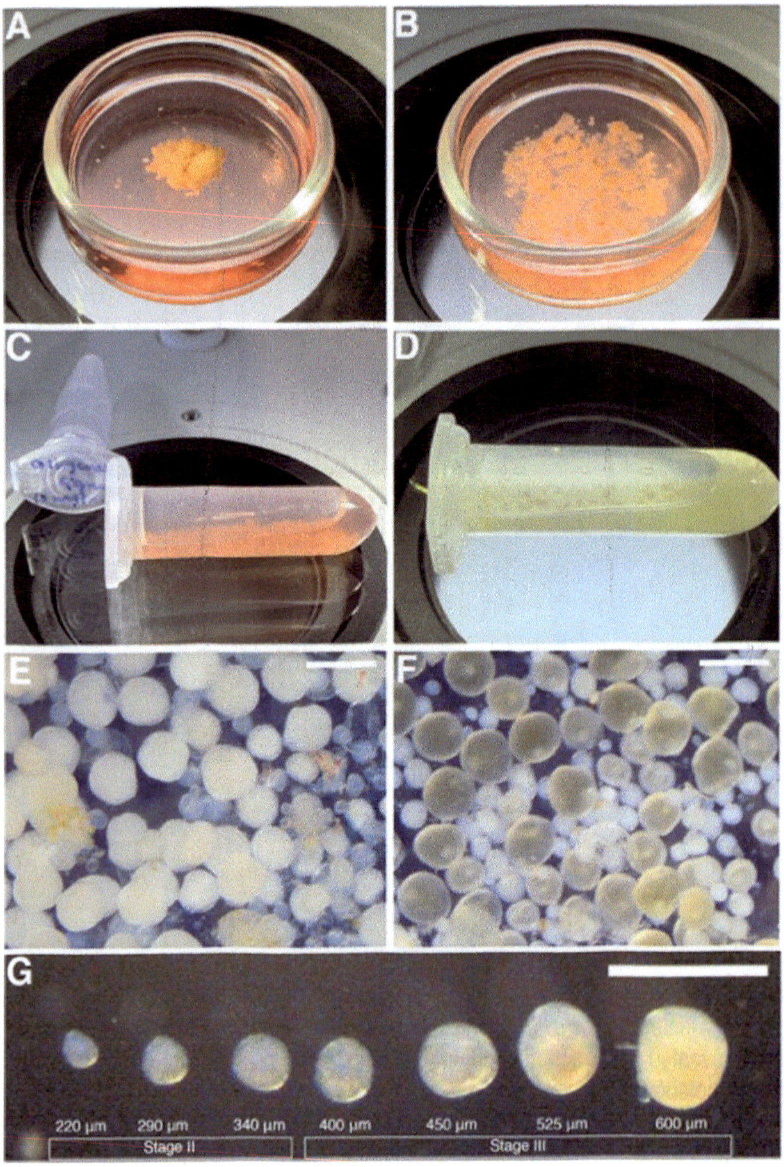

Fig. 3 Fixation for whole-mount immuno-staining followed by RNA in situ hybridization. (**a**, **b**) Dissected ovary was placed in L-15 medium (**a**) and the individual follicles were detangled (**b**). The follicles were collected in a 2-ml tube with 1 ml of L-15 medium and 50 µl of 10 mg/ml collagenase was added (**c**). After the collagenase treatment, the follicles were fixed and permeabilized using 5% formaldehyade and glacial acetic acid (**d**). (**e**–**f**) The follicles after fixation and permeabilization turned translucent (**f**) in comparison to the ones before fixation and permeabilization (**e**). The follicles were staged and used for further experiments (**g**). (Scale bar: 1 mm in **e**–**g**)

2. Remove the blocking solution carefully with 200 µl pipette tip.
3. Add 200 µl of diluted primary antibody solution.

4. Incubate the follicles either for 4 h at room temperature or overnight at 4 °C on a shaker.
5. Wash the follicles six times for 20 min each with PBDT-I solution.
6. Incubate the follicles with 200 μl of diluted secondary antibody solution either for 4 h at room temperature or overnight at 4 °C on a shaker.
7. Wash the follicles six times for 20 min each with PBDT-I solution (*see* **Note 15**): Both the primary and the secondary antibodies are diluted in 2% NGS-PBDT-I solution.
8. Wash the follicles briefly twice with PBSTw and post-fix for 10 min with 4% PFA solution without shaking.
9. Wash follicles again three times for 5 min with PBSTw on a shaker.

3.9 Whole-Mount In Situ Hybridization (WISH)

1. Replace PBSTw in each of these tubes with 350 μl of pre-hyb buffer pre-warmed at 60 °C and incubate at least for 1 h at 60 °C with gentle shaking.
2. Shortly before the end of the pre-hybridization, dilute 1 μl of fresh *cyclin b1* probe in 200 μl of pre-hyb solution and denature at 80 °C for 10 min.
3. Replace the pre-hyb buffer with the denatured probe solution and hybridize overnight at 60 °C in a hybridization oven under gentle shaking.
4. On the next day, remove the probe solution and briefly wash with 2 ml of Wash I solution pre-warmed at 60 °C. This is followed by:
   - Two washes of 30 min each in pre-warmed Wash I at 60 °C.
   - One wash of 15 min in pre-warmed Wash II at 60 °C.
   - Two washes of 30 min each in pre-warmed Wash III at 60 °C (*see* **Note 16**).
5. Wash the follicles twice 5 min with TNT buffer on a shaker at RT.
6. Block for 1 h with 350 μl of 2% BBR blocking solution on a shaker at RT.
7. Replace the blocking solution with 200 μl of diluted anti-DIG-AP antibody solution and incubate for 4 h on a shaker at RT or overnight at 4 °C on a shaker.
8. Remove the antibody solution by a brief wash with TNT buffer followed by six washes of 20 min with TNT buffer.
9. Wash the follicles twice for 5 min with the equilibration buffer.
10. Transfer the follicles with the buffer to a 12-well plate.

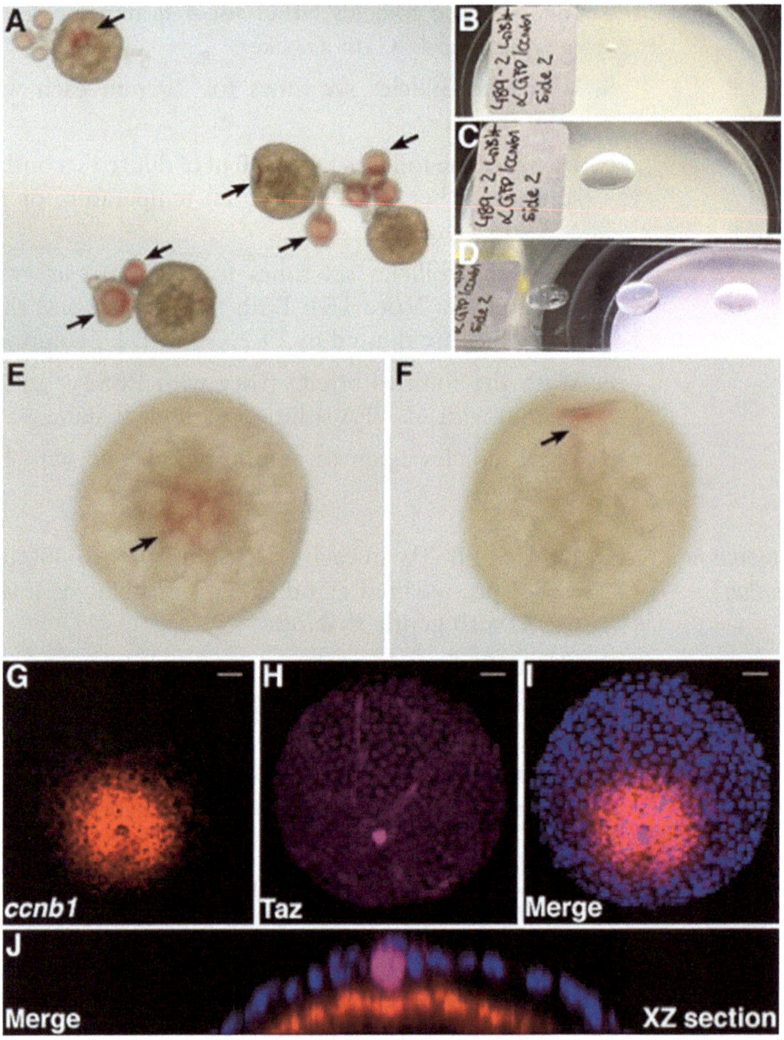

Fig. 4 Labeling the animal pole domain with *cyclinb1* mRNA and the micropylar cell with Taz. (**a–f**) Follicles stained for *cyclinb1* mRNA (**a**, black arrows) were mounted in a small drop of agarose on a 24 × 60 coverslip (**b–d**). The follicles were mounted in a way so that the animal pole domain was in contact with the coverslip (black arrow in **e**) for imaging on an inverted confocal microscope. (**f**) An example of follicle mounted with the animal pole up (black arrow in **f**). (**g–j**) Confocal images showing the maximum intensity projections of the *cyclinb1*-marked animal pole domain (**g**). Taz marks the micropylar cell (**h**) and appeared centrally placed in the animal pole domain on a merge panel (**i**). This was confirmed in the *X–Z* section passing through the follicle cell layer and the animal pole of the oocyte (**j**) DAPI was used to mark the nuclei of the micropylar cell and the follicle cells (Scale bar: 20 μm in **e–g**)

11. Replace the equilibration buffer with the 500 μl of FastRed staining solution.
12. Observe the color development reaction after 20–30 min on a stereomicroscope with transmitted light and epifluorescence (Fig. 4a) (*see* **Note 17**).

3.10 Mounting and Imaging the Stained Whole-Mount Follicles

1. Sort out the follicles by stage based on their size.
2. Mount each follicle individually in a drop of 1% low-melting agarose on a coverslip (Fig. 4b–d).
3. While the agarose is still liquid, use a bent needle to orient the follicle so that the animal pole marked by *cyclinb1* faces the coverslip surface (Fig. 4e, f) (*see* **Note 18**).
4. Store the coverslips in a humidified chamber until imaging.
5. Image on a confocal microscope as per the users' settings (Fig. 4g, j) (*see* **Note 19**).
6. Process images with Fiji.

4 Notes

1. Choose a young, sexually mature female (3–5 months old) that has not been paired yet or that has been paired once 8–10 days before dissection. We observed that these females have relatively younger oocytes (stage II–IV) in comparison to old females that have been crossed several times.
2. Prepare the 35% sucrose solution and TissueTek mix (1:1) while the fixed tissue is equilibrating in 35% sucrose. Vortex until the resulting solution becomes completely homogenous and keep it overnight at 4 °C.
3. Mount a maximum of eight individual sections per ultra superfrost slides.
4. Slides can be store at −80 °C at that stage.
5. Rhodamine phalloidin and DAPI can be combined with the secondary antibodies to label actin filaments and nuclei, respectively.
6. Rhodamine phalloidin also proved to be an important marker to screen for the micropylar cell. It distinctly marks the accumulation of filamentous actin at the contact point (arrow in Fig. 2b, f).
7. The dissected ovary is a nest of follicles at different stages of oogenesis [9].
8. The tube containing follicles is kept tilted to ensure proper exposure to the enzyme.
9. The medium should turn cloudy.
10. Prepare the 5% formaldehyde solution during the collagenase treatment.
11. One drop is approx. 50 μl. It is very important to add the glacial acetic acid immediately following the formaldehyde. The follicles should turn translucent after the addition of

glacial acetic acid. Although the reaction is performed in a 2 ml tube, we have placed the ovary in a glass dish before and after addition of glacial acetic acid for better visualization (compare Fig. 3e, f).

12. In this procedure, fixation and permeabilization occur at the same time [10].
13. Stage V (i.e., mature) oocytes are discarded and only oocytes ranging from mid-stage II to early-stage IV are kept to ultimately visualize the micropylar cell.
14. A maximum of 50–70 follicles are taken in each tube to avoid over-crowding and poor penetration of the antibody and later of the ISH probe.
15. Both the primary and the secondary antibodies are diluted in 2% NGS-PBDT-I or 5% NGS-PBDT-II solution.
16. The hybridization and all the stringency washes are performed at 60 °C in a hybridization oven under gentle shaking. All the stringency wash solutions are pre-warmed at 60 °C. Reducing salt concentration at a defined temperature increases the stringency to get rid of the non-specific binding of the probe. After the stringency washes, all steps are carried out at room temperature.
17. The *cyclinb1* probe marks the animal pole domain surrounding the micropylar cell.
18. A maximum of 4–6 follicles are mounted on a coverslip (Fig. 4d).
19. The *cyclinb1* domain is also imaged as a reference to the overlying micropylar cell (Fig. 4g, i, j).

References

1. Elkouby YM, Mullins MC (2016) Methods for the analysis of early oogenesis in zebrafish. Dev Biol. https://doi.org/10.1016/j.ydbio.2016.12.014
2. Yan Y-L, Desvignes T, BreMiller R et al (2017) Gonadal soma controls ovarian follicle proliferation through Gsdf in zebrafish. Dev Dyn 246:925–945. https://doi.org/10.1002/dvdy.24579
3. Roth S (2001) Drosophila oogenesis: coordinating germ line and soma. Curr Biol 11: R779–R781. https://doi.org/10.1016/s0960-9822(01)00469-9
4. Korta DZ, Hubbard EJA (2010) Soma-germline interactions that influence germline proliferation in *Caenorhabditis elegans*. Dev Dyn 239:1449–1459. https://doi.org/10.1002/dvdy.22268
5. Dingare C, Niedzwetzki A, Klemmt PA et al (2018) The Hippo pathway effector Taz is required for cell morphogenesis and fertilization in zebrafish. Development. https://doi.org/10.1242/dev.167023
6. Xia P, Gütl D, Zheden V, Heisenberg C-P (2019) Lateral inhibition in cell specification mediated by mechanical signals modulating taz activity. Cell 176(6):1379–1392.e14. https://doi.org/10.1016/j.cell.2019.01.019
7. Yi X, Yu J, Ma C et al (2019) The effector of Hippo signaling, Taz, is required for formation of the micropyle and fertilization in zebrafish. PLoS Genet 15:e1007408. https://doi.org/10.1371/journal.pgen.1007408
8. Aleström P, D'Angelo L, Midtlyng PJ et al (2019) Zebrafish: housing and husbandry recommendations. Lab Anim

2019:23677219869037. https://doi.org/10.1177/0023677219869037

9. Elkouby YM, Mullins MC (2017) Coordination of cellular differentiation, polarity, mitosis and meiosis – new findings from early vertebrate oogenesis. Dev Biol 430:275–287. https://doi.org/10.1016/j.ydbio.2017.06.029

10. Fernández J, Fuentes R (2013) Fixation/permeabilization: new alternative procedure for immunofluorescence and mRNA in situ hybridization of vertebrate and invertebrate embryos. Dev Dyn 242:503–517. https://doi.org/10.1002/dvdy.23943

Chapter 15

Visualization of Transcriptional Activity in Early Zebrafish Primordial Germ Cells

Fabio M. D'Orazio, Aleksandra Jasiulewicz, Yavor Hadzhiev, and Ferenc Müller

Abstract

Here, we describe a fast and straightforward methodology to in vivo detect transcriptional activity in the early zebrafish germ line. We report how fluorescently labeled morpholinos, targeted to nascent early transcripts, can be used to track the onset of transcriptional events during early embryogenesis. This method could be applied to any tagged cell line in a developing early zebrafish embryo as long as the gene of interest is expressed at high enough level for morpholino detection and is expressed at the first and main wave of genome activation, for which the protocol has been verified. The protocol, in combination with genetic manipulation, allows studies of mechanisms driving zygotic genome activation (ZGA) in individual cells. The reported procedures apply to a broad range of purposes for zebrafish embryo manipulation in view of imaging nuclear molecules in specific cell types.

Key words Zebrafish, Germ cells, Germ plasm, Bucky ball, Transcription, ZGA, Morpholinos, Microinjections, Microscopy

1 Introduction

Imaging approaches are fundamental for the understanding of developmental dynamics and embryogenic pathways, and allow high-resolution, non-invasive studies. Increased technological advances offer accurate and sensitive tracking of single molecules, which has been largely exploited for transcript detection, protein-chromatin interactions, and DNA contacts [1–3].

The natural transparency and extrauterine development of zebrafish larvae have boosted its use in imaging experiments. In particular, extensive transcriptional regulation takes place during early zebrafish embryogenesis when the zygotic genome activates and when cellular commitment begins [4]. Despite the importance of

Fabio M. D'Orazio and Aleksandra Jasiulewicz contributed equally to this work.

Roland Dosch (ed.), *Germline Development in the Zebrafish: Methods and Protocols*, Methods in Molecular Biology, vol. 2218, https://doi.org/10.1007/978-1-0716-0970-5_15, © Springer Science+Business Media, LLC, part of Springer Nature 2021

transcriptional visualization during early zebrafish development, detailed spatio-temporal studies are challenged by the fast cell division pace experienced by the early embryo. Currently, several methods for transcriptional visualization exist; however, the totality of these methods relies on time-consuming hybridization of oligonucleotide probes after embryo fixation or use of transgenesis. In vivo imaging overcomes the need for genetic manipulation or tissue cross-link and offers a wide range of possibilities to spatially and temporally determine single-cell behavior.

While morpholinos (MOs) have been mainly associated with loss-of-function experiments, their use as molecular probes for visualization has not been fully explored. Due to the intrinsic ability of MOs of binding RNA by the complementarity of bases, they can be directed toward the desired RNA target and function as probes when fluorescently tagged [5, 6]. This methodology has been adapted to track the early expression of genes such as miR-430 and zinc finger genes in transcribing cells during ZGA, to determine the length of S-phase and monitor transcription dynamics [7–9]. In particular, MOs targeting the 5′ end of the transcripts ease visualization of nascent RNAs. During the morpholinos design, it is important to consider the existence of secondary structures. Online tools such as the RNAfold web server [10] can be used to predict RNA folding and select regions free from secondary structures. Once such regions are identified, Gene Tools customer support should be contacted for designing MOs with optimal hybridization parameters in these regions. For optimal results, 2–3 MOs should be tested. Note that evidence for gene expression detection has only been reported for multicopy genes active during genome activation [6, 7].

The germ line is one of the first specified cell types during zebrafish embryogenesis, depending on maternally inherited cytoplasmic aggregates [11, 12]. These germ aggregates (germ plasm) accumulate in the maturing oocyte and preserve germ fate-associated RNAs and proteins [13–15]. Upon fertilization, the embryonic germ plasm segregates with only one daughter cell at each asymmetrical cell division and the cells inheriting the germ plasm become the progenitors of the germ cells. While the requirement of the germ plasm for germ cell formation has been validated [16–18], detailed mechanisms of action and the effect of the germ plasm on transcription have not been explored.

The highly specialized, germ plasm-contained proteins have allowed the development of several tagging methods for germ cell visualization in the past [14, 19, 20]. However, due to the initial transcriptional quiescent period (pre-ZGA), zygotically expressed transgenes only allow germ plasm visualization after 3.5 h post fertilization (hpf) at the earliest [19]. The Tg(Buc-GFP) line is to date the only available transgenic fish line allowing germ plasm tracking from the beginning of embryogenesis [21].

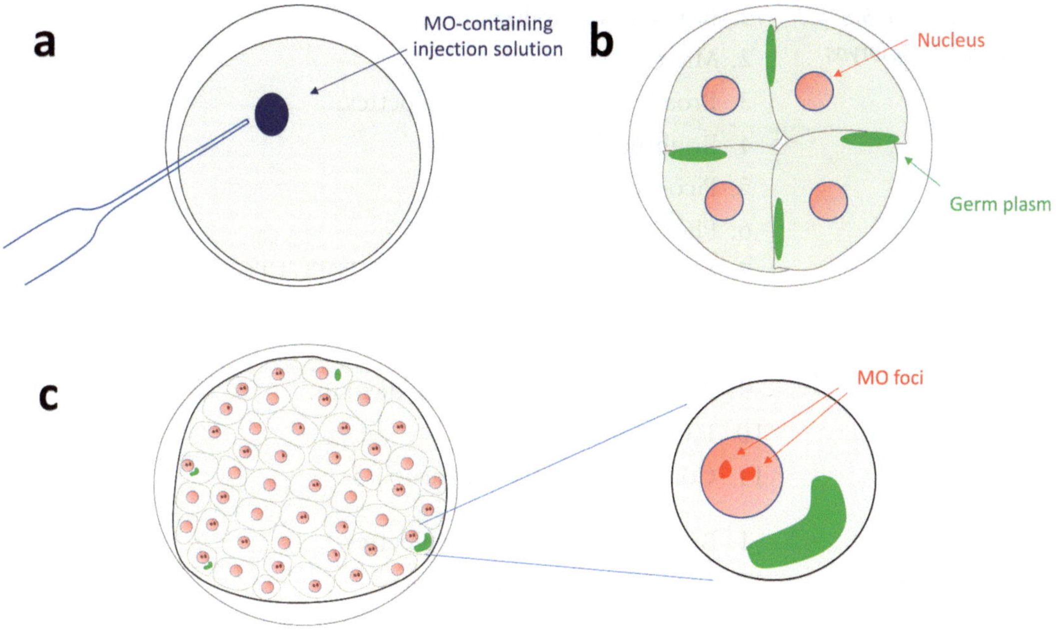

Fig. 1 Schematic of the experiment. (**a**) Zebrafish embryos are injected with fluorescently labeled MOs at 1-cell stage and let grow. (**b**) The germ plasm (green) can be tracked in the offspring of Buc-GFP$^+$ females. At 4-cell stage, four main masses accumulate along the cleavage furrows. (**c**) Overlook of an early-stage zebrafish embryo and magnified view where transcription can be visualized in the germ cells. The germ plasm-carrying cell can be identified (green) and transcriptional activity visualized (red dots)

Here, we detail a simple strategy to visualize and track transcription in embryonic germ cells by injection of fluorescently labeled morpholinos in zebrafish zygotes where the germ plasm is detectable by GFP (Fig. 1). This method has great applicability and it can be easily transferred to other embryonic regions (e.g., IVL cells) or labeled cell types. Additionally, the ability of tagged morpholinos to function as molecular probes offers visualization tools to investigate the dynamics of early-expressed genes.

2 Materials

2.1 Adult Zebrafish Cross and Embryo Collection

1. Tg(Buc-GFP) adult zebrafish females.
2. Polycarbonate, polyethylene, or glass crossing tanks with mesh.
3. E3 medium for embryo growth. To prepare a 60× stock, add the following ingredients to 10 l of water: 172 g NaCl, 7.6 g KCl, 29 g $CaCl_2$, 49 g $MgSO_4{\cdot}7H_2O$. Store at room temperature for up to 1 month.
4. Small net for embryo collection.
5. Graduated Falcon™ conical tubes.
6. Pronase from *Streptomyces griseus.*
7. Agarose-coated Petri dishes.

2.2 Microinjection of Zebrafish Embryos

1. Borosilicate capillaries.
2. Micropipette puller.
3. Broad-/fine-tip Pasteur pipettes.
4. Brightfield microscope.
5. Pico-injector.
6. Phenol red.
7. Morpholinos (CACACGCATCTTGTTGTCTGCTGTT) (*see* **Note 1**).
8. Nuclease-free grade water.
9. Rhodamine isothiocyanate dextran.
10. E3 (described above).
11. Constant temperature incubator.

2.3 Embryo Preparation for Live Imaging

1. Broad-/fine-tip Pasteur pipettes.
2. 1.5 ml microcentrifuge tubes.
3. Phosphate-buffered saline (PBS).
4. Low melting point agarose.
5. Fluorescent stereomicroscope.
6. Zeiss Z1 Light Sheet microscope or fast confocal microscope (Lsm880 with fast Airyscan module).

2.4 Embryo Preparation for Fixed-Embryo Imaging

1. Pasteur pipettes.
2. 1.5 ml microcentrifuge tubes.
3. 1× PBS.
4. HEPES (2-[4-(2-hydroxyethyl)piperazin-1-yl] ethanesulfonic acid).
5. 4% (w/v) formaldehyde (PFA). PFA is hazardous and should be handled with care using protective gloves and lab coat. PFA powder should be handled under the hood exclusively (*see* **Note 2**).
6. Vectashield Antifade Mounting Medium with DAPI or alternative mounting medium.
7. Confocal microscope.

3 Methods

3.1 Embryo Collection and Dechorionation

1. Set pairs of adult zebrafish in crossing tanks with separators the evening before the experiment. Although procedures applied in your animal facility should be followed, we recommend to use males and females in a one-to-two proportion (*see* **Note 3**).

2. The next morning, remove the separator and let the embryos pass through the mesh.
3. Use a net to collect embryos and transfer them to a Petri dish.
4. Add pronase solution (0.5 mg/20 ml) to the embryos and incubate until 10% of the embryos are out of the chorion (*see* **Note 4**).
5. Carefully wash the embryos in 1 l of fresh E3 or fish water three times.
6. Transfer the dechorionated embryos in agarose-coated Petri dishes and incubate at 28.5 °C.

3.2 Microinjection of Morpholinos in Zebrafish Embryos

1. Prepare the injection solution as follows:

Component	μl in a 10 μl reaction	Final concentration
Stock morpholinos (30 μM)	3	10 μM
Phenol red solution (0.5%)	2	0.1%
Rhodamine isothiocyanate dextran (5%)	1	0.5%
Nuclease-free water	To 10	–

2. Break the tip of the needle capillary with a sharp scalpel or fine forcipes.
3. Load the capillary with the desired volume of injection solution.
4. Pipette the embryos in an empty Petri dish and aspirate the water with a broad-tip Pasteur pipette. Remove any left-over water with a fine-tip Pasteur pipette.
5. Arrange the embryos in a monolayer to facilitate microinjections.
6. Perform microinjections manually or with a micromanipulator under a stereomicroscope.
7. Inject about 1 nl of injection solution per embryo (*see* **Note 5**).
8. When all the embryos in the Petri dish have been injected, transfer them in a new Petri dish with E3 medium and grow them at 28.5 °C until needed.

3.3 Sample Preparation and Imaging Acquisition

The offspring of transgenic females expressing GFP-labeled germ plasm (green) were collected and injected at 1-cell stage with 5 μM miR-430 morpholino-Cy5 (red). Images were acquired using a

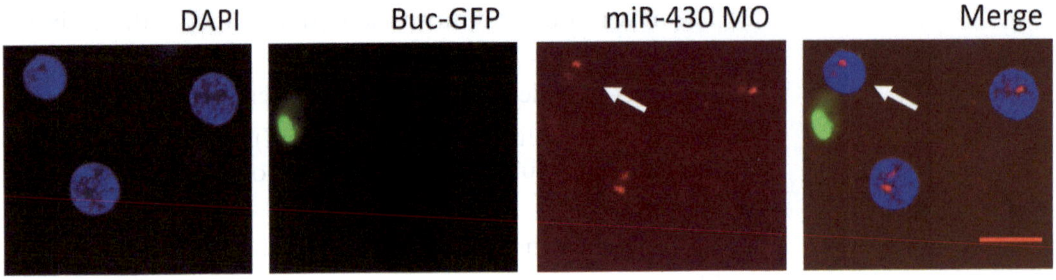

Fig. 2 Confocal microscopy of fixed embryos injected with labeled morpholinos allows single-cell detection of transcriptional events in time-course experiments. White arrows indicate the nucleus of the germ plasm-carrying cell. Scale bar is 10 μm

Zeiss Light Sheet Z1 microscope or a Zeiss 780 Confocal microscope and processed using Zen Blue and Zen Black software.

3.3.1 Embryo Fixation and Imaging

1. Collect about 20 dechorionated embryos in an Eppendorf tube and set them on ice.
2. Wash the embryos in cold PBS for three times.
3. Add 500 μl of Hepes 1 M to 4 ml of 4% PFA.
4. Add 1 ml of cold 3.5% PFA and Hepes to the embryos and fix on ice for 40 min (*see* **Note 6**).
5. Remove all the PFA solution and wash the embryos three times in cold PBS.
6. Aspire the liquid almost entirely, ensuring a thin layer of PBS is left to avoid exposing the embryos to the air.
7. Cover the embryos with about 100 μl of Vectashield Antifade Mounting Medium with DAPI and incubate for 30 min to overnight.
8. Transfer all the embryos to a glass-bottom imaging dish or microscope slide prior to image acquisition (Fig. 2).

3.3.2 Live Embryo Imaging via Light Sheet Z1 Microscope

1. Prior to mounting embryos for imaging, prepare 1% low melt point agarose. The agarose should be well mixed using a magnetic stirrer for 20 min at 70–80 °C. A clear and particle-free agarose helps optimal light sheet propagation and auto adjustment in the device during imaging acquisition. It is not recommended to use regular agarose, as it does not preserve a liquid phase at mild temperatures.
2. Preheat the microscope chamber filled with E3 medium to 28 °C (*see* **Note 7**).
3. Pre-select injected embryos by using a fluorescent microscope. Dispose any malformed, dead, and non-injected embryos (*see* **Note 8**).
4. Preheat low melt agarose pre-setting a magnetic mixer 120 °C or an Eppendorf thermo block 70–80 °C. The use of an

Eppendorf thermo block shortens the time needed to melt the agarose. After melting, keep the agarose at 42 °C to prevent its setting.

5. By using a Pasteur pipette pour a drop of hot agarose on a plate and wait until the agarose cools down to around 33–35 °C to avoid harm to the embryos caused by high temperatures.
6. Choose up to four embryos and transfer them into the liquid agarose. Hold a pipette with embryos vertically and let the embryos settle in the agarose by gravity. During the procedure, minimize the carry-over of medium to avoid excessive dilution of the agarose.
7. By using a color-coded capillary and plunger, uptake about 1 cm of agarose from the plate and then uptake the embryos. Embryos should not touch each other, but should not also be farther apart than 3–5 mm (*see* **Note 9**).
8. Wait for the agarose to solidify in the capillary and place it in the microscope chamber (Fig. 3).

3.4 Image Acquisition

1. Use a 20× lens for whole embryo image acquisition. Use 40× for higher magnification (*see* **Note 10**).

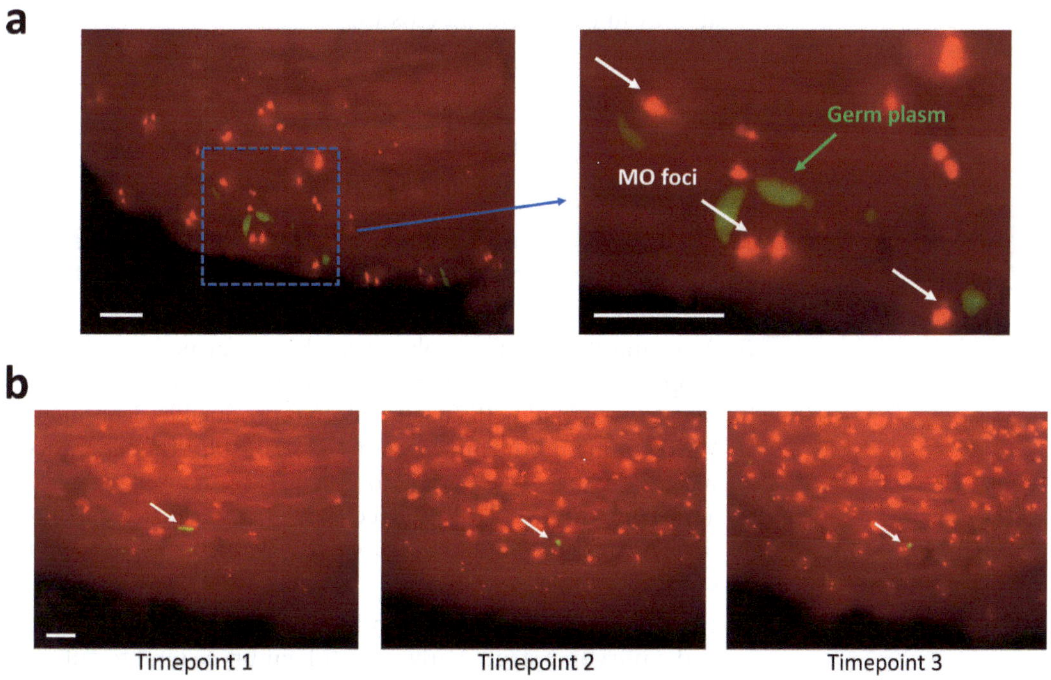

Fig. 3 Live imaging of transcriptional activity in a whole zebrafish embryo where the germ cells are labeled. (**a**) Dual laser acquisition of a zebrafish embryo allows visualization of transcriptional foci in the germ cells at blastula stages. White arrows indicate transcriptional foci in the germ plasm-carrying cells. Green arrow indicates the germ plasm. Scale bar is 30 μm. (**b**) Time course image acquisition by Light Sheet Z1 showing transcription occurrence in the whole embryo and in the labeled germ cells (white arrows). Scale bar is 30 μm

2. Set lasers depending on the required wavelength. Lasers used in these experiments are: 488 (green) and 568 (red).
3. If imaging zebrafish embryos, it is recommended to start the acquisition prior to the appearance of the fluorescent signal. If interested in expression over time, a time-lapse experiment can be set (*see* **Note 11**).
4. After image acquisition, data analysis was performed in Zen (Blue or lite 2.1) software.

4 Notes

1. Upon arrival, resuspend the morpholinos to the desired concentration. We recommend 300 μM for stocking. Minimize light exposure of the morpholinos throughout the procedure.
2. To prepare 4% PFA, warm up the required amount of water to 60 °C and gently pour it on PFA powder under a fume hood. Stir thoroughly on a warm, magnetic stirrer until the solution clears completely. Adjust the pH to 7.4 and freeze it in aliquots.
3. In our experience, keeping the Tg(Buc-GFP) line in heterozygosity helped egg yield. Buc-GFP-positive embryos are generated from the following crosses:
   - Homozygous adults.
   - Heterozygous adults.
   - Homozygous females × heterozygous males.
   - Heterozygous females × wild type males.
4. Dechorionation helps fast and reproducible fixation in PFA. It also eases the injection procedure. If embryos are used for live imaging, dechorionation is not necessary.
5. Assessing the injection volume can be performed based on the drop sizes. Accurate estimation of injection volume is important to avoid side effects and allow reproducible experiments. We recommend the injection of 1–2 nl of solution per embryo, which can be estimated with the use of a μm-scale. As an approximation, 1–2 nl of injections corresponds to a sphere diameter of 150 μm or 30% of the diameter of a zebrafish embryo.
6. Fixation time can vary depending on the downstream needs and embryonic stage. Early zebrafish embryos tend to fall apart if fixation is mild; however, the softness of the embryos allows short incubations in PFA. On the other hand, over-fixation can prevent antibody or probe accessibility and it is not

recommended. Over-fixation may also interfere with morpholino fluorescence.

7. If you are using any drug treatment you can prepare E3 with the soluble chemical of your choice.
8. The incubator and the microscope should be located in the same room to shorten the time the embryos spend at room temperature. Beware the room temperature should not be lower than 20 °C. This is particularly important at this stage as the ambient temperature in microscope rooms tend to be low.
9. Excess of space between embryos in the capillary will cause the agarose sausage to be too long which will make it hard to image (long sausage will swing away from the imaging pole) and additionally generate longer image acquisition time, lowering the time resolution.
10. The use of high magnification increases the laser exposure of the specimen, resulting in fast signal bleaching.
11. If imaging in time lapse, bear in mind that more embryos you have, longer imaging frame it will generate, lowering the imaging time resolution.

Acknowledgments

This work was supported by the Horizon 2020 programme Zencode-ITN, Wellcome Trust Investigator award (106955/Z/15/Z) to F.M., Horizon 2020 FET project to F.M. We thank Roland Dosch at the University Medical Center in Göttingen for providing the Tg(Buc-GFP) line and the BMSU facility at the University of Birmingham for the maintenance.

References

1. Reisser M, Palmer A, Popp AP, Jahn C, Weidinger G, Gebhardt JCM (2018) Single-molecule imaging correlates decreasing nuclear volume with increasing TF-chromatin associations during zebrafish development. Nat Commun 9(1):5218
2. Fukaya T, Lim B, Levine M (2016) Enhancer control of transcriptional bursting. Cell 166 (2):358–368
3. Stapel LC, Broaddus C, Vastenhouw NL (2018) Detection and automated analysis of single transcripts at subcellular resolution in zebrafish embryos. Methods Mol Biol Clifton NJ 1649:143–162
4. Schulz KN, Harrison MM (2019) Mechanisms regulating zygotic genome activation. Nat Rev Genet 20(4):221–234
5. Chen J, Wu J, Hong Y (2016) The morpholino molecular beacon for specific RNA visualization in vivo. Chem Commun Camb Engl 52 (15):3191–3194
6. Umemoto N et al (2013) Fluorescent-based methods for gene knockdown and functional cardiac imaging in zebrafish. Mol Biotechnol 55(2):131–142
7. Hadzhiev Y et al (2019) A cell cycle-coordinated polymerase II transcription compartment encompasses gene expression before global genome activation. Nat Commun 10 (1):691

8. Chan SH et al (2019) Brd4 and P300 confer transcriptional competency during zygotic genome activation. Dev. Cell 49(6):867–881.e8
9. Sato Y et al (2019) Histone H3K27 acetylation precedes active transcription during zebrafish zygotic genome activation as revealed by live-cell analysis. Dev Camb Engl 146:19
10. Hofacker IL (Jul. 2003) Vienna RNA secondary structure server. Nucleic Acids Res 31 (13):3429–3431
11. Eddy EM (1975) Germ plasm and the differentiation of the germ cell line. Int Rev Cytol 43:229–280
12. Seydoux G, Braun RE (2006) Pathway to totipotency: lessons from germ cells. Cell 127 (5):891–904
13. Yoon C, Kawakami K, Hopkins N (1997) Zebrafish vasa homologue RNA is localized to the cleavage planes of 2- and 4-cell-stage embryos and is expressed in the primordial germ cells. Dev Camb Engl 124(16):3157–3165
14. Köprunner M, Thisse C, Thisse B, Raz E (2001) A zebrafish nanos-related gene is essential for the development of primordial germ cells. Genes Dev 15(21):2877–2885
15. Knaut H, Pelegri F, Bohmann K, Schwarz H, Nüsslein-Volhard C (2000) Zebrafish vasa RNA but not its protein is a component of the germ plasm and segregates asymmetrically before germline specification. J Cell Biol 149 (4):875–888
16. Gross-Thebing T et al (2017) The vertebrate protein dead end maintains primordial germ cell fate by inhibiting somatic differentiation. Dev. Cell 43(6):704–715.e5
17. Tada H, Mochii M, Orii H, Watanabe K (2012) Ectopic formation of primordial germ cells by transplantation of the germ plasm: direct evidence for germ cell determinant in Xenopus. Dev Biol 371(1):86–93
18. Buehr ML, Blackler AW (1970) Sterility and partial sterility in the South African clawed toad following the pricking of the egg. J Embryol Exp Morphol 23(2):375–384
19. Blaser H et al (2005) Transition from non-motile behaviour to directed migration during early PGC development in zebrafish. J Cell Sci 118(Pt 17):4027–4038
20. Krøvel AV, Olsen LC (2002) Expression of a vas::EGFP transgene in primordial germ cells of the zebrafish. Mech Dev 116(1–2):141–150
21. Riemer S, Bontems F, Krishnakumar P, Gömann J, Dosch R (2015) A functional Bucky ball-GFP transgene visualizes germ plasm in living zebrafish. Gene Expr Patterns GEP 18(1–2):44–52

Chapter 16

Experimental Methodologies to Visualize Early Cell Biology in Zebrafish Embryos

Asfa Sabrin Borbora and Sreelaja Nair

Abstract

For organisms to function normally, biological molecules must work at the correct time in the right cells and within the right intracellular compartments of cells. Biological research relies heavily on discovering the cellular locations at which such molecular interactions occur. A mainstay technique in this process of discovery is the visualization of locations of proteins in cells and tissues, known as immunocytochemistry and immunohistochemistry, respectively. If performed correctly, these techniques can provide detailed information regarding the endogenous locations of proteins and their ectopic locations or absence in mutants and in disease states.

Key words Zebrafish, Antibody staining, Immunofluorescence, Immunohistochemistry, Immunocytochemistry, Early development

1 Introduction

Nearly 130 years ago, Emil Adolf von Behring and Shibasaburo Kitasato reported their seminal findings in which sera from animals infected with *C. tetani* or *C. diphtheriae* protected naïve animals against tetanus and diphtheria bacilli and toxins [1, 2]. These results laid the foundation for serum therapy in the early 1900s, in which sera from animals inoculated with disease-causing bacteria were used as cures for a variety of bacterial infections, including tetanus, which had emerged as a major cause for fatality in wounded soldiers during World War I. The underlying mechanism which made serum therapy a prophylactic success is now known as antigen-antibody interaction and is the foundation for the success of vaccine therapy. In the 1940s, Albert Coons used the antigen-antibody interaction as a diagnostic tool for detecting infections. He first attempted to visualize pneumococci by conjugating a fluorescent moiety to the anti-pneumococcal antibodies present in the immune serum [3]. Coons then tested the conjugated antibodies in mice infected with pneumococci and was able to see bright

Roland Dosch (ed.), *Germline Development in the Zebrafish: Methods and Protocols*, Methods in Molecular Biology, vol. 2218, https://doi.org/10.1007/978-1-0716-0970-5_16, © Springer Science+Business Media, LLC, part of Springer Nature 2021

fluorescent patches in infected tissues, which were absent in tissues from uninfected mice [4–6]. The basic principles used by Coons to generate the fluorescent-antibody conjugate and bind the conjugated antibody to an antigen in biological tissue remains largely unchanged to this day. In the years that followed, improvements in synthesis of fluorophores with unique spectra and microscopy techniques have refined the basic methodology described by Coons to allow visualization of antigens inside single cells and even in subcellular regions. This technique is known as immunofluorescence (IF) and can be used on whole tissues (immunohistochemistry (IHC)) and on single cells (immunocytochemistry (ICC)). The fundamentally conserved nature of antigen-antibody interactions and protein sequences makes it possible to detect proteins in a species using antibodies generated against homologous proteins in a different species, with careful use of controls for validation of the pattern seen [7].

Though initially conceptualized as a diagnostic tool, IF has become the mainstay technique in all studies that require spatial visualization of a protein, RNA-protein, or DNA-protein conjugate in single cells or in tissues. For embryonic development studies, in the absence of transgenic lines that express a fluorescently tagged version of a protein or RNA of interest, IF is an invaluable tool to understand the expression pattern during specific stages of embryogenesis. *Danio rerio* (zebrafish) debuted as a model organism in the 1960s and is now widely used to study vertebrate embryonic development and disease conditions [8, 9]. Zebrafish embryos are optically transparent and free of yolk granules, which make biological interactions within single cells or in entire cell layers accessible to visualization techniques till late larval stages [8]. In this chapter, we describe IF procedures to visualize the dynamic cell biology of developing zebrafish embryos [10, 11]. Though the protocols described here are particularly useful to obtain immunofluorescence labeling during the early cleavage stages of zebrafish development, the same procedures can also be used for larval stages of zebrafish.

2 Materials

2.1 Obtaining Synchronously Fertilized Zebrafish Embryos

2.1.1 Manual Extrusion of Mature Eggs from Female Zebrafish

1. Zebrafish mating cages with mesh bottom inserts.
2. Fish nets.
3. 90 mm plastic Petri dishes.
4. Tissue paper roll.
5. Glass beaker.
6. Plastic spoon.
7. Flat spatula.

8. Embryo medium, store at room temperature (RT): 5 mM sodium chloride (NaCl), 0.17 mM potassium chloride (KCl), 0.33 mM calcium chloride ($CaCl_2 \cdot 2H_2O$), 0.33 mM magnesium sulfate ($MgSO_4 \cdot 7H_2O$), 33 μl/L of methylene blue from a 5 mg/ml stock made in distilled water and stored at 4 °C.
9. Adult zebrafish females primed for manual egg extrusion.
10. Stock MESAB (0.4% in 1× phosphate-buffered saline, pH 7.4 (PBS)).
11. MESAB working solution: 30 ml MESAB stock diluted to 200 ml with fish facility water.
12. Stock BSA (10% BSA in autoclaved MilliQ water).
13. BSA working solution: 0.5% BSA stock in Hank's premix (without $NaHCO_3$).

2.1.2 Preparation of Sperm Solution from Male Zebrafish

1. Adult male zebrafish.
2. Fish nets.
3. 90 mm plastic Petri dishes.
4. Tissue paper roll.
5. Glass beaker.
6. Plastic spoon.
7. Dissection scissors.
8. Forceps.
9. 1.5 ml tubes.
10. Stock MESAB (0.4% in 1× phosphate-buffered saline, pH 7.4 (PBS)).
11. MESAB working solution: 30 ml MESAB stock diluted to 200 ml with fish facility water.
12. Hank's buffer: Hank's stock solutions (store at 4 °C):
    - Stock 1: 8.0 g NaCl +0.4 g KCl in 100 ml distilled water.
    - Stock 2: 0.358 g Na_2HPO_4 anhydrous 0.6 g KH_2PO_4 in 100 ml distilled water.
    - Stock 4: 0.72 g $CaCl_2$ in 50 ml distilled water.
    - Stock 5: 1.23 g $MgSO_4 \cdot 7H_2O$ in 50 ml distilled water.
    - Stock 6: 0.3 5 g of $NaHCO_3$ in 10 ml of distilled water, prepared fresh.

    Hank's premix (store at 4 °C):

    Prepare by mixing Hank's stock solutions in this order: 10 ml stock 1 + 1 ml stock 2 + 1 ml of stock 4 + 86 ml of distilled water + 1 ml of stock 5.

    Hank's buffer: 900 μl Hank's premix + 10 μl Hank's stock 6. Prepare immediately before experiment and keep on ice.

2.1.3 In Vitro Fertilization of Zebrafish Eggs

1. Manually extruded mature eggs in a dish.
2. Sperm solution on ice.
3. Timer.
4. Embryo medium.

2.2 Immunofluorescence Labeling of Zebrafish Embryos

2.2.1 Dechorionation of Live Embryos

1. 55 mm glass Petri dishes.
2. Pronase stock (20 mg/ml in autoclaved MilliQ water).
3. Embryo medium.
4. Long glass Pasteur pipette with rubber bulb.

2.2.2 Fixation and Storage of Fixed Embryos

1. Long glass Pasteur pipette with rubber bulb.
2. 90 mm plastic Petri dishes.
3. 4% PFA in PBS (bring PBS to a boil and then let it cool a bit. In the fume hood, add the required amount of PFA powder into hot PBS and stir till all PFA goes into solution. Make up the volume as required and store at −20 °C in 10 ml aliquots. Do not refreeze after thawing).
4. MT fixative: 10 ml 4% PFA + 110 μl glutaraldehyde (25% w/v) + 100 μl 500 mM EGTA, pH 8.0 + 20 μl TritonX-100.
5. Methanol.

2.2.3 Preparation of Embryos for Immunofluorescence Labeling

1. Long glass Pasteur pipette with rubber bulb.
2. 55 mm glass Petri dishes.
3. Fine Forceps #55.
4. Methanol rehydration series:
   - 75% Methanol solution in 1× PBS.
   - 50% Methanol solution in 1× PBS.
   - 25% Methanol solution in 1× PBS.
5. Proteinase K: 10 mg/ml in autoclaved MilliQ water, store at −20 °C.
6. $NaBH_4$ solution (0.5 mg/ml in 1× PBS, store at 4 °C and use within 1 week or prepare fresh).
7. PBS-Triton solution: PBS with 0.1% TritonX-100.

2.2.4 Incubation of Embryos with Primary Antibodies

1. Horizontal rocker at 4 °C and RT.
2. Blocking stock solution: 10% bovine serum albumin (BSA) in autoclaved MilliQ water.
3. Blocking solution: 1% BSA (10% BSA stock diluted in PBS-Triton).

4. Primary antibody of appropriate dilution in blocking solution (typically used at 1:500–1:2500).
5. PBS-Triton.

2.2.5 Incubation of Embryos with Secondary Antibodies and Nuclei Labeling

1. Horizontal rocker at 4 °C and RT.
2. PBS-Triton.
3. Blocking solution: 1% BSA (10% BSA stock diluted in PBS-Triton).
4. Secondary antibody at 1:100 dilution in blocking solution.
5. DAPI stock solution: 250 μg/ml in dimethylformamide, store at −20 °C in aliquots.
6. DAPI working solution: 1:500 dilution of DAPI stock solution in PBS.
7. 25% v/v glycerol in 1× PBS.
8. 50% v/v glycerol in 1× PBS.

2.2.6 Mounting Immunolabeled Zebrafish Embryos for Imaging

1. Hole reinforcement stickers.
2. Glass slide.
3. Fine forceps #55.
4. Coverslips.
5. Clear nailpolish.
6. Mounting media: 1 ml 75% v/v glycerol in 1× PBS + 20 μl 1 M Tris–HCl, pH 8.0 + a pinch of DABCO. Heat the solution to 65 °C until DABCO dissolves. Store in dark at 4 °C.

3 Methods

3.1 Obtaining Synchronously Fertilized Zebrafish Embryos

The early cell biology of zebrafish embryos is very dynamic with successive cell divisions occurring every 15 min [8]. Since fish lay eggs in short temporally spaced bursts, to obtain clutches of embryos undergoing comparable phases of early development, it is necessary to perform in vitro fertilization to obtain synchronously fertilized eggs.

3.1.1 Manual Extrusion of Mature Eggs from Female Zebrafish

1. Place an adult male and a gravid female into a mating cage with a mesh bottom insert in the evening. Set up ~10 such pairs before the fish room lights turn off for the day.
2. In the morning, watch the adult pairs for mating behavior and initiation of egg expulsion from the female.
3. As soon as the female begins to extrude eggs, separate the male and female and place the female into a new tank containing fish facility water (*see* **Note 1**). Do this for all the pairs that were set

up the night before. This process is known as priming and will ensure that the female will expel mature eggs upon manual extrusion for in vitro fertilization.

4. After obtaining primed females, anesthetize a female fish in a 250 ml glass beaker containing ~200 ml of MESAB working solution.
5. When the gill movements of the fish have slowed down, gently pick up the anesthetized fish using a plastic spoon, dip the spoon into a beaker containing fish facility water to rinse, and place the fish on a wad of tissue paper (*see* **Note 2**).
6. Pat dry the fish gently, especially around the anal fin area from where the eggs will get extruded (*see* **Note 3**).
7. Using a flat spatula, lift the body of the fish from the tissue paper wad and place gently on a dry plastic Petri dish. Turn the fish to its side and spread out the anal fin away from the body.
8. Using a flat spatula, gently press the gravid abdomen in a sloping downward motion toward the anal opening to extrude the eggs onto the Petri dish (*see* **Notes 4** and **5**).
9. Extruded eggs can be held inactive in ~300 μl of 0.5% BSA for up to 20 min, though fertilization efficiency decreases with time.

3.1.2 Preparation of Sperm Solution from Male Zebrafish

1. Transfer 1–2 adult male zebrafish into a 250 ml beaker containing ~200 ml of MESAB working solution.
2. Euthanize the fish by ensuring that all gill movements completely stop, usually takes ~5 min.
3. Gently pick up the euthanized fish using a plastic spoon, dip the spoon into a beaker containing fish facility water to rinse and place the fish on a wad of tissue paper.
4. Pat dry and decapitate the fish using a pair of dissection scissors.
5. Transfer the fish onto a clean, dry plastic Petri dish and make an incision along the ventral side of the abdomen, from the gills to the pelvic fin.
6. Place the fish on its back and gently pull away the abdominal wall on either side.
7. Under an incident light microscope, the testis will be visible as a pair of elongated white organs resting on either side of the abdominal walls (*see* **Note 6**).
8. Using a pair of forceps, gently hold one end of the testes, tug it away from the abdominal wall and pull it free along its length.
9. Transfer both testes into ~500 μl cold Hank's buffer kept on ice. Triturate the testes on ice by pipetting with a 1 ml pipette to release the sperm into the buffer. Let the solution settle for 5 min on ice.

10. Check the supernatant for the presence of motile sperm and use for fertilizing the extruded eggs (*see* **Note** 7).

3.1.3 In vitro Fertilization of Zebrafish Eggs

1. Using a spatula gently spread the pool of extruded eggs in the Petri dish so that the eggs are not on top of each other.
2. If using eggs held in BSA, blot almost all the BSA solution with tissue paper before adding sperm.
3. To the pool of eggs, add ~200 μl sperm solution and mix well by swirling the Petri dish vigorously.
4. Ten seconds after adding the sperm solution add 500 μl embryo medium and start a timer while mixing the egg-sperm pool by swirling the Petri dish gently.
5. One minute after adding embryo medium, flood the plate with embryo medium and incubate the plate at 28 °C until dechorionation and fixation.
6. Check for fertilization efficiency and polyspermy ~30–40 mins after fertilization (mpf, *see* **Note 8**).

3.2 Immunofluorescence Labeling of Zebrafish Embryos

The protocol described below is applicable for direct as well as indirect immunofluorescence (Fig. 1). Though the focus of this chapter is to enable readers to obtain immunofluorescent labeling of early zebrafish embryos, the incorporation of a short permeabilization step after fixation should enable immunofluorescent labeling of larvae as well (Fig. 2).

3.2.1 Dechorionation of Live Embryos

1. After ~4 min of the addition of sperm solution, transfer the embryos into a glass dish using a plastic 3 ml pipette (*see* **Note 9**). There should be ~4 ml of embryo medium in the glass dish.
2. Add 100 μl of Pronase solution and gently swirl the plate to mix the enzyme.
3. Leave the plate undisturbed, in ~5 min embryos will begin to squeeze through the weakened chorion.
4. Swirl the plate gently to facility the dechorionation process. In ~15 mins, most embryos should be dechorionated.
5. If embryos need to be dechorionated for very early stages (fixations less than 5–10 mpf), remove excess embryo medium and add ~200 μl Pronase solution. Keep the glass dish in a tilted position to facilitate dechorionation in a concentrated solution of Pronase. After a couple of minutes add embryo medium and swirl the plate gently to complete the dechorionation process.
6. Tilt the plate gently to pool the dechorionated embryos toward one side of the dish and discard the Pronase containing medium (*see* **Note 10**).
7. Gently add fresh embryo medium to rinse the embryos and discard the medium. Repeat the rinse two more times to remove excess enzyme (*see* **Note 10**).

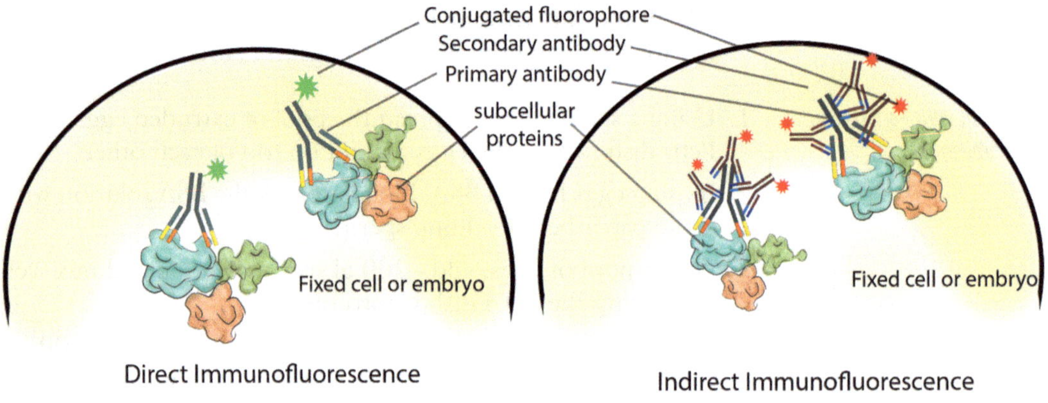

	Direct Immunofluorescence	Indirect immunofluorescence
Time/complexity	Direct immunofluorescence uses primary antibodies conjugated to fluorophores. In zebrafish, it reduces experiment time by at least a day.	Indirect immunofluorescence uses unlabeled primary antibodies, which are detected by fluorophore conjugated secondary antibodies. Due to the two-step detection, this requires more time.
Specificity	Specificity of antibody is high as the fluorophore-conjugated antibody directly binds to the protein of interest.	Though the primary antibody may be specific, the fluorophore conjugated secondary antibody may show non-specific binding to endogenous immunoglobulins in cells.
Cross-reactivity	Fluorophore-conjugated primary antibodies are specific to a protein in a species, and will detect homologs from other species only if the antigenic epitope is conserved.	Primary antibody specificity remains the same as in Direct IF. Secondary antibodies may cross react with immunoglobulins from closely related species. This can be overcome by using pre-adsorbed secondary antibodies.
Sensitivity	Primary antibodies, though specific, fail to amplify the signal.	Several conjugated secondary antibodies bind to the primary antibody, resulting in signal amplification.
Multiplexity	In principle, multiple fluorophore conjugated primary antibodies can be used simultaneously, even if derived from the same species.	If using multiple primary antibodies, each primary antibody must be derived from different species. This ensures that the secondary antibodies used will detect the primary antibodies from only one species each
Cost	Fluorophore tagged primary antibodies are costlier than both untagged primary antibodies, as well as tagged secondary antibodies.	Conjugated secondary antibodies are cheaper as they are not raised against a specific protein, but against the immunoglobulin of a particular species.

Fig. 1 Comparison of direct and indirect immunofluorescence labeling. Schematics depict general principles of the two techniques and the table lists some comparisons between the two techniques for practical use

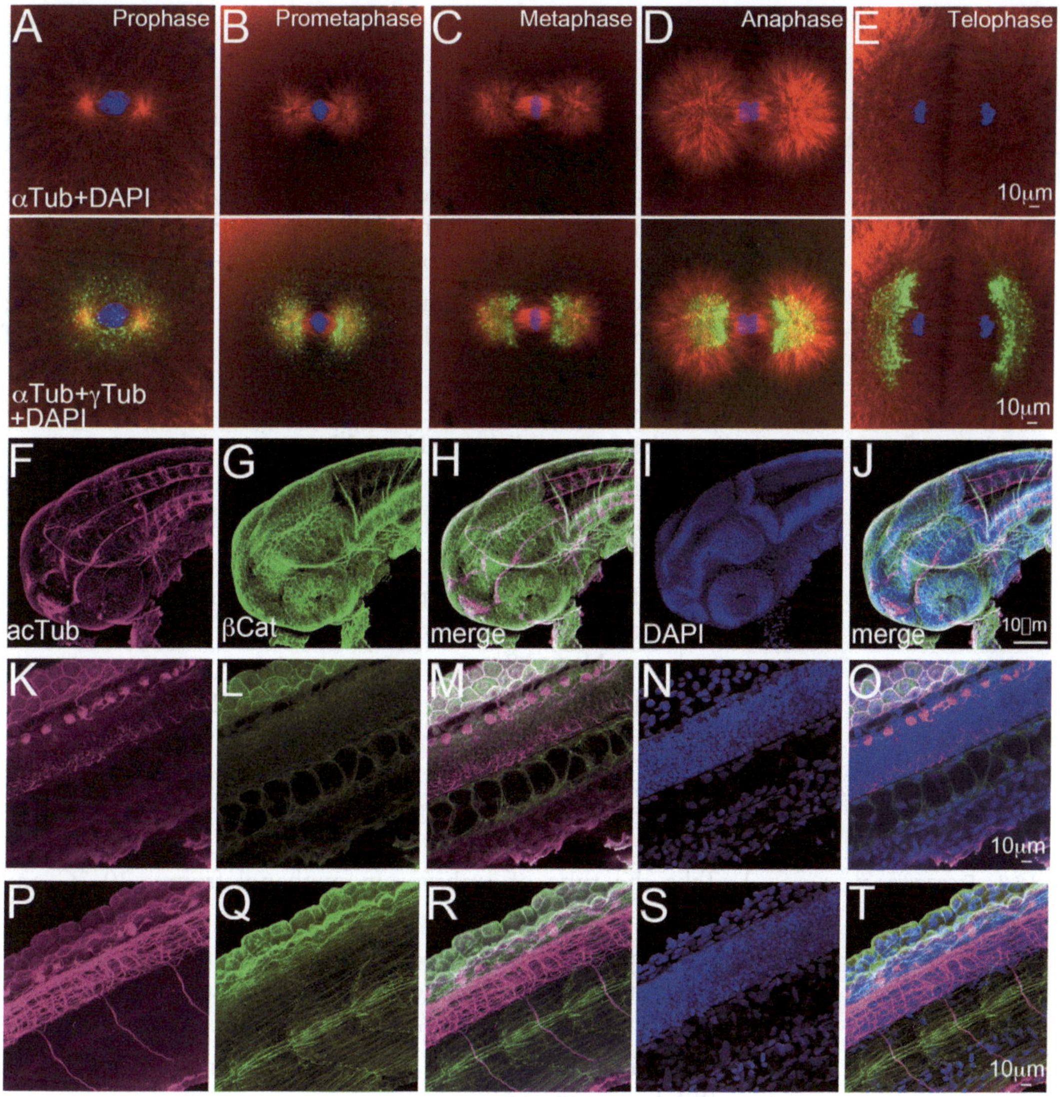

Fig. 2 Immunolabeling in early- and late-stage whole zebrafish embryos. Panels (**a–e**) show mitosis phases during cleavage stages (examples shown are from the first two cell division cycles) in early zebrafish embryos. Embryos are labeled for αTubulin (red), γTubulin (green), and DAPI (blue). Panels F to T show late-stage (~24 hpf) zebrafish embryos stained for acetylated Tubulin (acTub, magenta), βCatenin (green), and DAPI (blue). Panels (**f**, **k**, and **p**) show axonal tracts in the head (**f**), spinal cord nuclei and axonal tracts (**k**, **p**). In panels (**g**, **l**, and **q**) adhesion complex marked by βCatenin shows cell borders and bundled axonal tracts in the head (**g**), notochord (**l**), and muscles (**q**)

8. Dechorionated embryos can be kept in embryo medium till fixation time.

3.2.2 Fixation and Storage of Fixed Embryos

1. Transfer the required number of dechorionated embryos into a 1.5 ml tube using a glass pipette (*see* **Note 11**).
2. Take the pipette tip to the bottom of the tube and release the embryos into the bottom of the tube directly; do not drop embryos into the tube from the top.
3. Remove excess embryo medium as much as possible by tilting the tube. Gently add ~1 ml of 4% PFA from the sides of the tube.
4. Keep the tube horizontal, close the lid, and place it horizontal in a plastic Petri dish (*see* **Note 12**).
5. If fixing for microtubules, use fixative that contains a mixture of PFA and glutaraldehyde (MT fixative). For all other proteins, generally PFA only works well.
6. Leave the embryos in fixative for 6–8 h at RT and then transfer the plate with tubes containing fixed embryos to 4 °C overnight (*see* **Note 13**).
7. Discard the fixative solution and wash embryos twice in 1× PBS for 5 min each.
8. Replace PBS with 100% methanol and incubate the tubes overnight at −20 °C (*see* **Note 13**).

3.2.3 Preparation of Embryos for Immunofluorescence Labeling

1. Take the tube of embryos in methanol from −20 °C and discard the methanol.
2. Rehydrate embryos by adding 75% methanol in PBS at RT for 10 min, discard 75% methanol and do sequential 10 min washes with 50% and 25% methanol in PBS.
3. After discarding the 25% methanol, wash embryos twice with PBS for 5 min each.
4. If embryos are 24 h post fertilization or older, add 2 μl of proteinase K to 1 ml of PBS and incubate embryos in this for ~10 min for additional permeabilization before proceeding to **step 9**.
5. If the fixative contains glutaraldehyde, perform the following $NaBH_4$ treatment. If only PFA was used, proceed directly to **step 8**.
6. Discard PBS and add 1 ml of $NaBH_4$ solution to the embryos and incubate at RT for 30–45 min with the tube open (*see* **Note 14**).
7. Tap the sides of the tube to release the bubbles and add ~50 μl of PBS-Triton and mix gently without pulling the embryos into the pipette (*see* **Note 15**).

8. After all the embryos have settled to the bottom of the tube, discard the $NaBH_4$ solution and wash embryos twice with PBS for 5 min each.
9. Transfer the embryos into a glass dish containing PBS-Triton. Using a pair of forceps, manually remove the yolk from the embryos, without damaging the embryos (*see* **Note 16**).
10. Transfer the deyolked embryos into 1.5 ml tubes and wash twice with PBS for 5 min each (*see* **Note 17**).

3.2.4 Incubation of Embryos with Primary Antibodies

1. Discard PBS-Triton from the tube containing deyolked embryos. Add ~500 μl of blocking solution and leave the tube on its side at RT for 2 h, gently rocking on a horizontal rocker (*see* **Notes 17–19**).
2. Discard the blocking solution and add 500 μl of primary antibody of appropriate dilution in blocking solution to the deyolked embryos. Leave the tubes on its side overnight at 4 °C, gently rocking on a horizontal rocker (*see* **Note 20**).
3. Carefully remove primary antibody and save for reuse. Add ~1 ml of PBS-Trition and wash embryos for ~30 mins.
4. Discard PBS-Triton and repeat the washes 6–8 times, replacing the PBS-Triton at ~30 min intervals.
5. Replace the PBS-Triton with ~500 μl of blocking solution and leave the tube on its side at RT for 2 h, gently rocking on a horizontal rocker (*see* **Note 19**).

3.2.5 Incubation of Embryos with Secondary Antibodies and Nuclei Labeling

1. From here onwards, the tubes must be protected from exposure to light. Discard the blocking solution and add 500 μl of secondary antibody of appropriate dilution in blocking solution. Leave the tube on its side at RT for 2 h, gently rocking on a horizontal rocker (*see* **Note 19**).
2. Discard the secondary antibody and add ~1 ml of PBS-Triton and wash embryos for ~30 min.
3. Discard PBS-Triton and repeat the washes 6–8 times, replacing the PBS-Triton at ~30 min intervals.
4. Replace PBS-Triton with PBS and wash once for 5 min.
5. Replace PBS with 100 μl of DAPI working solution and incubate at RT for 10 min.
6. Discard DAPI and add ~1 ml of PBS-Triton and wash embryos thrice for ~10 min each.
7. Replace PBS with 30% glycerol and allow embryos to sink to the bottom of the tubes (*see* **Note 21**).
8. After the embryos have settled to the bottom, discard the 25% glycerol. Add 50% glycerol and let the embryos settle to the bottom (*see* **Note 21**).
9. Store in the dark till ready for imaging.

3.2.6 Mounting Immunolabeled Zebrafish Embryos for Imaging

1. On a glass slide stick two-hole reinforcement stickers (paper punch hole guards) one on top of another to form a circular chamber for placing the embryo (Fig. 3).
2. Place the slide on the stage of a transmitted light microscope and carefully pipette out 2–3 embryos from the 50% glycerol solution into the circular chamber using a glass pipette.
3. Remove excess 50% glycerol using a 10 μl pipette and add ~2 μl of mounting media into the chamber.
4. Orient the embryo with the top facing the coverslip side using fine forceps.
5. Place four to five 2 μl drops of mounting media over the rim of the chamber (on the hole reinforcement stickers) and carefully place a clean coverslip over the chamber to semi-flat mount the embryos.
6. Seal the coverslip in place using clear nail polish, allow the seal to dry. Image immediately or store the slides at 4 °C in the dark until ready to image.

4 Notes

1. Zebrafish mate and release eggs in short bouts that last 2–3 min over a 30-min window when the facility lights turn on in the morning. The male and female fish must be physically separated as soon as egg-laying initiates to ensure that the female does not release all the mature eggs during natural mating.
2. Do not wait till gill movements subside completely, which would result in the death of the fish.
3. During extrusion of eggs, care should be exercised to prevent the eggs from coming in contact with water. Water activates zebrafish eggs resulting in the resumption of the arrested meiosis and leads to subsequent decline in fertilization efficiency.
4. Eggs from multiple females can be pooled into a single Petri dish before adding sperm solution.
5. Extruded eggs must be used within 5 min to prevent drying out of ovarian fluid, which would result in a decline in fertilization efficiency.
6. Sometimes it may be necessary to remove the swim bladder and gut to access the testes.
7. Place a small drop of the sperm supernatant in a Petri dish and observe under incident light. Motile sperm appear as shiny moving particles.
8. Normally fertilized embryos will become two cells by ~30–40 min. If polyspermy has occurred, embryos will become

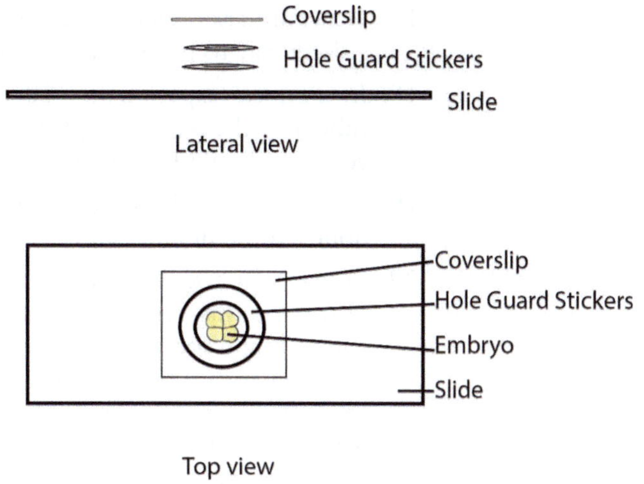

Fig. 3 Schematic for mounting zebrafish embryos for imaging. Zebrafish embryos can be semi-flat mounted in a circular chamber as shown. The embryos can be preserved in this sealed chamber till imaging

four-cell directly by 30–40 min and can be discarded from the analysis.

9. Embryos can be dechorionated after ~4 min of fertilization. Attempting to dechorionate earlier than this may interfere with the fertilization process.

10. To prevent lysis, embryos should not come in contact with air while adding and removing solutions.

11. Dechorionated embryos tend to stick to plastic and lyse, use glass pipettes and dishes after dechorionation.

12. Keeping the tubes horizontal during the addition of the fixative and during incubation ensures that the embryos do not come into contact with each other. This ensures minimal distortion in embryo shape during fixation, which may happen if embryos settle on top of each other when the tubes are held vertically.

13. Embryos can be left in fixative for a couple of days. For longer storage, it is best to transfer embryos into methanol. Embryos can be left in methanol at −20 °C for ~1 week.

14. Treatment with $NaBH_4$ causes the solution to release gas and the tubes can be propelled in air if the lids are kept closed.

15. During $NaBH_4$ treatment, embryos stick to the sides of the tubes. Do not pipette or swirl the solution to dislodge embryos. Addition of a small amount of PBS-Triton helps in dislodging the stuck embryos.

16. It is important to remove all traces of yolk granules to get the best quality immunolabeling and confocal images. Take care to not tear the edges of the embryo that are in contact with the yolk ball during deyolking.

17. Generally, after removal from methanol, it is best to proceed in a single day to primary antibody incubation.
18. In a typical experiment, ~10 embryos can be processed per tube for the volumes given. If using more embryos, the volumes should be scaled up.
19. Incubation in blocking solution and antibodies can also be done overnight at 4 °C.
20. Primary antibody incubation can also be done at RT for 2–3 h. However, this results in faster degradation of the primary antibody and precludes reuse of the primary antibody solution. The primary antibody solution should be monitored at regular intervals for microbial growth. If the solutions appear contaminated, discard and prepare a fresh batch.
21. Embryos can take ~1–2 h to settle to in glycerol at RT. These steps can also be done at 4 °C.

Acknowledgments

Our research is funded by a DBT/Wellcome Trust/India Alliance Intermediate Fellowship (IA/I/13/2/501042) to S.N. and by the Tata Institute of Fundamental Research, Mumbai.

References

1. Behring EV (1890) Ueber das Zustandekommen der Diphtherie-Immunität und der Tetanus-Immunität bei Thieren
2. Behring EV (1890) Untersuchungen über das Zustandekommen der Diphtherie-Immunität bei Thieren
3. Coons AH, Creech HJ, Jones RN (1941) Immunological properties of an antibody containing a fluorescent group. Proc Soc Exp Biol Med 47(2):200–202
4. Coons AH, Creech HJ, Jones RN, Berliner E (1942) The demonstration of pneumococcal antigen in tissues by the use of fluorescent antibody. J Immunol 45(3):159
5. Coons AH (1961) The beginnings of immunofluorescence. J Immunol 87:499–503
6. Coons AH, Kaplan MH (1950) Localization of antigen in tissue cells; improvements in a method for the detection of antigen by means of fluorescent antibody. J Exp Med 91(1):1–13
7. Saper CB (2009) A guide to the perplexed on the specificity of antibodies. J Histochem Cytochem 57(1):1–5
8. Alestrom P et al (2019) Zebrafish: Housing and husbandry recommendations. Lab Anim 2019:23677219869037
9. Kimmel CB et al (1995) Stages of embryonic development of the zebrafish. Dev Dyn 203 (3):253–310
10. Gard DL (1991) Organization, nucleation, and acetylation of microtubules in *Xenopus laevis* oocytes: a study by confocal immunofluorescence microscopy. Dev Biol 143 (2):346–362
11. Pelegri F et al (1999) A mutation in the zebrafish maternal-effect gene nebel affects furrow formation and vasa RNA localization. Curr Biol 9(24):1431–1440

Chapter 17

Observation of Medaka Larval Gonads by Immunohistochemistry and Confocal Laser Microscopy

Toshiya Nishimura and Minoru Tanaka

Abstract

The combination of immunohistochemistry and confocal laser microscopy enables the observation of cellular structures and protein localization within cells using whole-mount tissues. However, such high-resolution imaging requires several steps, such as proper dissection before fixation and antibody staining, and the appropriate positioning of tissues on a glass slide for observation. Here, we describe the method developed by our laboratory for the immunohistochemistry of medaka embryonic and larval gonads, focusing on the dissection and mounting of tissues for confocal laser microscopy. Positioning the gonad just beneath the coverslips is essential to obtain high-resolution images at a level where cellular components of germ cells, such as germ plasm and nuclear structures, can be clearly observed using an oil immersion objective lens.

Key words Medaka, Gonads, Germ cell, Immunohistochemistry, Dissection, Agar pads, Confocal laser microscopy

1 Introduction

Germ cells, precursors of sperm and eggs, are nursed by somatic cells in gonads. In teleosts, including medaka and zebrafish, sex determination and gonadal sex differentiation usually occur from the late embryonic stage to the early larval stage [1–3]. Observation of gonads at these stages is important for understanding sex determination and differentiation, but this is challenging because gonads are too small to dissect. Even if they are successfully dissected, they can easily be lost during handling for staining procedures, such as those for immunohistochemistry and in situ hybridization. Therefore, it is better to perform immunohistochemistry and in situ hybridization of gonads with whole medaka embryos and larvae.

Gonads are located between the hindgut and peritoneal wall in medaka (Fig. 1A, B) [4]. For better access by antibodies and observation via microscopy, the skin and gut of medaka should be removed from the body so that the gonad is exposed (Fig. 1C–E).

Roland Dosch (ed.), *Germline Development in the Zebrafish: Methods and Protocols*, Methods in Molecular Biology, vol. 2218, https://doi.org/10.1007/978-1-0716-0970-5_17, © Springer Science+Business Media, LLC, part of Springer Nature 2021

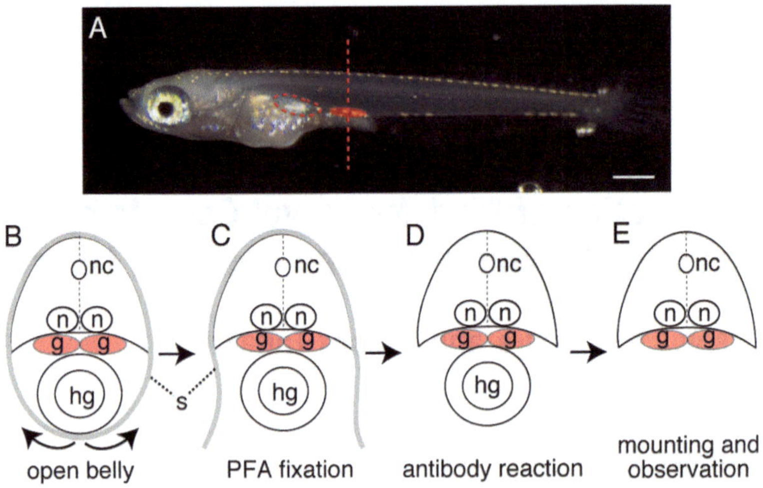

Fig. 1 Overview of immunohistochemistry procedures. (A) The medaka larva at 5 dph. The red spot indicates the gonad. The red dotted line encircles the position of the air bladder. The cross section at the red dotted line is illustrated in B. (B) Cross section of an intact medaka larva at a position where the gonad can be observed. Before PFA fixation, the skin is torn with tweezers to open the belly. (C) Cross section of the larva after opening the belly. PFA fixation is performed at this stage. (D) Cross section of the larva after removal of skin. Antibody reactions are performed at this stage. (E) Cross section of the larva after removal of the hindgut. The gonad contained in the upper half of the body is mounted on agarose pads for observation by confocal laser microscopy. *nc* notochord, *n* nephric duct, *g* gonad, *hg* hindgut, *s* skin. The bar indicates 200 μm

In this chapter, we will describe the method developed by our laboratory for immunohistochemistry using medaka larvae at 5 day-post-hatching (dph), a stage at which gonadal sex differentiation occurs [5], focusing, in particular, on required skills that include (1) opening the belly using tweezers before paraformaldehyde (PFA) fixation (Fig. 1B), (2) removal of the skin before antibody staining (Fig. 1C), (3) removal of the gut, and (4) mounting the gonads with bodies on agarose pads for observation by conventional confocal laser microscopy (Fig. 1D and E, *see* also Fig. 5A–C).

2 Materials

2.1 Dissection and Fixation

1. Fish: medaka (OK Cab line) larvae at 5 dph.
2. Balanced salt solution (BSS): 0.65% NaCl, 0.04% KCl, 0.02% $MgSO_4{\cdot}7H_2O$, 0.02% $CaCl_2{\cdot}2H_2O$, 0.001% phenol red. Sterilize BSS by autoclaving and adjust its pH to 8.3 with 5% $NaHCO_3$ [6].

3. Anesthetic solution: 160 μg/ml ethyl 3-aminobenzoate methanesulfonate (MS-222) in BSS. Prepare 25× stock solution (4 mg/ml MS-222 and 4 mg/ml $NaHCO_3$ in Milli-Q water). Dilute to 1× solution with BSS. The MS-222 solution should be kept at 4 °C.
4. 4% PFA: Dissolve paraformaldehyde in 1× PBS (pH 7.4) at 50 °C. After cooling, add Tween®20 (0.1% final concentration).
5. 100% Methanol.
6. Two pairs of fine tweezers (INOX 5, A. DUMONT&FILS, *see* **Note 1**).
7. 60 mm plastic dish (*see* **Note 2**).

2.2 Immunohistochemistry

1. 1× PTW: 1× PBS (pH 7.4) containing 0.1% Tween®20.
2. Blocking solution: 20% Blocking One (Nakalai Tesque, Kyoto, Japan) in 1× PTW.
3. Primary antibody reaction mix: anti-VASA (rat) [7] in blocking solution (1/200) (*see* **Note 3**).
4. Secondary antibody reaction mix: 10 μg/ml Alexa Fluor®488 goat anti-rat IgG (H + L) (Thermo Fisher Scientific, Waltham, MA, USA) and 25 μg/ml DAPI (4′,6-diamidino-2-phenylindole) in blocking solution.

2.3 Preparation of Agarose Pads

1. 1% agarose dissolved in Milli-Q water.
2. Glass slides (76 × 26 mm, thickness: 0.9–1.2 mm).
3. Coverslips (22 × 24 mm, thickness: 0.13–0.17 mm).
4. Vinyl tape (width: 19 mm, thickness: 0.2 mm, e.g. Nitto Denko, Osaka, Japan).

2.4 Mounting on the Agar Pad and Observation

1. Agarose pads.
2. 1× PTW: 1× PBS (pH 7.4) containing 0.1% Tween®20.
3. 27G × 3/4″ needle (0.40 × 19 mm).
4. Vectashield mounting medium.
5. Confocal laser microscope (e.g., FV1000 Olympus, Tokyo, Japan).
6. 60× oil immersion objective lens (e.g., UPlanSApo, NA = 1.35, Olympus, Tokyo, Japan).

3 Methods

3.1 Dissection and Fixation

For better access of PFA to the gonads, cut and open the belly of medaka with tweezers before fixation (Fig. 1B, C).

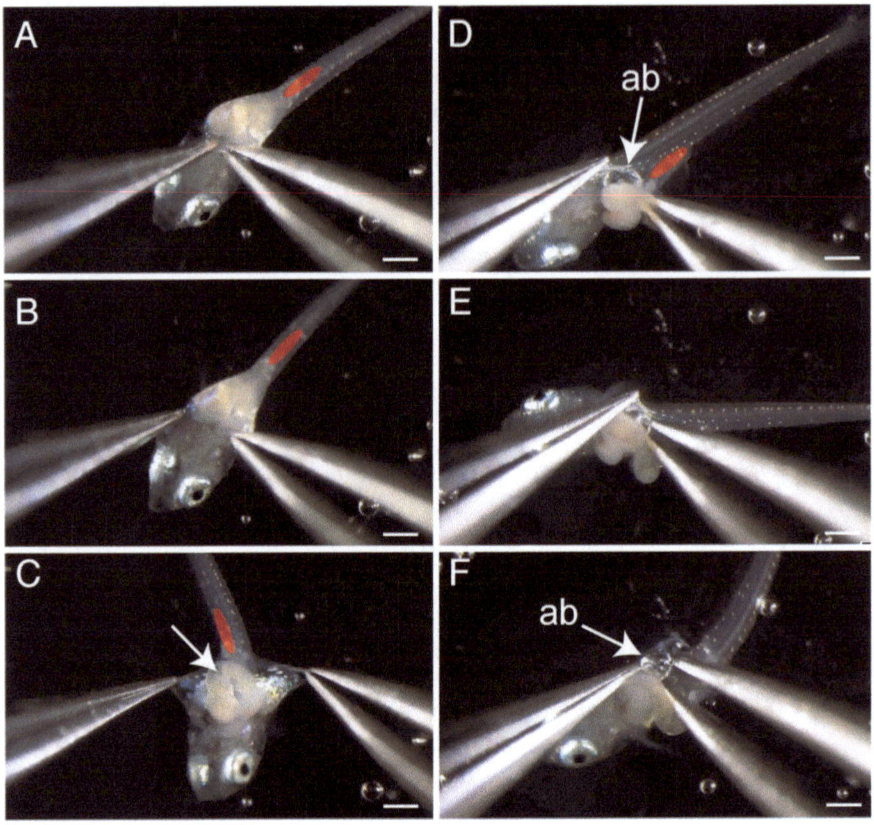

Fig. 2 Dissection before PFA fixation. (**a–c**) Opening the belly with tweezers. Grasp the skin over the heart with two tweezers (**a**) and tear the skin slowly apart, pulling one pair of tweezers away from the other (**b**). Open the belly until the gut (arrow) is exposed (**c**). (**d–f**) Removal of the air bladder (ab), which is located above the gut (*see* also Fig. 1A). Hold the torn skin with one pair of tweezers (**d**, left) and pick the air bladder with another pair (**e**, right). The air bladder can be removed by gently pulling it out using the right pair of tweezers (bf). Bars indicate 200 μm

1. Place medaka larvae in chilled anesthetic solution until movement stops.
2. Turn medaka larvae facing belly-up using tweezers.
3. Pick the skin above the heart with two fine tweezers and tear the skin to open the belly toward the posterior position where the gonad is located (Fig. 2a–c), without damaging the gonads.
4. Hold the torn skin near the heart with one pair of tweezers (left, Fig. 2d) and remove the air bladder located dorsally to the intestine with the other pair of tweezers (right, Fig. 2e, f) (*see* **Note 4**).
5. Optional: if genotyping is required, heads or tails should be cut and put into a lysis buffer for genomic DNA extraction.

6. Put the dissected samples into 2 ml tubes containing 4% PFA (2 ml) and incubate them at 4 °C with gentle shaking on a shaker overnight (12 h <).
7. Replace the 4% PFA with 100% methanol. The samples can be stored at −30 °C until use (*see* **Note 5**).

3.2 Immunohistochemistry

For better access of antibodies to the gonads, remove the skin covering the whole body prior to incubation with antibodies (Fig. 1D). This process is especially essential when using medaka larvae at or beyond the hatching stage (*see* **Note 6**).

1. Replace 100% methanol with 2 ml of 1× PTW. Incubate for 5 min with gentle shaking on a shaker (washing step).
2. Repeat the washing step five times.
3. Place the samples on the 60 mm plastic dish filled with 1× PTW.
4. Removal of skin: hold the anterior side of the body with one pair of tweezers (left) and pick the anterior skin with the other pair of tweezers (right, Fig. 3a, b). Peel the skin away by slowly moving the right pair of tweezers posteriorly to the end of tail or until the gonad is exposed (Fig. 3b–f) (*see* **Note 7**).
5. Transfer the skin-peeled samples into 2 ml tubes containing 1× PTW.
6. Replace the 1× PTW with 2 ml of blocking solution and incubate with gentle shaking on a shaker at room temperature for 2 h.
7. Replace the blocking solution with 50–100 μl of the primary antibody solution and incubate with gentle shaking on a shaker at 4 °C overnight (12 h <). Make sure that the samples are immersed in the antibody solution.
8. Replace the primary antibody solution with 2 ml of 1× PTW. Incubate with gentle shaking on a shaker for 30 min at 4 °C (washing step).
9. Repeat the washing step three times.
10. Replace the blocking solution with 50–100 μl of the secondary antibody solution and incubate with gentle shaking on a shaker at 4 °C overnight (12 h <).
11. Remove the secondary antibody solution and add 2 ml of 1× PTW. Incubate with gentle shaking on a shaker at 4 °C for 30 min.
12. Replace 1× PTW with 2 ml of 4% PFA. The sample can be stored at 4 °C for at least 1 week in the dark. If you wish to immediately proceed to mounting and observation, skip this step.

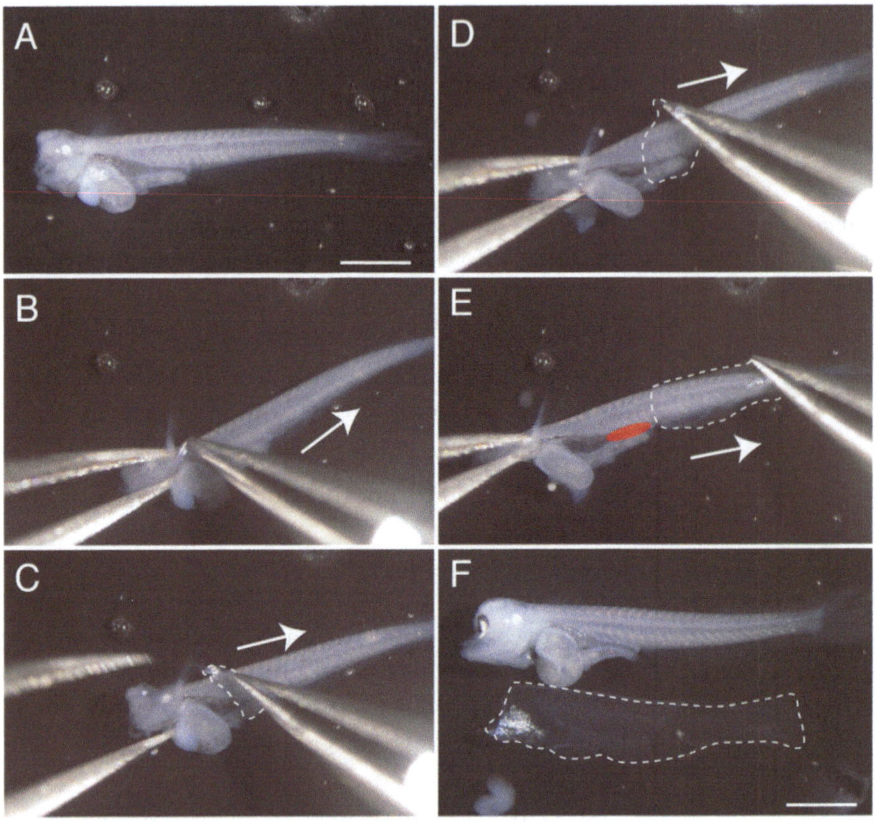

Fig. 3 Removal of the skin before antibody reactions. (**a**) The fixed larva before removal of the skin. (**b–e**) Procedures for removal of the skin. (**b**) Hold the anterior part with one pair of tweezers (left) and pick the anterior torn skin with another pair (right). (**c–e**) Gently, peel the skin by moving the right tweezers posteriorly (arrows) until the gonad (red) is exposed (**e**). (**f**) The larva after removal of the skin. Dotted lines indicate the skin. Bars indicate 1 mm

3.3 Preparation of Agarose Pads

Agarose pads are useful for stably fixing samples at certain positions for microscopic observation (Fig. 4).

1. Layer two pieces of vinyl tape (approximately 19 × 24 mm) onto a glass slide. One should be approximately 10 mm away from the other so that medaka larvae can be placed between them. The number of layered vinyl tapes can be changed according to the size of the larvae. For medaka larvae at 5 dph, three layers of vinyl tape are enough.
2. Apply 1% agarose gel between the vinyl tapes and immediately place the coverslip on them before the gel is polymerized.
3. Cool the gel for 1 min at room temperature.
4. Remove the coverslip just before use.

3.4 Mounting and Observation

So far, gonads have been successfully immunostained. However, the hindgut still prevents the observation of gonads from the

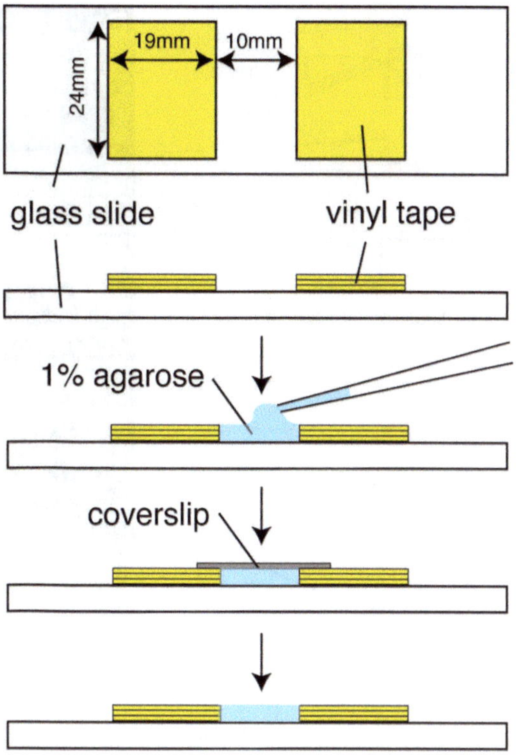

Fig. 4 Preparing the agarose pad. A pair of cut vinyl tapes (approximately 24 × 19 mm) are layered onto a glass slide, separated by a distance of approximately 10 mm (the length between two layered vinyl tapes depends on the size of the larvae). Pour 1% agarose gel (100–200 μl) into the space between the vinyl tapes. Before setting of the agarose gel, a coverslip is placed on it, which makes the surface of the agarose flat. After cooling the agarose for at least 1 min at room temperature, the coverslip can be removed

ventral side (Fig. 1D). In order to observe a whole gonad clearly from the ventral side, any obstacles, such as the gut and, if present, skin, should be removed (Fig. 1E) before mounting and observing using a 60× oil immersion objective lens (Fig. 5A–C).

1. Place the immunostained samples in a 60 mm dish filled with 1× PTW.
2. Hold the anterior part with a pair of tweezers (left) and cut the connective tissue between the gut and body with a 27 G needle (Fig. 5A, D, E). When the posterior end of the gut is detached, the gut can be removed by sliding the needle between the gut and gonad from the posterior to the anterior (Fig. 5A, F). Alternatively, you can pick the posterior end of the gut using tweezers and pull slowly toward anterior end (*see* **Note 8**).

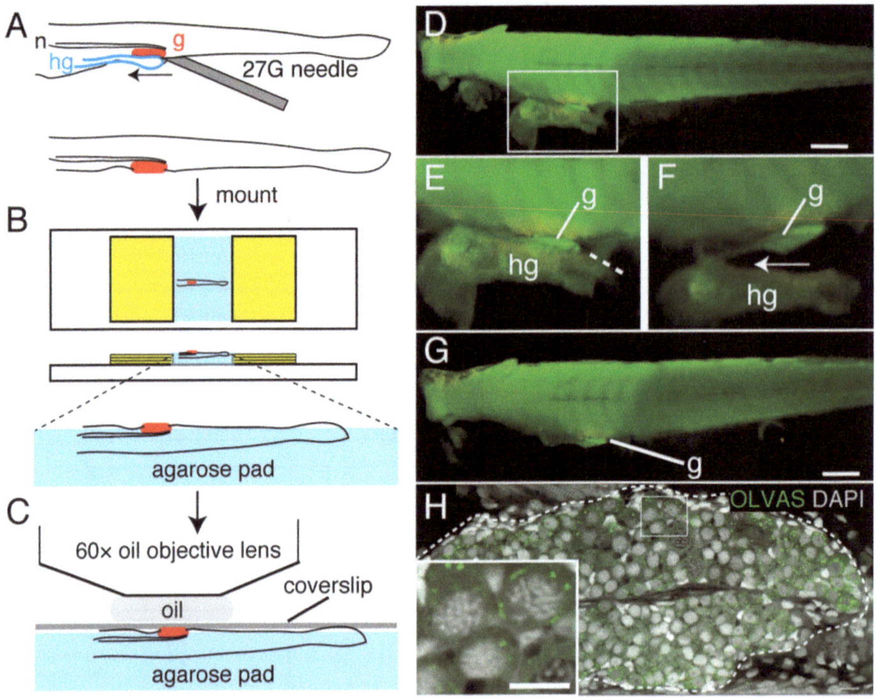

Fig. 5 Removal of the gut and mounting the gonad on the agarose pad. (A–C) Schematic representation of the removal of the gut (A), mounting on the agarose pad (B), and observation with an oil immersion objective lens (C). (A) The larva is viewed from the lateral side. The hindgut (hg) is removed using a 27 G needle by sliding anteriorly (arrow) between the hindgut and gonad. (B) The larva is mounted on an agarose pad and embedded in the space that is made by scratching with tweezers. Note that the larva should be embedded at the level where the gonad just appears at the surface of the agarose pad. (C) After a drop of mounting medium was added, the coverslip was placed on the samples. Samples can be observed using a 60× oil immersion objective lens. (D) Lateral side view of the immunostained larva before removal of the gut. The white rectangle is magnified in E. (E) The gonad (g) is located just above the hindgut (hg). The connective tissue that should be cut with a 27 G needle is shown by a white dotted line. (F) After cutting the connective tissue, the gut is removed by sliding the needle anteriorly (arrow) between the gonad and gut. (G) Lateral side view of the immunostained larva after removal of the gut. The sample is ready for mounting. (H) An example of the gonad at 5 dph observed from the ventral side using a 60× oil immersion objective lens and an FV1000 confocal laser microscope (Olympus). The magnified image in the white rectangle is shown in the inset. Germ plasm (green) surrounding nuclei and thread-like chromosomes (gray) can be observed in meiotic germ cells. Bars represent 200 μm (D and G) and 10 μm (H). *n* nephric duct, *hg* hindgut, *g* gonad

3. Pick the tail of the larva with a pair of tweezers and transfer it to the agarose pad. Five to six samples at maximum can be mounted on one agarose pad (Fig. 5B).
4. A scratch is made with the tip of tweezers in the agarose. The length and the depth of scratch are similar to the body length and height of medaka larvae, respectively.
5. Place the sample into the space created by the scratch, with the gonad facing upward. Do not embed the whole larvae deeply

into the agarose gel; the gonads should appear just at the surface of the agarose (Fig. 5B).

6. Put a drop of mounting medium Vectashield on the samples and place a coverslip on them (Fig. 5C). Nail polish can be used to adhere the coverslip to vinyl tape.
7. Observe the mounted samples through confocal laser microscopy. The 60× oil immersion objective lens allows the observation of individual gonads and the cellular components of germ cells, such as germ plasm and nuclear structures (Fig. 5C, H) (*see* **Note 9**). Other examples of confocal images obtained by the present methods can be seen in [8–10].

4 Notes

1. Keeping the tips of tweezers sharp is essential for conducting delicate procedures, such as the removal of the skin and gut.
2. To avoid the tips of tweezers from bending, which is often caused by compression and/or putting pressure to the bottom of the dish, use a plastic dish with a layer of PDMS (Polydimethylsiloxane, SILPOT® 184 W/C, DOW CORNING, Midland, MI, USA) at the bottom.
3. The concentration of antibodies depends on the titer of each antibody. You may need to optimize the concentration for your experiment by serial dilutions.
4. Once the air bladder is removed, the samples sink to the bottom of the tube. This makes handling easier during the replacements of buffers and antibody solutions.
5. This process involves permeabilization for better access of an antibody to its cognate epitope. However, depending on the type of antibody or epitope (i.e., lipoprotein or membrane protein), this process can be skipped.
6. When you want to observe medaka embryos before hatching, the removal of skin can be performed after antibody reactions (before removal of the gut). The skin of medaka embryos is thin enough for antibodies to penetrate the gonads.
7. The skin is often torn during peeling and remains on the body. As long as the skin around the gonads is removed, antibodies can gain access. Therefore, it might not be necessary to remove the entire skin.
8. This is a difficult procedure. Gonads are often detached from the body together with the hindgut. In this case, the gonad with the gut is mounted on agarose pads, but make sure that the gonads are facing up.

9. As the working distance of an objective lens is inversely proportional to its numerical aperture [11], to obtain a high-resolution image with a 60× or 100× oil immersion objective lens, gonads should be located just beneath the coverslip. If you cannot obtain clear images, the samples are likely embedded too deeply in the agarose. In such a case, you will need to re-mount the samples.

Acknowledgments

This work was supported by JSPS KAKENHI Grant Numbers JP18K19332 (T.N.) and JP18KK0200 (T.N.).

References

1. Matsuda M, Nagahama Y, Shinomiya A, Sato T, Matsuda C, Kobayashi T, Morrey CE, Shibata N, Asakawa S, Shimizu N, Hori H, Hamaguchi S, Sakaizumi M (2002) *DMY* is a Y-specific DM-domain gene required for male development in the medaka fish. Nature 417:559–563
2. Nanda I, Kondo M, Hornung U, Asakawa S, Winkler C, Shimizu A, Shan Z, Haaf T, Shimizu N, Shima A, Schmid M, Schartl M (2002) A duplicated copy of *DMRT1* in the sex-determining region of the Y chromosome of the medaka, *Oryzias latipes*. Proc Natl Acad Sci U S A 99:11778–11783
3. Tong SK, Hsu HJ, Chung BC (2010) Zebrafish monosex population reveals female dominance in sex determination and earliest events of gonad differentiation. Dev Biol 344:849–856
4. Nishimura T, Tanaka M (2014) Gonadal development in fish. Sex Dev 8:252–261
5. Kobayashi T, Matsuda M, Kajiura-Kobayashi-H, Suzuki A, Saito N, Nakamoto M, Shibata N, Nagahama Y (2004) Two DM domain genes, *DMY* and *DMRT1*, involved in testicular differentiation and development in the medaka, *Oryzias latipes*. Dev Dyn 231:518–526
6. Kinoshita M, Murata K, Naruse K, Tanaka M (2009) A laboratory manual for medaka biology. Wiley-Blackwell, Ames, IA, p 397
7. Aoki Y, Nagao I, Saito D, Ebe Y, Kinjo M, Tanaka M (2008) Temporal and spatial localization of three germline-specific proteins in medaka. Dev Dyn 237:800–807
8. Nishimura T, Sato T, Yamamoto Y, Watakabe I, Ohkawa Y, Suyama M, Kobayashi S, Tanaka M (2015) *foxl3* is a germ cell-intrinsic factor involved in sperm-egg fate decision in medaka. Science 349:328–331
9. Nishimura T, Yamada K, Fujimori C, Kikuchi M, Kawasaki T, Siegfried KR, Sakai N, Tanaka M (2018) Germ cells in the teleost fish medaka have an inherent feminizing effect. PLoS Genet 14:e1007259
10. Kikuchi M, Nishimura T, Saito D, Shigenobu S, Takada R, Gutierrez-Triana JA, Cerdán JLM, Takada S, Wittbrodt J, Suyama M, Tanaka M (2019) Novel components of germline sex determination acting downstream of *foxl3* in medaka. Dev Biol 445:80–89
11. Paddock SW (2000) Principles and practices of laser scanning confocal microscopy. Mol Biotechnol 16:127–149

Chapter 18

Methods for Visualization of RNA and Cytoskeletal Elements in the Early Zebrafish Embryo

Christina L. Hansen and Francisco Pelegri

Abstract

Zebrafish embryos, with their large size (>0.5 mm) and accessibility, are valuable tools for investigating core cellular processes. Many of those processes, such as cell division, asymmetric inheritance of cellular components, and structural dynamics involved in cell motility and morphology rely on cytoskeletal rearrangements and associated macromolecules. In addition to the protein-rich cytoskeleton, the early embryo is packed with maternally deposited RNA, which serves essential roles in establishing cell polarity, cell fate, and cell organization. Here, we present methods for visualizing endogenous RNA along with cytoskeletal structures, including microtubules and filamentous actin (F-actin) in the context of an intact vertebrate embryo. Each of the four protocols described herein (embryo fixation, RNA probe design/synthesis, double fluorescent in situ hybridization with tubulin immunofluorescence, and fluorescent in situ hybridization with phalloidin labeling of F-actin) are intended for optimal preservation and visualization of both the cytoskeleton and RNAs of interest. These methods can also be modified and applied to a broad range of other uses.

Key words Zebrafish, Embryo, Cytoskeleton, F-Actin, Microtubules, In situ hybridization, RNA localization, Phalloidin, Immunofluorescence

1 Introduction

Interactions between RNA and the cytoskeleton are critical for a number of cellular processes, with notable examples including mRNA transport [1], germ granule aggregation [2], and specific RNA localization necessary for a subcellular organization [3]. Strategies that permit simultaneous visualization of both RNA and protein components of cytoskeletal structures, such as tubulin and actin, are therefore useful and necessary experimental tools. The methods described in this chapter are all vital steps for the successful completion of combinatorial RNA and cytoskeletal protein labeling, including fixation, RNA probe design and synthesis, and in situ hybridization with either microtubule immunofluorescence or F-actin labeling.

Roland Dosch (ed.), *Germline Development in the Zebrafish: Methods and Protocols*, Methods in Molecular Biology, vol. 2218, https://doi.org/10.1007/978-1-0716-0970-5_18, © Springer Science+Business Media, LLC, part of Springer Nature 2021

These protocols can be readily adapted to a number of RNA-protein combinations. Subheading 3.1 describes a fixation procedure optimized for the preservation of intracellular cytoskeletal structures (Fig. 1a). Several key aspects differentiate this from a typical fixation, including (1) removal of chorions prior to fixation, (2) glutaraldehyde and EGTA as additives in standard 4% PFA, and (3) subsequent reduction of glutaraldehyde with sodium borohydride ($NaBH_4$). These steps may all be omitted for other concurrent RNA and non-cytoskeletal protein labeling purposes. For example, our lab frequently uses beta-catenin protein labeling as a membrane marker while simultaneously visualizing mRNA localization by in situ hybridization, for which the standard fixation procedures are sufficient.

Subheading 3.2 (Figs. 1b and 2), which presents a PCR-based strategy for de novo RNA probe design and synthesis using T7-promoter containing primers [4, 5], is similarly adaptable. Although some aspects of probe design are largely a matter of preference, several important considerations, such as probe length and target region [6], are also discussed (*see* **Note 1**).

In Subheading 3.3, we describe the use of Tyramide Signal Amplification (TSA™) [7, 8] with fluorescein (green) and digoxigenin/Cy3 (red) haptens to visualize two different RNA transcripts, and utilize a Cy5 (far-red) secondary antibody to visualize tubulin proteins (Fig. 1c). However, this aspect of the protocol is easily amendable to changes, such as visualizing only one RNA by omitting the second day of in situ labeling and proceeding directly to protein immunofluorescence (Fig. 3) or adding a third day of in situ procedure with another fluorophore to visualize three RNAs of interest (*see* **Note 2**). Indeed, the fluorescence in situ portion of the methods described below (Subheading 3.3, **steps 1–32**) is largely based on an excellent triple in situ protocol shared by the Talbot lab (Triple Fluorescent In Situ; zfin.org).

We have also included a robust, one-step protocol using phalloidin for visualization of filamentous actin (F-actin) in the early embryo. Fluorophore-conjugated versions of phalloidin, a toxic peptide originally isolated from death cap mushrooms [9], is now commonly used to stain actin structures in cells due to its specificity and efficacy in binding F-actin [10, 11]. Here, we present a modified, one-step version of standard phalloidin staining protocols by recommending the addition of fluorophore-conjugated phalloidin to the fixation procedure rather than post-fixation. This optimized procedure provides two major benefits over the traditional method: (1) reduced total time involved as it eliminates the need for a later staining step, and most importantly, (2) improved preservation of F-actin structural details within the sample. The latter is likely achieved due to phalloidin's ability to stabilize F-actin through its binding site, located in-between F-actin subunits and effectively preventing depolymerization [12, 13]. Demonstrating the power

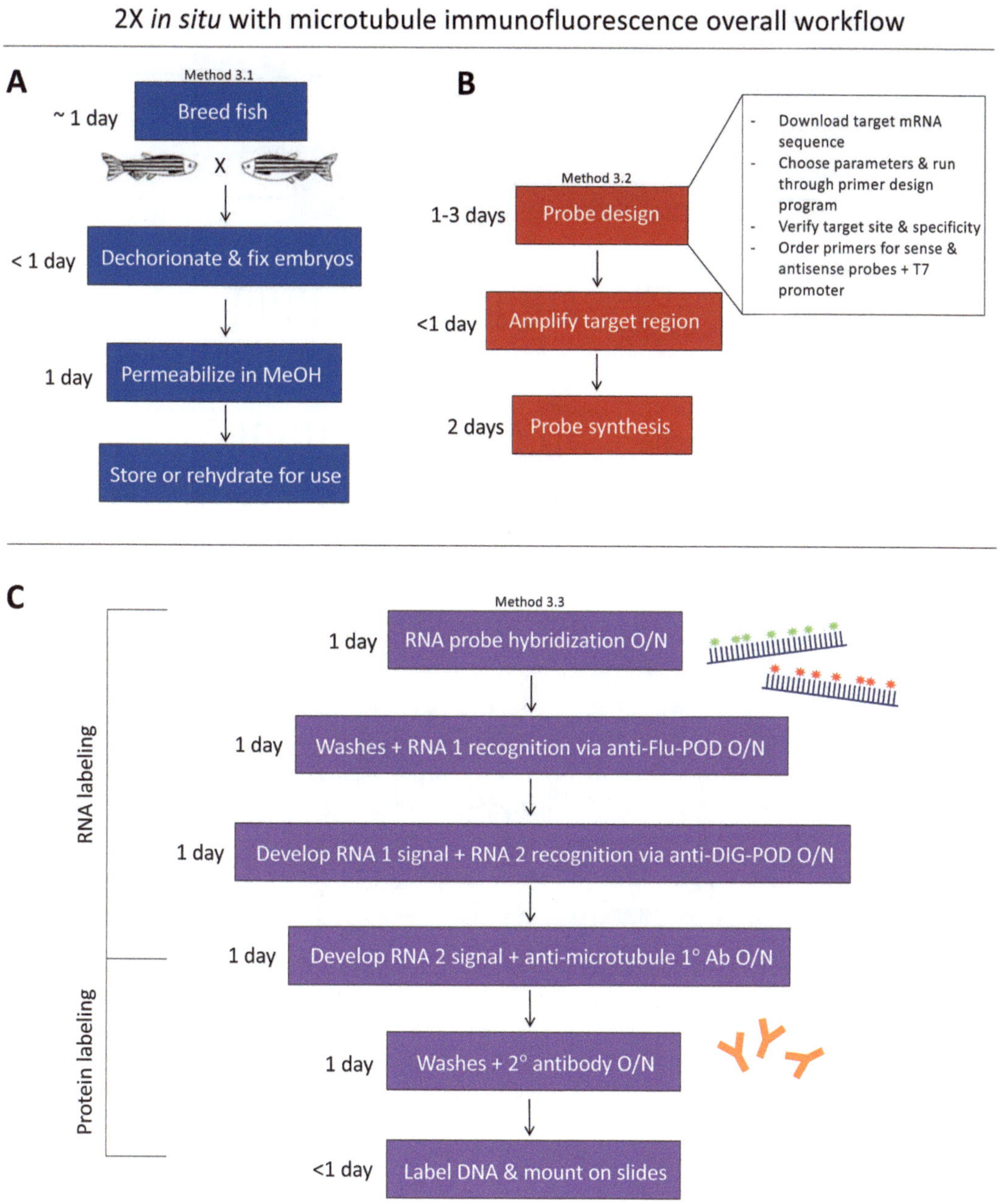

Fig. 1 (**a–c**) A generalized diagram and approximate timeline of Subheadings 3.1–3.3, representing the overall workflow required for two-color fluorescence in situ hybridization with microtubule labeling. Note that embryo fixation and RNA probe preparation (Subheadings 3.1 and 3.2) can be performed well in advance of Subheading 3.3 (or other downstream uses, such as Subheading 3.4)

of this technique, previously unobserved aspects of F-actin structure in the early embryo, such as micron-scale actin trenches in the cortex and cleavage furrows, were able to be visualized [14].

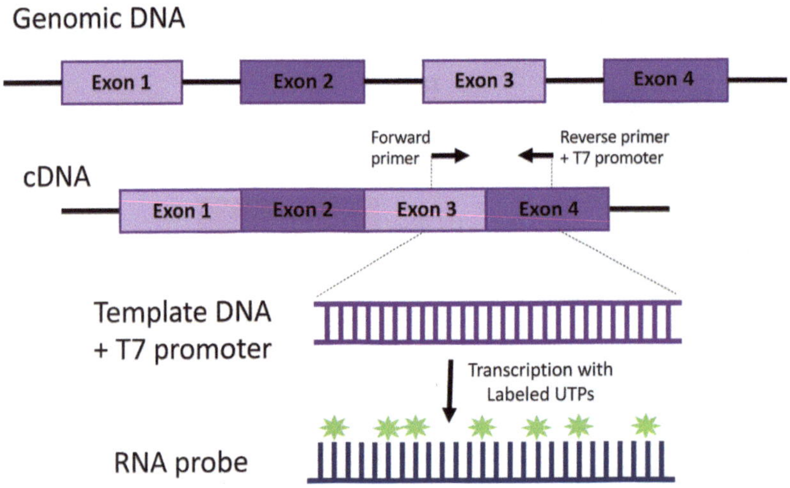

Fig. 2 Generalized schematic of PCR-based RNA probe synthesis. This diagram depicts an RNA probe designed to target a region spanning an exon-exon junction, but other design strategies, such as including portions of the 5′ or 3′ UTR, can be employed (*see* **Note 1**)

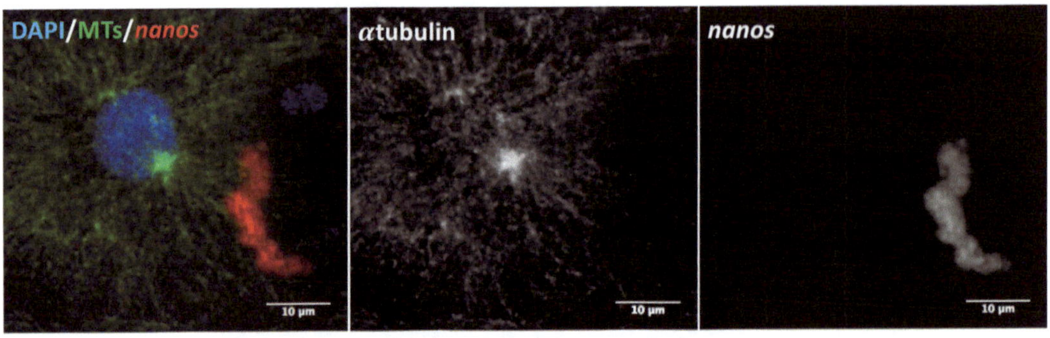

Fig. 3 Micrograph of a germ plasm aggregate (*nanos*) and microtubules surrounding a nucleus in a fixed 2-h post-fertilization zebrafish embryo. Using the combined in situ/immunofluorescence protocol described in Subheading 3.3, digoxigenin-labeled probes targeting *nanos* RNA and antibodies targeting alpha-tubulin were used to simultaneously visualize an RNA-rich structure (germ plasm) with microtubules and a DAPI-stained nucleus. Image is a 2D projection of a Z-stack acquired on a Zeiss LSM510 confocal microscope

Other methods of F-actin labeling that do not require phalloidin, such as anti-actin-antibodies or transgenic fluorescent actin reporter lines, provide other potential avenues for simultaneous visualization of actin with RNA, but often have significant additional drawbacks (reviewed nicely in [11]). However, a longstanding limitation of phalloidin staining is its incompatibility with many common in situ hybridization methods, such as the one described here in Subheading 3.3. There are several aspects of in situ protocols that could lead to this incompatibility, including methanol treatment [15], treatment with high concentrations of formamide, and extended exposure to high temperatures. By addressing each of these key points, we have developed a functional

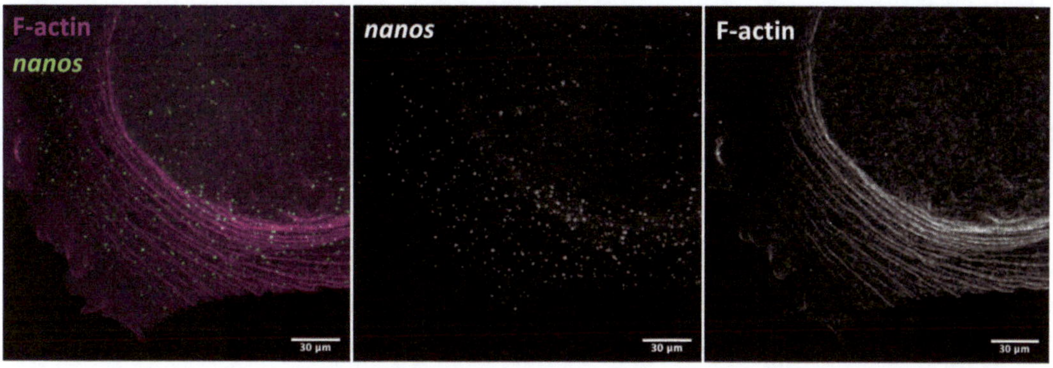

Fig. 4 Micrograph of F-actin arcs and germ plasm RNPs (*nanos*) in the blastodisc periphery of a fixed 2-cell (45 min post-fertilization) zebrafish embryo. As described in Subheading 3.4, fluorophore-conjugated phalloidin was incorporated with the fixation step to label F-actin (pseudocolored magenta here) followed by in situ hybridization to visualize fluorescein-labeled probes bound to *nanos* RNA. Image is a 2D projection of a Z-stack acquired on a Zeiss LSM780 confocal microscope

combinatorial phalloidin/in situ protocol (Subheading 3.4, also *see* **Note 3**). This protocol provides a unique opportunity to concurrently visualize both RNAs and F-actin structures at a sub-micron scale in fixed vertebrate embryos (Fig. 4), and we are pleased to share what we hope will be a valuable resource for the research community.

2 Materials

2.1 Fixation of Zebrafish Embryos

1. 1× PBS-Tween (PBST): 1× PBS, pH 7.0–7.4, 0.1% Tween-20. For 1 l stock solution, mix 100 ml 10× PBS, 12.5 ml 20% Tween-20, and add nuclease-free H_2O to 1 l. Store at room temperature.
2. Paraformaldehyde (PFA) fix solution: 4% paraformaldehyde (PFA) in PBST. Under a fume hood and in a flask of at least a 250 ml volume capacity, dissolve 4 g paraformaldehyde in 100 ml 1× PBS-Tween by bringing to a boil on a hotplate. Be careful to not overboil. Remove from heat and let cool to room temperature. Aliquot into desired amounts (we typically prepare 1 and 10 ml aliquots) and store at −20 °C.
3. Cytoskeleton Fix: 4% PFA, 0.25% glutaraldehyde, 5 mM EGTA, 0.2% Triton X-100. Per 1 ml of 4% PFA, add 10 µl 25% glutaraldehyde, 10 µl 500 mM EGTA, and 2 µl Triton X-100. Should be prepared freshly each day it is used.
4. 100% Methanol (MeOH).
5. Pronase: 30 mg/ml stock solution.
6. 1× embryo medium (E3): 5 mM NaCl, 0.17 mM KCl, 0.33 mM $CaCl_2$, 0.33 mM $MgSO_4$, and 10^{-5}% methylene

blue. Prepare 10× stock by dissolving 23.4 g NaCl, 1.0 g KCl, 3.9 g $CaCl_2 \cdot 2H_2O$, 6.5 g $MgSO_4 \cdot 7H_2O$ and fill with H_2O to a total volume of 8 l. For 1× working dilution, dilute 1:10 and add methylene blue to inhibit microbial growth (800 ml 10× stock +8 ml 0.01% methylene blue and fill with H_2O to 8 l total).

7. Forceps (Dumont #5).
8. Plastic transfer pipettes (~2 mm opening, for washes and transferring embryos pre-dechorionation).
9. Glass Pasteur pipettes (for transferring dechorionated embryos).
10. Glutaraldehyde deactivation solution: 0.5 mg/ml sodium borohydride ($NaBH_4$).

2.2 De Novo RNA Probe Design and Synthesis

1. Primer design program (e.g., Primer3 http://bioinfo.ut.ee/primer3/ [12]).
2. Custom DNA oligo primers (we order IDT single-stranded 25 nmol DNA Oligos).
3. Nuclease-free H_2O.
4. Thermocycler.
5. dNTPs and PCR buffer (*see* **Note 4**).
6. cDNA (10–100 ng per reaction) from the embryonic stage of interest (*see* **Note 5**).
7. EconoTaq DNA polymerase: 5 units/μl.
8. Materials for diagnostic gel: agarose, Tris/Borate/EDTA (TBE) buffer, gel box, DNA marker size ladder (suggested size range 100–1 kbp), loading dye.
9. Optional: DNA Clean & Concentrator kit (Zymo or equivalent).
10. Optional: RNase AWAY (ThermoFisher or equivalent).
11. Incubator or water bath to produce a constant temperature of 37 °C.
12. RNA labeling mix: DIG (Roche) and/or Fluorescein (Roche).
13. T7 RNA polymerase (50,000 units/ml) and 10× reaction buffer (NEB or equivalent).
14. RiboLock RNase inhibitor (40 units/μl) (ThermoFisher or equivalent).
15. TURBO DNase (2 units/μl) (Ambion or equivalent).
16. 7.5 M Lithium chloride (ThermoFisher or equivalent).
17. 100% Ethanol (also diluted in nuclease-free water to make 70% EtOH).
18. Centrifuge.

19. 100% formamide.
20. Hybridization buffer (hyb): 50% formamide, 5× SSC, 100 μg/ml yeast tRNA, 50 μg/ml heparin, 0.25% Tween-20, 1 M citric acid to pH 6.0 (~0.02 M citric acid final concentration). For a 50 ml stock solution, mix 25 ml formamide, 12.5 ml 20× SSC, 1 ml of 25 mg/ml yeast tRNA, 50 μl of 50 mg/ml heparin, 250 μl 20% Tween-20, ~460 μl 1 M citric acid, 10.74 ml DEPC-treated H_2O. Store hyb stock at −20 °C.
21. PCR strip tubes.
22. 1.5 ml Eppendorf tubes.
23. Incubator or water bath to produce a constant temperature of 37 °C.

2.3 Two-Color Fluorescence In Situ Hybridization with Microtubule Immunofluorescence

1. 1× PBS-Tween (PBST): 1× PBS, pH 7.0–7.4, 0.1% Tween-20. For 1 l stock solution, mix 100 ml 10× PBS, 12.5 ml 20% Tween-20, and add nuclease-free H_2O to 1 l. Store at room temperature.
2. 100% Methanol (MeOH).
3. Methanol/PBS-Tween series: 2:1, 1:1, and 1:2 mixes of MeOH:PBST can be prepared in advance in 15 ml conical tubes.
4. Forceps (Dumont #5).
5. Prehybridization buffer (prehyb): 50% formamide, 5× SSC, 0.1% Tween-20. For a 250 ml stock solution, mix 125 ml formamide, 62.5 ml 20× SSC, 1.25 ml 20% Tween-20, and 62.5 ml nuclease-free H_2O. Store prehyb stock at −20 °C.
6. Hybridization buffer (hyb): 50% formamide, 5× SSC, 100 μg/ml yeast tRNA, 50 μg/ml heparin, 0.25% Tween-20, 1 M citric acid to pH 6.0 (~0.02 M citric acid final conc.). For a 50 ml stock solution, mix 25 ml formamide, 12.5 ml 20× SSC, 1 ml of 25 mg/ml yeast tRNA, 50 μl of 50 mg/ml heparin, 250 μl 20% Tween-20, ~460 μl 1 M citric acid, 10.74 ml DEPC-treated H_2O. Store hyb stock at −20 °C.
7. RNA probe stock: To be added to the hybridization buffer at 150–500 ng/ml (*see* Subheading 3.2, **step 22**). Can be made as described in Subheading 2.2.
8. 2× SSC: 2× SSC, 0.25% Tween-20. For a 400 ml stock solution, mix 40 ml 20× SSC (sodium citrate buffer), 5 ml 20% Tween-20, and nuclease-free H_2O to 400 ml. Store at room temperature. 50 ml of 0.2× SSC can be prepared by mixing 5 ml 2× SSC with 45 ml nuclease-free H_2O.
9. Prehyb/2× SSC wash solutions: 2:1, 1:1, and 1:2 mixes of Prehyb:2× SSC should be prepared in advance in 15 ml conical tubes and can be stored at room temperature.

10. Hydrogen peroxide (H_2O_2): to make 2% dilutions in PBS-Tween or TNT.
11. TNT: 0.1 M Tris–HCl pH 7.5, 0.15 M NaCl, 0.5% Tween-20. For a 1 l stock solution, mix 100 ml Tris pH 7.5, 30 ml 5 M NaCl, 25 ml 20% Tween-20, add nuclease-free H_2O up to 1 l. Store at room temperature.
12. Blocking solution/TBSTB: TNT with 0.5% PerkinElmer Blocking Powder. Measure 0.25 g PerkinElmer Blocking Powder or equivalent (*see* **Note 6**) into a 50 ml conical or glass bottle and fill to 50 ml with TNT. Invert multiple times to mix and heat to 69 °C for at least 1 h (ideally while rocking and/or rotating) to fully dissolve powder into solution. Pipet 1 ml aliquots into Eppendorf tubes and store at −20 °C. Once thawed, never refreeze.
13. TSA Cyanine 3 and Fluorescein System: PerkinElmer Kit includes 1× Amplification Diluent, Tyr-Fluorescein Amplification Reagent, and Tyr-Cy3 Amplification Reagent. The amplification reagents will need to be resuspended in 150 μl DMSO.
14. Anti-Fluorescein-POD: Roche or equivalent.
15. Anti-Digoxigenin-POD: Roche or equivalent.
16. DAPI: 100 μg/ml stock solution.
17. Glycerol: used to make 30% and 50% Glycerol in PBS-Tween.
18. 1.5 ml Eppendorf tubes.
19. Plastic transfer pipettes (convenient for washes).
20. Glass Pasteur pipettes (for transferring embryos).
21. Incubator or water bath to produce a constant temperature of 69 °C.

2.4 Phalloidin Labeling for Preservation and Visualization of F-Actin Structures in the Early Zebrafish Embryo (with Optional In Situ Hybridization Steps for RNA Labeling)

1. Paraformaldehyde (PFA) fix solution: 4% paraformaldehyde in PBS-Tween. Under a fume hood and in a flask of at least 250 ml capacity, dissolve 4 g paraformaldehyde in 100 ml 1× PBS-Tween by bringing to a boil on a hotplate. Be careful to not overboil. Remove from heat and let cool to room temperature. Aliquot into desired amounts (we typically prepare 1 and 10 ml aliquots) and store at −20 °C.
2. Cytoskeleton Fix: 4% PFA, 0.25% glutaraldehyde, 5 mM EGTA, 0.2% Triton X-100. Per 1 ml of 4% PFA, add 10 μl 25% glutaraldehyde, 10 μl 500 mM EGTA, and 2 μl Triton X-100. Should be prepared freshly each day it is used.
3. Conjugated phalloidin [e.g., rhodamine-phalloidin (Cytoskeleton) or Alexa-488-Phalloidin (Thermo Fisher)]; to be added to Cytoskeleton Fix at 0.2 units/ml.
4. Pronase: 30 mg/ml stock solution.

5. 1× embryo medium (E3): 5 mM NaCl, 0.17 mM KCl, 0.33 mM $CaCl_2$, 0.33 mM $MgSO_4$, and 10^{-5}% methylene blue. Prepare 10× stock by dissolving 23.4 g NaCl, 1.0 g KCl, 3.9 g $CaCl_2 \cdot 2H_2O$, 6.5 g $MgSO_4 \cdot 7H_2O$ and fill with H_2O to a total volume of 8 l. For 1× working dilution, dilute 1:10 and add methylene blue to inhibit microbial growth (800 ml 10× stock + 8 ml 0.01% methylene blue and fill with H_2O to 8 l total).
6. Forceps (Dumont #5).
7. Plastic transfer pipettes (for washes and transferring embryos pre-dechorionation).
8. Glass Pasteur pipettes (for transferring dechorionated embryos).
9. 0.5 mg/ml sodium borohydride ($NaBH_4$).
10. DAPI: 100 μg/ml stock solution.
11. 1× PBS-Tween (PBST): 1× PBS, pH 7.0–7.4, 0.1% Tween-20. For 1 l stock solution, mix 100 ml 10× PBS, 12.5 ml 20% Tween-20, and add nuclease-free H_2O to 1 l. Store at room temperature.
12. Glycerol: used to make 30% and 50% Glycerol in PBS-Tween.
13. 1.5 ml Eppendorf tubes.
14. Plastic transfer pipettes (convenient for washes).
15. Glass Pasteur pipettes (for transferring embryos).
16. Incubator or water bath to produce a constant temperature of 37 °C.
17. 2× SSC: 2× SSC, 0.25% Tween-20. For a 400 ml stock solution, mix 40 ml 20× SSC (sodium citrate buffer), 5 ml 20% Tween-20, and nuclease-free H_2O to 400 ml. Store at room temperature. 50 ml of 0.2× SSC can be prepared by mixing 5 ml 2× SSC with 45 ml nuclease-free H_2O.
18. Minimal prehybridization buffer (minimal prehyb): 10% formamide, 2× SSC. For a 100 ml stock solution, mix 10 ml formamide and 90 ml 2× SSC. Store minimal prehyb at 4 °C.
19. Hybridization buffer (hyb): 50% formamide, 5× SSC, 100 μg/ml yeast tRNA, 50 μg/ml heparin, 0.25% Tween-20, 1 M citric acid to pH 6.0 (~0.02 M citric acid final conc.). For a 50 ml stock solution, mix 25 ml formamide, 12.5 ml 20× SSC, 1 ml of 25 mg/ml yeast tRNA, 50 μl of 50 mg/ml heparin, 250 μl 20% Tween-20, ~460 μl 1 M citric acid, 10.74 ml DEPC-treated H_2O. Store hyb stock at −20 °C.
20. RNA probe stock: To be added to hybridization buffer at 150–500 ng/ml (*see* Subheading 3.2, **step 22**). Can be made as described in Subheading 2.2.

21. Hydrogen peroxide (H_2O_2): to make 2% dilutions in PBS-Tween or TNT.
22. TNT: 0.1 M Tris–HCl pH 7.5, 0.15 M NaCl, 0.5% Tween-20. For a 1 l stock solution, mix 100 ml Tris pH 7.5, 30 ml 5 M NaCl, 25 ml 20% Tween-20, add nuclease-free H_2O up to 1 l. Store at room temperature.
23. Blocking solution/TBSTB: TNT with 0.5% PerkinElmer Blocking Powder. Measure 0.25 g PerkinElmer Blocking Powder or equivalent (*see* **Note 6**) into a 50 ml conical or glass bottle and fill to 50 ml with TNT. Invert multiple times to mix and heat to 69 °C for at least 1 h (ideally while rocking and/or rotating) to fully dissolve powder into solution. Pipet 1 ml aliquots into Eppendorf tubes and store at −20 °C. Once thawed, never refreeze.
24. TSA Cyanine 3 and/or Fluorescein System: PerkinElmer Kit includes 1× Amplification Diluent, Tyr-Fluorescein Amplification Reagent, and Tyr-Cy3 Amplification Reagent. The amplification reagents will need to be resuspended in 150 μl DMSO.
25. Anti-Fluorescein-POD or Anti-Digoxigenin-POD: Roche or equivalent.

3 Methods

3.1 Fixation of Zebrafish Embryos for Cytoskeleton Visualization

Preservation of dynamic networks such as the cytoskeleton requires rapid fixation. Here, chorion removal and the addition of glutaraldehyde to the fixation solution increase the speed of fixation, thus improving the ability to visualize intact cytoskeletal structures in the early embryo.

1. Collect embryos and let develop in Petri dishes containing embryo medium (E3). Shortly prior to the desired stage, carefully treat with a pronase solution (2 mg/ml in E3) to facilitate chorion removal. While being treated with pronase, swirl the dish and periodically (every ~30 s, no longer than 2 min) press into chorions with forceps to test their integrity. Once you see chorions start to easily buckle or dimple inwards from gentle pressure, carefully pour off the majority of the pronase/E3 solution, taking care not to expose embryos to the air:water interface (which can cause lysis), and quickly replace with E3. The force of the E3 wash will likely cause the majority of embryos to fall out of their chorions, and remaining chorions can be removed manually with forceps or by carefully pipetting through a glass Pasteur pipette (*see* **Note 7**).
2. Wash the embryos two more times by carefully pouring off the E3 (again, taking care not to expose embryos to the air:water interface) and replacing it with fresh, then use a glass Pasteur

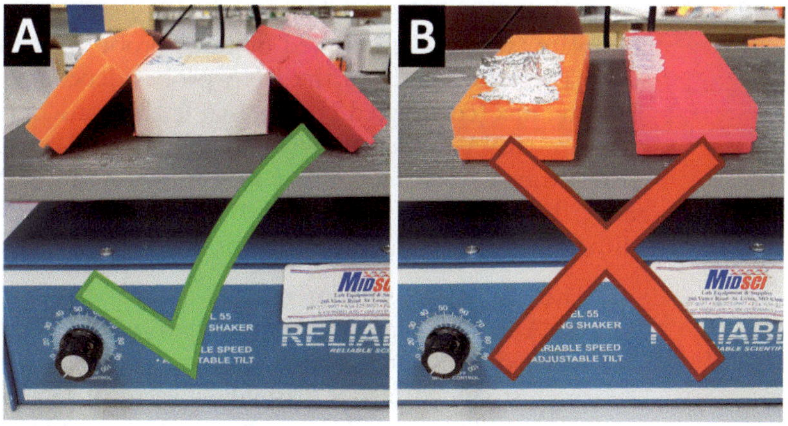

Fig. 5 During washes, we recommend that tube racks should be tilted at an angle (**a**) rather than sitting flat (**b**) on the rocker in order to help the wash solutions fully immerse the embryos

pipette to transfer embryos into 1.5 ml Eppendorf tubes that have been pre-filled with a small amount of E3 (~300 μl) to reduce risk of direct contact with the tube walls and air:water interface. We recommend fixing no more than 45 embryos at once in a single tube.

3. Carefully remove as much of the E3 as possible without drying out the embryos, then fill the tube with Cytoskeleton Fix. Lay the tube on its side and let fixation occur at room temperature for at least 4 h. After 4 h, you can let the embryos rock gently (*see* **Note 8**, Fig. 5) overnight at 4 °C or proceed immediately to the next steps.
4. Wash four times for 5 min each in PBS-Tween while rocking gently at room temperature.
5. Treat embryos with 500 μl/tube 0.5 mM sodium borohydride ($NaBH_4$) for 30 min while rocking gently at room temperature.
6. Dehydrate the embryos into methanol through a PBS-Tween/Methanol series (**steps 6–9**). *Important: Omit these steps if planning to visualize actin (see* ***Note 3****).*
7. Wash embryos in 2:1 PBS-Tween:MeOH for 5 min while rocking gently at room temperature.
8. Wash embryos in 1:1 PBS-Tween:MeOH for 5 min while rocking gently at room temperature.
9. Wash embryos in 1:2 PBS-Tween:MeOH for 5 min while rocking gently at room temperature.
10. Add 1 ml of 100% methanol to each tube of embryos and leave at least overnight at −20 °C. Embryos can be stored in MeOH at −20 °C for several months.

3.2 De Novo RNA Probe Design and Synthesis

This method includes the design and PCR amplification of a probe template (**steps 2–7**) followed by RNA probe synthesis (**steps 8–22**). The fluorescence of the RNA probes depends on the RNA labeling mix (conjugated UTPs) used in the transcription reaction mixture (RNA probe synthesis reaction; **step 9**). If planning to perform two-color fluorescence in situ hybridization (Subheading 3.3), the probes for each target RNA will need to incorporate a different marker (e.g., fluorescein or digoxigenin).

1. These initial steps are intended to guide the design of template DNA for RNA probes from scratch using a PCR-based, T7 method [5], but if you already have the DNA template of interest cloned into a plasmid, linearize it (*see* **Note 9**) and proceed directly to RNA probe synthesis in **step 9**.
2. Upload the exonic sequence of your target gene to a primer design resource, such as Primer3 [16]. Most settings can be left as default, but you will need to select product size ranges. This is up to your discretion, but we typically choose relatively short probes (100–300 nt, *see* **Note 1** and Fig. 2).
3. Select "Pick Primers" and decide which suggested set is most appropriate (*see* **Note 1**). For an antisense probe template, add the T7 promoter sequence (*see* **Note 10**) to the 5′ end of the reverse primer. For a sense probe template, add the T7 promoter sequence to the 5′ end of the forward primer.
4. Order primers and resuspend in nuclease-free H_2O to a 100 mM stock concentration. For maximum longevity, store at −20 °C and minimize freeze-thaw cycles if possible by aliquoting. However, we find that most primers do well even when stored at 4 °C for several months.
5. Prepare the following reaction mix for PCR amplification of the template DNA:

PCR amplification mix (per reaction)	
	Amount (μl)
Nuclease-free H_2O	3.85
dNTP/buffer mix	17.9
10 mM Fwd primer + 10 mM Rev. primer mix	0.15
cDNA (~10–100 ng)	3.00
EconoTaq DNA polymerase	0.10
Total reaction volume	25.0

And use the following PCR settings:

Step	Temperature, °C	Time
Initial denaturation	95	2 min
45 Cycles	95 45–65 (*see* **Note 11**) 72	30 s 15–60 s 1 min/kb
Final extension	72	5 min
Hold	4	–

6. Run 3 μl of PCR product + 5 μl loading dye vs. a ladder on a 1% agarose gel at 95 V to verify that template was produced and is at the expected size.
7. Optional: use a Clean & Concentrate kit to increase the concentration of PCR product. Only a few microliters of the DNA template will be used for the transcription reaction, so we recommend storing the excess at −20 °C for probe synthesis in the future.
8. From this point forward, be careful about avoiding potential RNase contamination. We find that standard laboratory cleanliness practices and spraying lab bench area, equipment, and gloves with a decontamination reagent such as RNase AWAY is sufficient.
9. Set up transcription reaction for RNA probe synthesis in RNase-free 1.5 ml Eppendorf tubes as shown below:

RNA probe synthesis mix (per reaction)	
	Amount (μl)
Nuclease-free H_2O	12.0
10× transcription buffer	2.00
RNA labeling mix (conjugated UTPs)	2.00
Template DNA (~1–2 μg from PCR or plasmid)	1.00
Ribolock RNase inhibitor	1.00
T7 RNA polymerase	2.00
Total reaction volume	20.0

Incubate tubes at 37 °C for 2–4 h.

10. To remove leftover template DNA at the end of the transcription reaction, add 1 μl of Turbo DNase per tube and incubate at 37 °C for 15 min.
11. Remove from incubator and precipitate the newly synthesized RNA probes by adding 1.3 μl LiCl and 75 μl chilled 100%

ethanol per tube and leave overnight at −20 °C. You will likely observe the solution immediately becomes cloudy as you add the LiCl and EtOH, which is a positive sign that RNA is beginning to precipitate.

12. Isolate the RNA probes out of solution by centrifugation at 13,000 rpm (12,000–16,000 × *g*) for 10 min at 4 °C. There should be a visible RNA pellet at the bottom of the tube.
13. Discard the supernatant without disturbing the RNA pellet and replace with 100 μl 70% EtOH.
14. Centrifuge again at 13,000 rpm (12,000–16,000 × *g*) for 5 min at 4 °C.
15. Discard as much of the supernatant as possible without disturbing the RNA pellet, then air dry for no more than 5 min at room temperature. One way of doing this is to place several KimWipes on the benchtop surface, and prop up the tubes open and upside down on top of the KimWipe. All ethanol and visible liquid should evaporate.
16. Add 20 μl DEPC-treated H_2O and leave at room temperature for 2 min.
17. Gently vortex at a low setting and/or flick the tube several times to resuspend, then briefly spin down to the bottom of the tube.
18. Transfer 2 μl of the resuspension to a PCR tube for later use on a diagnostic gel.
19. Add 20 μl 100% formamide and store the RNA probe stock, protected from light, at −20 to −80 °C.
20. Run the 2 μl of resuspended RNA that was set aside in **step 19** + 5 μl loading dye vs. a ladder on a 1% agarose gel at 120 V to verify that the transcription reaction worked (should visualize a band, not a large smear).
21. For use in in situ hybridization (colorimetric or fluorescent), prepare a working mix of probes + hybridization buffer, to be stored at −20 °C and which can be reused at least three times. If a specific concentration is desired (suggested ranges are 150–500 ng/ml), the RNA concentration can be estimated from the gel or on a Nanodrop and then diluted accordingly. We find that a standard 5 μl of each probe stock per 1 ml of hybridization buffer is sufficient to fully saturate our target transcripts for multiple experiments.

3.3 Two-Color Fluorescence In Situ Hybridization with Microtubule Immunofluorescence

In order to visualize two different target RNAs, this method uses two probes—one made with fluorescein and the other with digoxigenin. Hybridization of both probes is carried out simultaneously, followed by fluorescence development for each color via Tyramide Signal Amplification (TSA™) performed on subsequent days (fluorescein-labeled RNA detection: **steps 17–24**, digoxigenin-labeled RNA detection: **steps 25–31**). Next, microtubules are recognized by anti-tubulin antibodies and secondary antibody conjugated to Cy5 (**steps 32–38**). A suggested DNA stain with DAPI is included in **steps 39** and **40**.

1. Rehydrate the embryos into PBS-Tween through a PBST/methanol series (**steps 2–5**).
2. Wash embryos in 2:1 PBS-Tween:MeOH for 5 min while rocking gently at room temperature (*see* **Note 12**, Fig. 5).
3. Wash embryos in 1:1 PBS-Tween:MeOH for 5 min while rocking gently at room temperature.
4. Wash embryos in 1:2 PBS-Tween:MeOH for 5 min while rocking gently at room temperature.
5. Wash at least 2 × 5 min in PBST. If you are only interested in staining structures in the embryonic blastodisc, you can use this time to remove the yolk using forceps while still in PBST.
6. Wash once in prehybridization buffer (Prehyb) for 5 min, then add 300 μl of fresh prehyb to each tube of embryos. The embryos can be stored in prehyb at −20 °C or immediately incubated at 69 °C for at least 4 h. We recommend that each tube used for RNA and/or protein labeling contain only 10–20 embryos so that each embryo is fully saturated by all reagents and wash solutions. Transferring fixed embryos between tubes can be performed with a standard plastic transfer pipette that has an opening wide enough to not damage the embryos as they pass through.
7. While the embryos are incubating, prepare the desired probe (s)/hyb dilution if it was not already made. We typically use 5 μl of probe stock per 1 ml of hybridization buffer, and aim to have at least 250 μl of probe mix prepared for each tube of embryos. For dual labeling, probes for each target RNA will be added to the same probe hybridization mixture (e.g., 5 μl RNA_1-fluorescein probes + 5 μl RNA_2-digoxigenin probes in 1 ml hybridization buffer). Once prepared, allow the probe hybridization mix to heat up in the same incubator as the embryos for the remainder of their incubation period.
8. After the incubation period, remove the prehyb from embryos and replace with at least 250 μl of prepared probe mix. Allow to hybridize in the incubator overnight (*see* **Note 8**) at 69 °C.
9. 9. Preheat the SSC wash solutions to 69 °C for at least 10 min. Remove probe mix and store at −20 °C for reuse, then begin a prehyb:2× SSC wash series (detailed below).
10. Wash with a 2:1 solution of prehyb: 2× SSC for 5–10 min at 69 °C.

11. Wash with a 1:1 solution of prehyb: 2× SSC for 5–10 min at 69 °C.
12. Wash with a 1:2 solution of prehyb: 2× SSC for 5–10 min at 69 °C.
13. Wash with 2× SSC for 5–10 min at 69 °C.
14. Wash three times with 0.2× SSC for 15 min each at 69 °C.
15. Wash with PBS-Tween for 10 min while rocking gently at room temperature (*see* **Note 12**).
16. Freshly prepare a 2% H_2O_2 in PBS-Tween solution, enough for ~500 μl/tube. Add the 2% peroxide solution to each tube and let gently rock for 60 min. This step is critical to remove potential background from native peroxidase activity but can damage the embryos if left longer than an hour. Be aware that the tubes may pop open due to pressure building up from the peroxide reaction, so be careful not to spill their contents when handling.
17. Wash four times with TNT while rocking gently at room temperature.
18. Block for at least 3–4 h in ~400 μl TBSTB or similar blocking solution (*see* **Note 6**) while rocking gently at room temperature. The embryos can be left rocking overnight in blocking solution at 4 °C if needed, but if possible it is best to proceed to the antibody step before leaving overnight.
19. Remove block and replace with ≥250 μl 1:1000 Anti-Fluorescein-POD in blocking solution, and leave rocking gently overnight at 4 °C. If needed, the embryos can be left in antibody solution at 4 °C for multiple nights (3 maximum) before proceeding to next steps.
20. Discard antibody solution and wash eight times in TNT while gently rocking at room temperature over the course of 1–2 h.
21. Prepare 1:50 Tyr-Fluorescein Amplification Reagent in Amplification Diluent for 25 μl × (n + 1) tubes to ensure there will be enough mix for all samples. Add 25 μl of the 1:50 amplification mix to each tube, and cover with foil to protect from light. From this point on, the embryos should be protected from light during all wash steps because the fluorescence has been developed/activated.
22. Shake upright at room temperature for 1 h. If you do not have a shaker, a benchtop rocker will also work.
23. Wash two times for 5 min each in TNT while rocking gently at room temperature.
24. Wash 1 h in freshly prepared 2% H_2O_2 in TNT while rocking gently at room temperature.

25. Wash three times for 5 min each in TNT while rocking gently at room temperature.
26. Block for 1–2 h in ~400 μl blocking solution while rocking gently at room temperature. The embryos can be left rocking overnight in blocking solution at 4 °C if needed, but if possible it is best to proceed to the antibody step before leaving overnight.
27. Remove block and replace with ≥250 μl 1:1000 anti-DIG-POD in blocking solution, and leave rocking gently overnight at 4 °C. If needed, the embryos can be left in antibody solution at 4 °C for multiple nights (3 maximum) before proceeding to next steps.
28. Discard antibody solution and wash eight times in TNT while gently rocking at room temperature over the course of 1–2 h.
29. Prepare 1:50 Tyr-Cy3 in amplification diluent for 25 μl × $(n + 1)$ tubes to ensure there will be enough mix for all samples. Add 25 μl of the 1:50 amplification solution to each tube.
30. Shake upright at room temperature for 1 h. If you do not have a shaker, a benchtop rocker will also work.
31. Wash two times for 5 min each in TNT while rocking gently at room temperature.
32. Wash 1 h in freshly prepared 2% H_2O_2 in TNT while rocking gently at room temperature.
33. Wash 5 min in TNT, then 5 min in 1:1 TNT:PBS-Tween while rocking gently at room temperature.
34. Wash three times for 10 min each in PBS-Tween, then replace with blocking solution and let rock gently at room temperature for at least 2 h.
35. Replace blocking solution with at least ≥250 μl/tube of prepared anti-microtubule primary antibody diluted in blocking solution, and leave rocking gently overnight at 4 °C. If needed, the embryos can be left in antibody solution at 4 °C for multiple nights (3 maximum) before proceeding to next steps.
36. Wash four times for 30 min each in PBS-Tween while rocking gently at room temperature.
37. Add ~400 μl of blocking solution per tube and let block for at least 1 h while rocking gently at room temperature.
38. Replace blocking solution with appropriate secondary antibody dilution in blocking solution and let rock gently overnight at 4 °C. It is possible to reduce the time for this step to 2–4 h, but we find that leaving the embryos in secondary overnight produces better labeling.

39. Wash two times for 30 min each in PBS-Tween while rocking gently at room temperature. Wash two additional times if not proceeding to DAPI labeling.
40. In order to visualize nuclei, add 200 μl per tube of 1:200 DAPI in PBS-Tween and let rock gently at room temperature for 10–20 min.
41. Wash three times for 5 min each in PBS-Tween while rocking gently at room temperature.
42. To prepare for mounting, remove the PBS-Tween and replace with ~500 μl of a 30% glycerol in PBS-Tween solution. Leave the tubes on a flat, stationary surface so the embryos can settle to the bottom, then remove the 30% glycerol and replace with 50% glycerol in PBS-Tween. Again, let the embryos settle to the bottom. They can be left in the glycerol solution for multiple nights (maximum 3) at 4 °C before being mounted if necessary.
43. Mount embryos as desired for imaging (for mounting tips, *see* **Note 13**).

3.4 Rapid Phalloidin Labeling for Preservation and Visualization of F-Actin Structures in the Early Zebrafish Embryo (with Optional In Situ Hybridization Steps for RNA Labeling)

In this method, embryo fixation and F-actin labeling (**steps 1–6**) take place simultaneously due to the addition of fluorophore-conjugated phalloidin to the fixation solution, which promotes better preservation of F-actin structure. For RNA visualization in combination with the actin cytoskeleton, optional in situ hybridization steps are included (**steps 8–25**). A suggested DNA stain with DAPI is included in **steps 26** and **27**.

1. Freshly prepare the combined cytoskeleton fixative/phalloidin labeling mix the day of the experiment by mixing the following:

Cytoskeleton fixative/phalloidin labeling mix (per tube of embryos)		
	Amount (μl)	**Final concentration**
4% paraformaldehyde (PFA)	965.5	–
25% glutaraldehyde	10.0	0.25%
500 mM EGTA	10.0	5 mM EGTA
Triton X-100	2.0	0.2%
1:200 conjugated phalloidin	12.5	0.2 units/ml

Important: Whenever possible, protect all phalloidin-containing solutions and phalloidin-labeled samples from light.

2. Collect embryos and let develop in Petri dishes containing embryo medium (E3). Shortly prior to the desired stage, carefully treat with a pronase solution (2 mg/ml in E3) to facilitate chorion removal. While being treated with pronase, swirl the dish and periodically (every ~30 s) press into chorions with forceps to test their integrity. Once you see chorions start to easily buckle or dimple inwards from gentle pressure, carefully pour off the majority of the pronase/E3 solution and quickly replace with E3. The force of the E3 wash will likely cause the majority of embryos to fall out of their chorions, and any remaining chorions can be removed manually with forceps.
3. Wash the embryos two more times by carefully pouring off the E3 and replacing it with fresh, then use a glass Pasteur pipette to transfer embryos into 1.5 ml Eppendorf tubes. We recommend fixing no more than 60 embryos at once in a single tube.
4. Carefully pipette off as much of the E3 as possible without drying out the embryos, then fill the tube with cytoskeleton fixative/phalloidin labeling mix. Lay the tube on its side, protect it from light (e.g., covered with foil or left in a drawer) and leave at room temperature for at least 4 h. After 4 h, you can let the embryos rock gently overnight at 4 °C or proceed immediately to the next steps. We find that phalloidin labeling is slightly improved if allowed to continue overnight.
5. Wash four times for 5 min each in PBS-Tween while rocking gently at room temperature (*see* **Note 12**, Fig. 5). Continue protecting embryos from light during all of the wash steps.
6. Treat embryos with 500 μl/tube 0.5 mM sodium borohydride ($NaBH_4$). This step is necessary for deactivating/reducing the glutaraldehyde.
7. Wash three times for 5 min each in PBS-Tween while rocking gently at room temperature.
8. If additional protein labeling is desired, the embryos can immediately go into blocking buffer at this point to begin an immunofluorescence assay (e.g., for combined F-actin and microtubule labeling, here proceed to **step 34** of Subheading 3.3 and follow through to end). For RNA labeling via in situ hybridization, proceed to the next steps. Otherwise, skip directly to **step 26** for DNA staining and mounting instructions.
9. Ensure that the embryos are still protected from light whenever possible. Wash twice in a minimal prehybridization buffer (10% formamide in 2× SSC) for 5 min each at room temperature, then add fresh minimal prehyb to each tube of embryos and let incubate at 37 °C for at least 2 h (*see* **Note 3**).

10. While the embryos are incubating, prepare the desired probe/hyb dilution if it was not already made. We typically use 5 μl of probe stock per 1 ml of hybridization buffer, and aim to have at least 250 μl of probe mix prepared for each tube of embryos. Once prepared, allow the probe mix to heat up to 37 °C for the remainder of their incubation period.
11. After the incubation period, remove the minimal prehyb from embryos and replace with at least 250 μl of prepared probe mix. Allow to hybridize at 37 °C in the incubator overnight (*see* **Note 8**).
12. Preheat the SSC wash solutions to 37 °C for at least 10 min. Remove probe mix and store at −20 °C for reuse, then begin wash series (detailed below).
13. Wash with a 1:1 solution of minimal prehyb:2× SSC for 5–10 min at 37 °C.
14. Wash with 2× SSC for 5–10 min at 37 °C.
15. Wash three times with 0.2× SSC for 15 min each at 37 °C.
16. Wash with PBS-Tween for 10 min while rocking gently at room temperature (*see* **Note 12**).
17. Freshly prepare a 2% H_2O_2 in PBS-Tween solution, enough for ~500 μl/tube. Add the 2% peroxide solution to each tube and let gently rock for 60 min. This step is critical to remove potential background from native peroxidase activity, but can damage the embryos if left longer than an hour. Be aware that the tubes may pop open due to pressure building up from the peroxide reaction, so be careful not to spill their contents when handling.
18. Wash four times for 5 min each with PBS-Tween while rocking gently at room temperature.
19. Block for at least 3–4 h in ~400 μl TBSTB or similar blocking solution (*see* **Note 6**) while rocking gently at room temperature. The embryos can be left rocking overnight in blocking solution at 4 °C if needed, but if possible it is best to proceed to the antibody step before leaving overnight.
20. Remove block and replace with ≥250 μl 1:1000 Anti-Fluorescein-POD in blocking solution, and leave rocking gently overnight at 4 °C.
21. Discard antibody solution and wash 8 times in PBS-Tween while gently rocking at room temperature over the course of 1–2 h.
22. Prepare 1:50 Tyr-Fluorescein Amplification Reagent in Amplification Diluent for 25 μl × (n + 1) tubes to ensure there will be enough mix for all samples. Add 25 μl of the 1:50 amplification mix to each tube, and cover with foil to protect from light.

23. Shake upright at room temperature for 1 h. If you do not have a shaker, a benchtop rocker will also work.
24. Wash two times for 5 min each in TNT while rocking gently at room temperature.
25. Wash 1 h in freshly prepared 2% H_2O_2 in TNT while rocking gently at room temperature.
26. Wash three times for 5 min each in TNT while rocking gently at room temperature.
27. In order to visualize nuclei, add 200 μl per tube of 1:200 DAPI in PBS-Tween and let rock gently at room temperature for 10–20 min.
28. Wash three times for 5 min each in PBS-Tween while rocking gently at room temperature.
29. To prepare for mounting, remove the PBS-Tween and replace with ~500 μl of a 30% glycerol in PBS-Tween solution. Leave the tubes on a flat, stationary surface so the embryos can settle to the bottom, then remove the 30% glycerol and replace with 50% glycerol in PBS-Tween. Again, let the embryos settle to the bottom. They can be left in the glycerol solution for multiple nights (max 3) at 4 °C before being mounted if necessary.
30. Mount embryos as desired for imaging (for mounting tips, *see* **Note 13**).

4 Notes

1. There are numerous strategies that can be applied to probe design, and here we will only briefly discuss a few key considerations. Depending on your needs and/or preferences, both sequence length and target region can be extensively varied. Increasing probe length and/or including stretches of the 5′ or 3′ UTR are strategies that increase target specificity. Designing probes that span exon-exon junctions ensures that the probes will only detect spliced transcripts, rather than genomic sequence, although this is less of a concern since RNA:RNA hybrids are much more stable than RNA:DNA hybrids [17]. In older embryos, probes that are longer than ~600 nt may have difficulty fully penetrating the sample; however, we have used full-length probes ~1–1.5 kB in length without difficulty in the early embryo (up to 6 h post-fertilization). Another potential benefit of longer probes is increased fluorescent signal due to greater numbers of conjugated-UTPs versus the quantity incorporated into short probes. However, for most purposes, short to medium length probes (~100–300 nt) work well because they are short enough to avoid potentially problematic

sequence (e.g., repetitive sequence, regions of self-complementarity, extreme G-C content), can easily penetrate whole embryos, typically provide ample fluorescent signal, and the template sequence can be reliably amplified with standard DNA polymerases such as EconoTaq. For additional information, probe design considerations are described in further depth by Aquino de Muro, 2008 [6].

2. In order to visualize three RNAs in a single sample rather than two, one would need to synthesize an additional RNA probe stock that incorporated UTPs conjugated to a different fluorophore and/or hapten than fluorescein or digoxigenin, such as dinitrophenol (DNP-11-UTP, PerkinElmer), and simply add it to the probe hybridization mixture at the same concentration as the other two probes, as described in Subheading 3.3, **step 7**. Further modify Subheading 3.3 by adding another day of RNA recognition and signal development procedure using the TSA Cyanine 5 System (PerkinElmer) after **step 32**. For specific directions, see the Triple Fluorescent In Situ protocol shared by the Talbot lab: https://wiki.zfin.org/display/prot/Triple+Fluorescent+In+Situ.

3. In standard in situ protocols, methanol treatment is intended to increase penetration of RNA probes by permeabilizing the embryo; however, at early stages (cleavage to early blastula), the embryo is naturally quite permeable, so we find that this step (Subheading 3.1, **steps 5–9**) can be omitted without negative consequence. For older embryos, other permeabilization agents have been suggested, such as ethanol, glacial acetic acid, or detergents. To address the potential for high formamide concentration negatively impacting phalloidin labeling quality, we found that decreasing the formamide concentration to 10% in 2× SSC, rather than 50% in prehybridization buffer (Subheading 2.3, **item 5**), is sufficient to rescue the phalloidin staining quality. This alteration slightly decreases the stringency of post-hybridization washes, so a high-quality/high-specificity RNA probe is necessary. Lastly, simply decreasing the probe hybridization temperature to 37 °C prevents the negative impact of high-temperature exposure on phalloidin staining. As with decreasing the formamide concentration of wash buffers, significantly decreasing the hybridization temperature runs the risk of decreasing in situ efficacy, so we recommend first attempting it with well-established, high-quality RNA probes.

4. Many labs have their own preferred method of adding dNTPs and reaction buffer to their PCR master mix, and we encourage

readers to use the method that works best for them. We prepare and freeze ready-to-use 1 ml aliquots of combined reaction buffer and dNTPs in ~200 ml batches by mixing the following materials:

dNTP/buffer mix (aliquots stored at −20 °C)		
	Amount (ml)	**Final concentration**
1 M $MgCl_2$	0.393	2.05 mM
1 M pH 8.4 Tris–HCl	2.613	13.7 mM
1 M KCl	13.092	68.3 mM
BSA (100 mg/ml)	3.468	27.1 mM
100 mM dATP	0.262	137 μM
100 mM dCTP	0.262	137 μM
100 mM dGTP	0.262	137 μM
100 mM dTTP	0.262	137 μM
Nuclease-free H_2O	171.12	–

5. An in-depth protocol and instructional video for RNA isolation from embryonic zebrafish and cDNA synthesis can be found in Peterson and Freeman, 2009 [18].
6. The PerkinElmer Blocking Powder Reagent is recommended by the manufacturer when using the PerkinElmer TSA System, and is sometimes conveniently included when purchasing various PerkinElmer kits. Roche Blocking Reagent for nucleic acid hybridization and detection also works well when used at the same concentration as the PerkinElmer powder (0.5% in TNT), and both perform better than 1% BSA. Both the PerkinElmer and Roche powders are primarily comprised of purified casein protein, so we predict that using 0.5% casein powder in TNT as an in-house alternative would perform similarly, although we have not yet tried this formulation.
7. Embryos can also be manually dechorionated with a pair of fine forceps after a very brief (<1 min) exposure to pronase solution. It may be possible to omit dechorionation via pronase treatment prior to fixation if you do not plan to visualize the cytoskeleton. Pre-fixation dechorionation and the suggested modified 4% PFA solution are used to fix embryos faster and allow better preservation of intracellular structures than traditional methods. Fluorescent in situ hybridization and immunofluorescence of many proteins other than tubulin and/or actin can be performed on early-stage embryos (pre-epiboly) that were fixed with 4% PFA without any additional reagents

and while still in their chorions. However, especially for proteins that may have a subcellular localization, we recommend always testing whether conditions for fixation of cytoskeletal elements result in improved signal, localization, or resolution. If the cytoskeleton-specific steps are omitted, manually dechorionate post-fixation with forceps prior to permeabilization with 100% MeOH.

8. We recommend a "lengthy" overnight hybridization timeframe of at least 18 h. The embryos can also be allowed to hybridize for an additional day and overnight.
9. A detailed explanation of T7 transcription reactions using plasmid or PCR-amplified target DNA, including a protocol for linearizing plasmids, can be found in Brunelle and Green, 2013 [19].
10. We include a short spacer sequence at the beginning of the T7 promoter sequence, so that the total sequence added is: 5′-G AAATTAATACGACTCACTATAGG.
11. Determine annealing temperature based on primer pair T_m. For simplicity, our lab finds that a standard annealing setting of 55 °C for 30 s per cycle works in most cases.
12. Instead of leaving the tube of embryos fully upright in their racks while on a benchtop rocker, place a small object (e.g., box lid or another tube rack) on the rocker and tilt the rack with embryos up against it. This helps spread out the embryos inside the tube and fully immerses them in the wash solutions/ reagents (Fig. 5).
13. There are many mounting procedures and we encourage readers to use the one that works best for them and their samples. Our lab typically follows the following methods: To flat mount blastodiscs from cleavage to late blastula stage embryos, we find that stacking two binder hole reinforcement rings on a glass slide provides a conveniently sized "well" for holding the deyolked embryos in place while also serving as a spacer to prevent them from becoming squashed by the coverslip. After the embryos have settled in 50% glycerol, transfer several (≤8) to the well with as little liquid as possible. Add a drop of mounting medium of choice on top of the embryos. There should be a visible "bubble" of liquid slightly raised, but there should not be enough to break surface tension and spill over onto the page reinforcement label surrounding the well. Using forceps, gently move the deyolked embryos so that the blastodiscs are facing up (i.e., so that the animal cortex of the embryo—not the more vegetally located blastodisc edges—will be closest to the coverslip). Blastodiscs mounted facing down typically cannot be visualized properly at high magnifications due to the increased distance to the objective. To better

discern the orientation of the blastodiscs, use a dissection needle or forceps to tilt them sideways, then orient them in the desired orientation. Try to avoid placing any directly next to the edge of the well. Carefully place a glass coverslip over the embryos by holding the coverslip on two opposite edges and bringing it as close to the sample as possible before releasing. Very gently press down on the four corners of the coverslip, then paint around all of the edges with clear nail polish (or sealant of choice) to seal it in place and prevent dehydration.

Acknowledgments

This work was supported by NIH grant GM007133-44 to C.L.H. and GM065303 to F.P., with additional support from the College of Agricultural and Life Sciences, the School of Medicine and Public Health, and the Laboratory of Genetics at the University of Wisconsin – Madison.

References

1. Jansen RP (1999) RNA-cytoskeletal associations. FASEB J 13:455–466
2. Theusch EV, Brown KJ, Pelegri F (2006) Separate pathways of RNA recruitment lead to the compartmentalization of the zebrafish germ plasm. Dev Biol 292:129–141. https://doi.org/10.1016/j.ydbio.2005.12.045
3. Kloc M, Zearfoss NR, Etkin LD (2002) Mechanisms of subcellular mRNA localization. Cell 108:533–544. https://doi.org/10.1016/S0092-8674(02)00651-7
4. Milligan JF, Uhlenbeck OC (1989) Synthesis of small RNAs using T7 RNA polymerase. Methods Enzymol 180:51–62
5. David R, Wedlich D (2001) PCR-based RNA probes: a quick and sensitive method to improve whole mount embryo in situ hybridizations. BioTechniques 30:769–772
6. Aquino de Muro M (2008) Probe design, production, and applications. In: Walker JM, Rapley R (eds) Molecular biomethods handbook, 2nd edn. Humana, Totowa, NJ, pp 41–53
7. Clay H, Ramakrishnan L (2005) Multiplex fluorescent in situ hybridization in zebrafish embryos using tyramide signal amplification. Zebrafish 2:105–111. https://doi.org/10.1089/zeb.2005.2.105
8. Zaidi AU, Enomoto H, Milbrandt J, Roth KA (2000) Dual fluorescent in situ hybridization and immunohistochemical detection with tyramide signal amplification. J Histochem Cytochem 48:1369–1375. https://doi.org/10.1177/002215540004801007
9. Lynen F, Wieland U (1938) Über die Giftstoffe des Knollenblätterpilzes. IV Justus Liebigs. Ann Chem 533:93–117. https://doi.org/10.1002/jlac.19385330105
10. Wulf E, Deboben A, Bautz FA, Faulstich H, Wieland T (1979) Fluorescent phallotoxin, a tool for the visualization of cellular actin. Proc Natl Acad Sci U S A 76:4498–4502
11. Melak M, Plessner M, Grosse R (2017) Actin visualization at a glance. J Cell Sci 130:525–530. https://doi.org/10.1242/jcs.189068
12. Dancker P, Löw I, Hasselbach W, Wieland T (1975) Interaction of actin with phalloidin: polymerization and stabilization of F-actin. Protein Struct 400:407–414
13. Vandekerckhove J, Deboben A, Nassal M, Wieland T (1985) The phalloidin binding site of F-actin. EMBO J 4:2815–2818. https://doi.org/10.1002/j.1460-2075.1985.tb04008.x
14. Eno C, Pelegri F (2018) Modulation of F-actin dynamics by maternal Mid1ip1L controls germ plasm aggregation and furrow recruitment in the zebrafish embryo. Development 145:dev156596. https://doi.org/10.1242/dev.156596

15. Kellogg DR, Mitchison TJ, Alberts BM (1988) Behaviour of microtubules and actin filaments in living Drosophila embryos. Development 103:675–686
16. Untergasser A, Cutcutache I, Koressaar T, Ye J, Faircloth BC, Remm M, Rozen SG (2012) Primer3—new capabilities and interfaces. Nucleic Acids Res 40:e115. https://doi.org/10.1093/nar/gks596
17. Sugimoto N, Nakano S, Katoh M, Matsumura A, Nakamuta H, Ohmichi T, Yoneyama M, Sasaki M (1995) Thermodynamic parameters to predict stability of RNA/DNA hybrid duplexes. Biochemistry 34:11211–11216. https://doi.org/10.1021/bi00035a029
18. Peterson SM, Freeman JL (2009) RNA isolation from embryonic zebrafish and cDNA synthesis for gene expression analysis. J Vis Exp. https://doi.org/10.3791/1470
19. Brunelle JL, Green R (2013) In vitro transcription from plasmid or PCR-amplified DNA. In: Lorsch J (ed) Methods in enzymology. Academic, Cambridge, MA, pp 101–114

Chapter 19

Glyoxal Fixation as an Alternative for Zebrafish Embryo Immunostaining

Nadia Rostam and Roland Dosch

Abstract

Immunohistochemistry has been widely used as a robust technique to determine the cellular and subcellular localization of proteins. This information ultimately helps to understand the function of these proteins and how biological processes are regulated. Antibodies applicable for labeling in zebrafish are limited, making immuno-staining challenging. Recently glyoxal fixation was rediscovered in tissue culture, mouse, rat, and *Drosophila*, expanding the list of effective antibodies for these species. Here, we compare a protocol for zebrafish staining using glyoxal as a fixative agent with PFA. We demonstrate that glyoxal fixation improves the antigenicity of some epitopes thereby increasing the number of useful antibodies in zebrafish.

Key words Immunostaining, Zebrafish, Fixation method, PFA, Glyoxal, Germ plasm, Bucky ball

1 Introduction

Zebrafish has fast become an excellent vertebrate model for studying the function of proteins and complex biological processes in the context of an entire organism. To understand the function of a protein, it is crucial to know where and when it is expressed inside an organism or even inside a cell. Therefore, investigating localization and expression pattern of proteins using immunohistochemistry (IHC) is very important in identifying the function of genes, which can be achieved by immune labeling of fixed or frozen tissue samples or whole-mount samples. Up to today, IHC is a powerful and one of the most widely used tools in detecting cellular and subcellular localization of proteins [1–5]. Principally, IHC works for a broad range of biological tissues and it is based on antigen detection using specific antibodies raised against the antigens of interest, which recognizes specific epitopes from proteins [2, 6]. In addition, this technique can be used more specifically to detect the activity of proteins, such as in their phosphorylated state, which is not possible with other methods like RNA in situ hybridization [7].

Roland Dosch (ed.), *Germline Development in the Zebrafish: Methods and Protocols*, Methods in Molecular Biology, vol. 2218, https://doi.org/10.1007/978-1-0716-0970-5_19,

© Springer Science+Business Media, LLC, part of Springer Nature 2021

Whole-mount immunostaining of zebrafish embryos has significantly revealed the expression pattern of tissue-specific proteins. However, there is a shortage in the availability of antibodies for this organism, making IHC problematic in our field [8]. In addition, a major problem with this application is that detailed protocol optimization is needed based on the type of tissue investigated and the antibody used, especially because antibodies need different conditions to give the highest possible signal. Moreover, IHC is a multistep technique which takes different levels of optimization and various combination of reagents is needed until maximum signal intensity from the antibodies is detected [6, 9].

For example, fixation is a critical step in immunohistochemistry which needs to be performed in a way that it keeps the balance between the integrity of the sample and antigenicity [10]. Previous studies have examined different protocols and fixative methods such as formalin, methanol, formaldehyde, and paraformaldehyde (PFA) for immunostaining [11–13]. More recently, it was shown that glyoxal fixation improves immune staining in tissue culture, mouse, rat, and *Drosophila* [14], but zebrafish was not tested in this study.

Here, we provide an alternative protocol for zebrafish embryo staining (Fig. 1). We show that glyoxal fixation provides better resolution, less background, and more intense signal production for protein visualization in zebrafish embryos but not for every antibody. We compared glyoxal and paraformaldehyde (PFA) fixation for Bucky ball (Buc) and tight junction protein Zonula occludens 1 (ZO1) antibodies. The ZO1 antibody gave a much better signal using glyoxal fixation, whereas Buc provides a strong signal using PFA fixation. We believe that our method can be extended to the staining of other tissue types of the zebrafish embryo, providing that optimal conditions for the antibodies are used.

2 Materials

2.1 Zebrafish Embryo Collection and Handling

1. Embryos of corresponding stage from wild-type zebrafish line maintained at standard conditions [15].
2. Stereomicroscope.
3. Plastic Petri dish.
4. Plastic Pasteur pipette 2 ml.
5. Eppendorf tubes 2 ml.
6. E3 medium: 5 mM NaCl, 0.17 mM KCl, 0.33 mM $CaCl_2$, 0.33 mM $MgSO_4$, 0.00001% methylene blue.
7. DUMONT forceps (size 5).

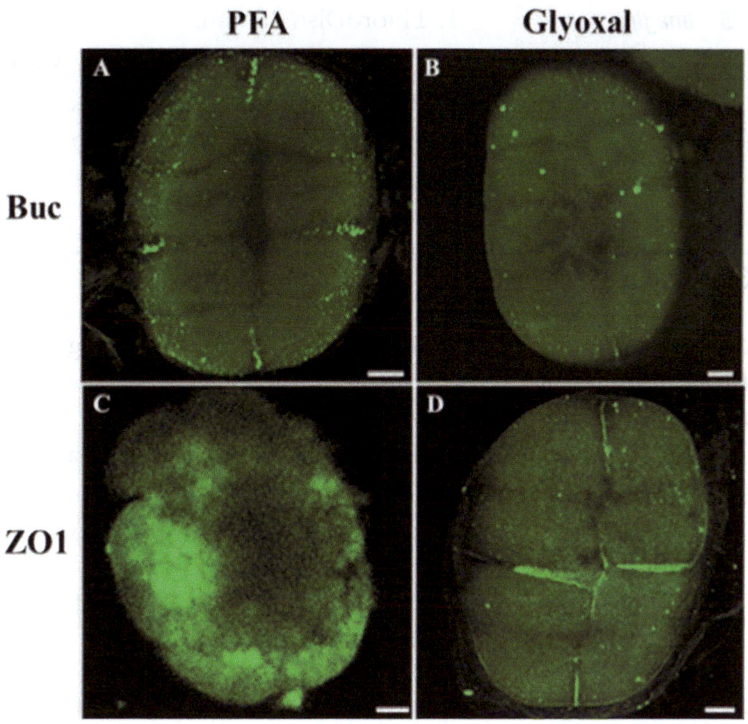

Fig. 1 Comparison of antibody staining of zebrafish embryos fixed with PFA and glyoxal. Labeling of PFA-fixed embryos (8-cell stage, animal view) is shown in the left column and of glyoxal-fixed embryos in the right column (*see* **Note 6**). The upper row shows Buc labeling (**A** and **B**) and the lower ZO1 (**C** and **D**). Bucky ball antibody showed almost no signal with glyoxal fixation and a very strong signal with PFA fixation, while ZO1 antibody showed a much stronger signal with glyoxal fixation. Scale bars: 50 μm

2.2 Embryo Fixation

1. PBS: 137 mM NaCl, 10 mM Na_2HPO_4, 2.7 mM KCl, 1.76 mM KH_2PO_4 (pH 7.4).
2. 4% PFA in 1× PBS: 4 mg PFA, 96 ml PBS.
3. 100% methanol (MeOH).
4. 100% ethanol (EtOH).
5. Glyoxal solution: 2.789 ml H_2O, 0.789 ml 100% EtOH, 0.313 ml 40% Glyoxal, 30 μl 100% acetic acid (pH 6.5) (*see* **Note 9**).
6. Rotator at 4 °C and room temperature (RT) conditions.

2.3 Blocking and Washings

1. 10% goat serum.
2. PBS/ 0.5% Triton X-100 (PBT) (*see* **Note 1**).
3. 2% Tween 20 in PBS: 2 ml Tween 20, 98 ml PBS (*see* **Note 1**).

2.4 Staining

1. Primary antibodies.
2. Secondary antibodies.

2.5 Imaging

1. FluoroDish 35 mm.
2. LSM780 confocal microscope (Carl Zeiss Microscopy, Jena).
3. ZEN 2011 software (Carl Zeiss Microscopy, Jena).
4. Murray's mounting medium: 2/3benzyl benzoate, 1/3 benzyl alcohol.

3 Methods

3.1 Zebrafish Embryo Collection and Handling

1. Set up fish pairs the evening before embryo collection by separating each male from his female partner in a mating box overnight (*see* **Note 11**).
2. The next morning, put fish pairs together and let them mate and lay eggs.
3. Collect the eggs, keep them at 28 °C to grow up to the desired cell stage.
4. Collect the embryos in a 2 ml Eppendorf tube containing E3 medium (*see* **Note 12**).

3.2 Embryo Fixation

Embryos were fixed using one of the following fixation methods:

3.2.1 Glyoxal-Based Embryo Fixation

1. Prepare fresh glyoxal solution.
2. Keep the solution at 4 °C.
3. Remove the E3 medium from the 2 ml Eppendorf tube containing embryos and replace it with glyoxal solution.
4. Keep the embryos on ice for 30 min.
5. Keep the embryos for another 30 min at RT.
6. Store at 4 °C.

3.2.2 PFA-Based Embryo Fixation

1. Prepare 4% PFA solution from frozen aliquots.
2. Replace the E3 medium from the 2 ml Eppendorf tube containing embryos with 4% PFA.
3. Rock the embryos on a rotator at 4 °C overnight.
4. Discard the PFA and replace it with PBS. Wash the embryos twice 10 min with PBS, each time for 10 min.
5. Gradually dehydrate the embryos in a MeOH/PBT series and store at least overnight or until used for staining at −20 °C.

3.3 Immunostaining

1. Rehydrate the embryos in a MeOH: 0.5% PBT series (50%, 75%, 87.5%, 100%) for PFA-fixed embryos and an EtOH: 0.5% PBT series (50%, 75%, 87.5%, 100%) for glyoxal-fixed embryos.
2. Manually dechorionate the embryos using a pair of forceps and remove damaged embryos (*see* **Note 2**).

3. Transfers embryos into a 24-well tissue culture plate (*see* **Note 8** and **Note 13**).
4. Replace the PBT with 500 µl fresh 0.5% PBT + 10% goat serum and rock for 2 hrs at RT (*see* **Note** 7).
5. Replace the blocking solution with 500 µl 0.5 PBT + 2% BSA + 10% goat serum and the appropriate dilution of the primary antibodies.
6. Rock embryos overnight at 4 °C.
7. Replace the primary antibody solution with 2 ml 2% Tween 20 and rock at RT for 15 min (repeat three times) and wash once with 2 ml of 0.5% PBT for 15 min.
8. Replace the PBT with 500 µl 0.5% PBT/10% goat serum and the appropriate dilution of the fluorescently labeled secondary antibodies.
9. Cover the dish with aluminum foil to protect from light and rock for 2 hrs at 4 °C.
10. Replace the primary antibody solution with 2 ml 2% Tween 20 and rock at RT for 15 min (repeat three times) and wash once with 2 ml of 0.5% PBT for 15 min.
11. Gradually dehydrate the embryos in a 0.5% PBT: MeOH series for PFA-fixed embryos and a 0.5% PBT: EtOH series for glyoxal-fixed embryos.
12. Embryos are ready to be imaged or stored at −20 °C until imaging.

3.4 Imaging

1. Transfer the embryos from the dish into an imaging FluoroDish.
2. Remove MeOH/EtOH.
3. Add a drop or more of Murray's Clearing Medium into the FluoroDish, covering the embryos (*see* **Note 10**).
4. Wait for about 5 min until the embryos become clear and image with confocal microscope.

4 Notes

1. We used a combination of two different detergents, Tween 20 and Triton, since this gave our samples a clearer and less damaged texture. Tween 20 is a soft detergent while Triton is more harsh, when used alone they are either not effective enough or harmful. Therefore, washing three times with Tween 20 preserves the texture of the embryos and one final wash with Triton effectively washes out the remaining Tween 20.

2. Manually dechorionating the embryos better preserves their morphology; therefore, we recommend avoiding dechorionation using chemicals like Pronase.
3. Glyoxal-fixed embryos have a much clearer morphology after fixation. This makes them easier to deyolk, since the dark embryos are easily distinguishable from the transparent yolk compared to the dark yolk after PFA fixation (Fig. 2a, b). We recommend glyoxal fixation method if it works for your antibodies.
4. It is much easier to deyolk glyoxal-fixed embryos than PFA-fixed embryos. In the case of glyoxal, the yolk is stickier and more elastic, which is easier to remove, compared to the brittle yolk after PFA fixation (Fig. 2c, d). Fragile PFA-fixed embryos are also problematic since autofluorescent sticky yolk globules cause unwanted background during imaging.
5. Glyoxal-fixed embryos are very easy to dechorionate and deyolk. Pressing the embryo with a forceps bursts the embryo out of the chorion. This procedure simultaneously deyolks the embryo (Fig. 2d).
6. Glyoxal fixation gives less background during imaging.
7. We highly recommend blocking with serum derived from the same species as the secondary antibodies to get the least possible background.
8. We recommend using an embryo sucker to transfer embryos during deyolking and mounting (Fig. 2e).
9. Glyoxal-fixed embryos can be stored in a 4 °C fridge for more than a week before staining. However, because of the easy and less time-consuming nature of this method, we prefer making fresh glyoxal always and staining embryos within few days after fixation.
10. Murray's mounting medium is toxic and should be handled very carefully. Wear gloves and work inside a hood while dealing with this substance.
11. Fish should be kept under 14 hrs light and 10 hrs dark condition before setting them up.
12. We prefer using 2 ml tube for embryo collection since these tubes have round bottoms, which make the movement and transfer of the embryos in them easy.
13. During embryo transfer with embryo sucker, it is important to suck some medium up before sucking up the embryos. This will keep the embryos in the fluid and prevents them from moving up and sticking to the rubber tube. It is better not to suck the embryos beyond the glass capillary, because they might get stuck at the rim at the capillary–tube transition.

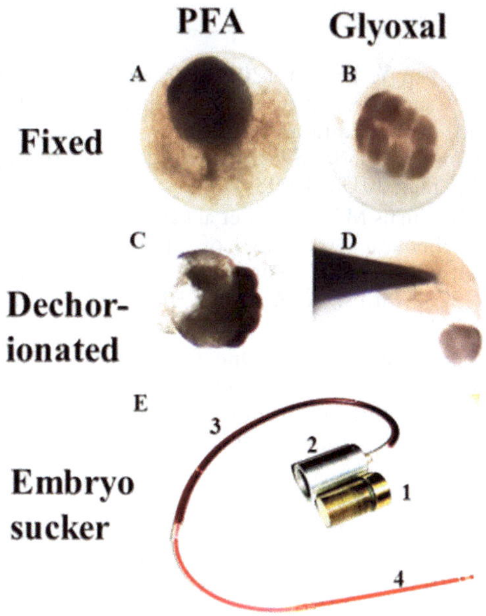

Fig. 2 Zebrafish embryo fixation using PFA and glyoxal. PFA fixation is shown in the left column and glyoxal fixation in the right. (**a**) After fixation with PFA embryos look dark and grainy. (**b**) A glyoxal-fixed embryo, which looks more transparent. The yolk becomes transparent with glyoxal fixation, which makes it easier to differentiate it from the embryo. Embryonic cells are clearly visible. (**c**) A manually dechorionated PFA-fixed embryo (*see* **Note 3**), showing aggregated granules of yolk around the broken yolk of the embryo. This commonly happens since embryos fixed with PFA are more fragile (*see* **Note 4**). (**d**) A glyoxal-fixed embryo after dechorionation, which is much more simple than with PFA. Careful pressing with forceps (dark triangle) on one chorion side, until it bursts, makes the embryo jump out without the yolk (*see* **Note 5**). As deyolking and dechorionation is a single-step process, autofluorescent yolk globules around the embryo are absent. (**e**) Embryo sucker. Using this instrument for embryo transfer preserves the texture of the embryos and prevents their destruction. Embryo sucker is composed of three parts. Part (1) is a metal screw which fits in part (2). The screw needs to be attached to part 2 during usage. Screwing and unscrewing the screw pushes and pulls pressure inside the rubber tube (3), which provides the needed force for sucking up and transferring out the deyolked or dechorionated embryos. To make this construction airtight, the screw is lubricated with grease, e.g., Vaseline. Part (3) is an extension rubber tube which can hold extra fluid medium for the embryos and provides surface area for producing enough pressure. Part (4) is a glass capillary tube. During embryo transfer, the screw should be locked first, and then the tip of the capillary tube should be held close to the embryos, unscrewing the screw sucks the embryos along with a minimal amount of embryonic medium up into the tube. Finally, the embryos can be pushed out by unscrewing

References

1. Han R, Li Z, Fan Y, Jiang Y (2013) Recent advances in super-resolution fluorescence imaging and its applications in biology. J Genet Genomics 40:583–595. https://doi.org/10.1016/j.jgg.2013.11.003
2. Waldvogel HJ, Curtis MA, Baer K et al (2007) Immunohistochemical staining of post-mortem adult human brain sections. Nat Protoc 1:2719–2732. https://doi.org/10.1038/nprot.2006.354
3. Inoue D, Wittbrodt J (2011) One for all-a highly efficient and versatile method for fluorescent immunostaining in fish embryos. PLoS One 6:1–7. https://doi.org/10.1371/journal.pone.0019713
4. Barlow AL, MacLeod A, Noppen S et al (2010) Colocalization analysis in fluorescence micrographs: verification of a more accurate calculation of Pearson's correlation coefficient. Microsc Microanal 16:710–724. https://doi.org/10.1017/S143192761009389X
5. Childs GV (2014) History of Immunohistochemistry. Elsevier Inc., Amsterdam
6. Zhang H, Wen W, Yan J (2017) Application of immunohistochemistry technique in hydrobiological studies. Aquacult Fish 2:140–144. https://doi.org/10.1016/j.aaf.2017.04.004
7. Perdana (2018) 済無No title. J Chem Inf Model 53:1689–1699. https://doi.org/10.1017/CBO9781107415324.004
8. Deflorian G, Cinquanta M, Beretta C et al (2009) Monoclonal antibodies isolated by large-scale screening are suitable for labeling adult zebrafish (*Danio rerio*) tissues and cell structures. J Immunol Methods 346:9–17. https://doi.org/10.1016/j.jim.2009.04.012
9. Yang J, Xu X (2012) Immunostaining of dissected zebrafish embryonic heart. J Vis Exp 59: e3510. https://doi.org/10.3791/3510
10. Thomask U (2003) Immunocytochemistry of the amphibian embryo – from overview to ultrastructure. Int J Dev Biol 383:373–383
11. Moon IS, Cho SJ, Jin IN, Walikonis R (2007) A simple method for combined fluorescence in situ hybridization and immunocytochemistry. Mol Cells 24:76–82
12. Novak AE, Ribera AB (2003) Immunocytochemistry as a tool for zebrafish developmental neurobiology. Methods Cell Sci 25:79–83. https://doi.org/10.1023/B:MICS.0000006894.43940.b1
13. Fernández J, Fuentes R (2013) Fixation/permeabilization: new alternative procedure for immunofluorescence and mRNA in situ hybridization of vertebrate and invertebrate embryos. Dev Dyn 242:503–517. https://doi.org/10.1002/dvdy.23943
14. Richter KN, Revelo NH, Seitz KJ et al (2018) Glyoxal as an alternative fixative to formaldehyde in immunostaining and super-resolution microscopy. EMBO J 37:139–159. https://doi.org/10.15252/embj.201695709
15. Aleström P, D'Angelo L, Midtlyng PJ, et al (2019) Zebrafish: housing and husbandry recommendations. Lab Anim 0:1–12. https://doi.org/10.1177/0023677219869037

Chapter 20

Histological Analysis of Gonads in Zebrafish

Kellee R. Siegfried and Jocelyn S. Steinfeld

Abstract

Zebrafish is an excellent system for the study of gonad development due to available genetic tools and its utilization as a human disease model. The zebrafish serves as an experimental system to model human disorders affecting the reproductive system, toxicological effects on fertility and sexual development, and hormonal regulation of fertility. Forward genetic screens have been used to uncover genetic causes of infertility and reverse genetic approaches have demonstrated that genes involved in germ cell development have similar functions in zebrafish and mammals. The most comprehensive picture of the gonad can be visualized by histology. There are a variety of methods that give excellent histology of zebrafish gonads. Below are methods for two staining approaches for the histology of paraffin-embedded zebrafish gonads.

Key words Ovary, Testis, Histology, Germ cells

1 Introduction

Histological analysis of tissues allows visualization of all cells and tissues within an organ or organism. Most methods of histology involve embedding tissue in either a hard resin or in paraffin wax providing a matrix to maintain tissue integrity for thin sectioning. The advantages of resin embedding are that thinner sections can be achieved as well as high resolution of subcellular structures and sometimes less tissue distortion. Paraffin-embedded tissue also gives high-quality tissue and cellular morphology yet is slightly less laborious and has the advantage that serial sections are more easily attained. The general procedure underlying paraffin sectioning involves processing through a series of solutions that facilitate infiltration of the tissue with paraffin wax. Paraffin is not miscible with aqueous solutions, and therefore the tissue first needs to be dehydrated then infiltrated with a non-toxic xylene substitute (e.g., Citrisolv), which is miscible with molten paraffin. To do this, the tissue is fixed, then dehydrated through a series of successively increasing percentages of ethanol. The gentle dehydration of tissue is important to avoid tissue damage that can occur with rapid

Roland Dosch (ed.), *Germline Development in the Zebrafish: Methods and Protocols*, Methods in Molecular Biology, vol. 2218, https://doi.org/10.1007/978-1-0716-0970-5_20, © Springer Science+Business Media, LLC, part of Springer Nature 2021

dehydration. Once the tissue is in a 100% ethanol solution, it can then be infiltrated with a xylene substitute and then with molten paraffin. Embedding of paraffin infiltrated tissue is accomplished by placing the tissue in a mold suitable for the microtome specimen holder and allowing the paraffin to cool to room temperature. Tissues are extremely stable once embedded in paraffin and can be stored indefinitely at room temperature.

After sectioning, a wide range of stains are available that label various cellular substructures. By far the most common staining method used on animal tissues is hematoxylin and eosin (H&E). Hematoxylin stains DNA whereas eosin Y generally stains the cytoplasm. An alternative staining combination is periodic acid Schiff and hematoxylin (PAS&H). The outcome is somewhat similar to H&E except that the basement membranes are also visible, which nicely outlines the testes tubules. Procedures for both staining methods are provided.

Fixatives are also numerous and are equally important to selecting as the embedding media and stains. When selecting a fixative one should consider the tissue type, embedding media, and downstream applications. Bouin's fixative gives excellent morphology of the gonads, is easy to use and prepare, and can be purchased as a premade solution. Due to the acid content in Bouin's fixative, the bones become adequately softened negating the need for decalcification. The main disadvantage of Bouin's fixative is the presence of picric acid, which is highly toxic, explosive when dry, and difficult to wash out of the tissue. An alternative to Bouin's fixative that also gives excellent gonad morphology, and is more suitable if antibody labeling may be carried out, is Davidson's fixative. Davidson's fixative is similar to Bouin's fixative without the picric acid. The lower acid content does not adequately soften bones of adult fish, therefore a decalcification step should be included after fixation if bones are present in the sample.

The zebrafish gonads can be readily visualized using histological methods. In this chapter, we focus on paraffin-embedded samples. An excellent description of testis morphology of resin-embedded zebrafish testis has been previously reported [1]. The methods and equipment provided in this protocol require minimal specialized equipment and can be set up readily in any lab provided a rotary microtome is available.

Histological analysis reveals the organization of somatic gonadal cells and germ cell differentiation. Descriptions of adult zebrafish gonad histology can be found in print [2] or online at http://zfin.org/hh_atlas/. Stages of oogenesis were characterized in detail by Selman et al. [3]. Histology of testicular germ cells has been described by Leal et al. [1]. H&E staining of paraffin-embedded gonads (Bouin's or Davidson's fixed) gives a clear picture of spermatogenic/oogenic stages as well as somatic cells and general gonad organization (Figs. 1 and 2). There are also several

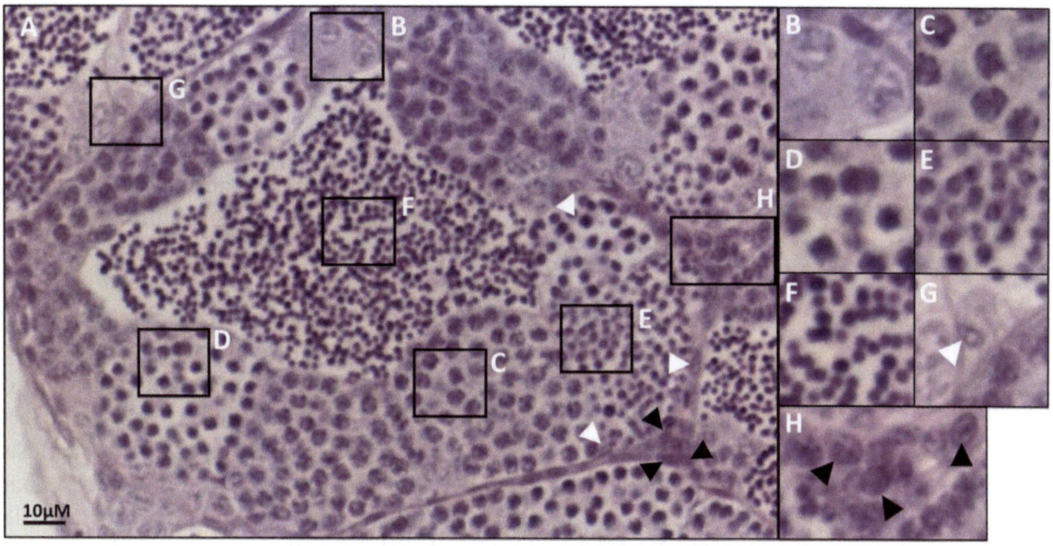

Fig. 1 Histology of mature zebrafish testis visualized with Hematoxylin and Eosin staining. (A) Zebrafish testes are organized as tubules containing cysts of developing germ cells and spermatozoa in the center. Within each cyst germ cells divide and differentiate synchronously. (B) Spermatogonia have large nuclei that stain weakly with hematoxylin. (C) Leptotene stage primary spermatocytes. (D) Zygotene stage primary spermatocytes. (E) Secondary spermatocytes. (F) Spermatozoa. (G) A Sertoli cell (white arrowhead) next to a spermatogonia cell. (H) A Leydig cells (black arrowheads) outside the tubule. Black arrowheads = Leydig cells; white arrowheads = Sertoli cells. Scale bar = 10 μM

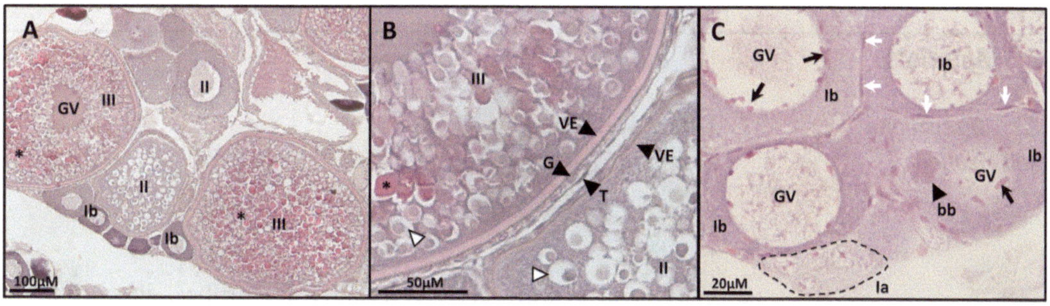

Fig. 2 Histology of Zebrafish Ovaries Visualized with Hematoxylin and Eosin Staining. (**a**) Meiotic oocytes (Stages Ib–III) and somatic cells are visible in varying stages of development. Oocytes grow drastically in size as they progress through meiosis I. (**b**) Higher resolution image of stage II and III oocytes showing the vitellin envelope (VE) and granulosa (G) and theca (T) cell layers. (**c**) Pre-follicle (Ia, dotted outline) and early follicle stage oocytes (Ib) are shown. Follicle stages are characterized by the following histological features: Stage 1a = Pre-follicle, oogonia are present in nests and are not yet separated by follicle cells. Stage Ib = Early follicles have a large lightly stained germinal vesicle (GV) with perinuclear granules (black arrow). Somatic follicle cells envelop each oocyte (white arrowheads). Stage II = Corticle alveolus stage: corticle alveoli (i.e., corticle granules) are apparent (white arrowhead). Stage III = Vitellogenic stage: yolk vesicles form and stain a bright pink color (asterisk). The vitelline envelope (VE) becomes more prominent and stains light pink. *GV* germinal vesicle, *asterisk* yolk vesicle, *VE* vitelline envelope, *G* granulosa cell, *T* theca cell, *white arrowhead* corticle granule, *bb* balbiani body, *dotted line* follicle nest containing Ia oocytes, *white arrow* follicle/granulosa cells, *black arrows* perinuclear granules

publications that describe histological gonad development and mutant defects (e.g., [4–10]). The zebrafish testis is organized into tubules each containing cysts of developing sperm cells. The germ cells within each cyst are derived from a single spermatogonia cell and proceed through spermatogenesis synchronously. The Sertoli cells are stretched thinly to encase the germ cell cysts and therefore Sertoli cell nuclei appear sparse and are difficult to find in sectioned testes (Fig. 1, white arrowheads). The Leydig cells are found in interstitial regions (i.e., outside of the testis tubules) (Fig. 1, black arrowheads). Spermatogonial cells are relatively large cells and have large nuclei with little hematoxylin staining due to decondensed chromatin (Fig. 1b). As cells proceed into meiosis, the nuclei become more strongly stained (Fig. 1a, c–e). Nuclear and cell size continues to decrease as spermatogenesis proceeds (Fig. 1a–f). The ovary is composed of oocytes that increase in size as they continue to mature and accumulate yolk (Fig. 2). Each oocyte is found within a follicle surrounded by a layer of granulosa and theca cells (Fig. 2b). Between this somatic cell layer and the oocyte is the zona pellucida, which stains pink from the eosin, and will later form the chorion (Fig. 2b). Prior to follicle formation, the oogonia are present in nests—these cells are extremely small relative to later-stage oocytes and therefore difficult to find in sections of large ovaries containing many late-stage oocytes (Fig. 2c).

2 Materials

2.1 Dissection and Fixation

1. Dissection tools.
   - No. 4 size forceps.
   - Small spring scissors.
   - Scalpel.
2. Glass scintillation vials or 2 ml microfuge tubes, depending on the sample size.
3. Bouin's fixative.
4. Davidson's fixative (215 ml): 50 ml 37% formalin, 75 ml 95% EtOH, 25 ml glacial acetic acid, 75 ml H_2O.
5. Phosphate-buffered saline (PBS).
6. Decalcification solution: 0.5 M EDTA, pH 8.
7. 190 proof EtOH.
8. 200 proof EtOH.
9. Paraplast plus.
10. Citrisolv, or other xylene substitute.
11. Modified Harris Hematoxylin (Fisher SH26-500D).

12. Eosin Y stock solution: 10 g eosin Y, 200 ml distilled H_2O, 800 ml 95% EtOH (store at room temperature, protected from light).
13. Eosin Y working solution: 55 ml eosin stock solution, 165 ml 80% EtOH, 1.25 ml acetic acid.
14. Acid alcohol: 2 ml 12 N HCl, 200 ml 70% EtOH.
15. Bluing solution: 1 ml 1 N ammonium hydroxide, 200 ml H_2O.
16. 0.5% Periodic Acid: 0.5 g Periodic Acid, 0.5 g Periodic Acid (make fresh before use).
17. Shiff's solution.
18. Mayer's Hematoxylin.
19. Mounting medium (e.g., permount).
20. Plain slides (optional).
21. Charged Slides (e.g., Colorfrost plus).
22. Coverslips.
23. Incubator capable of holding temperature at 60 °C.
24. Rotary microtome.
25. Slide warmer.
26. Tissue floatation bath.
27. Alcohol burner.
28. Staining dishes/jars.

3 Methods

Wash step time is dependent on the sample size. For whole trunks from adults, washes are carried out minimally for 30 min with gentle agitation at room temperature unless otherwise noted. Wash times can be decreased depending on the sample size.

3.1 Fixation

1. Prepare glass scintillation vials with the desired fixative (*see* **Note 1**). The volume of fixative should be minimally 10× the sample volume.
2. Euthanize fish using institute approved method [11].
3. Remove the head and tail by cutting with a scalpel just caudal to the pectoral fins and at or slightly rostral to the anal pore. Immediately place the entire trunk into fixative in a glass scintillation vial (*see* **Note 2**). Fix overnight up to 24 h at room temperature (*see* **Note 3**).
4. Proceed to decalcification step or, if decalcification is not necessary, wash samples 3× in 70% EtOH (diluted with water) (*see* **Note 4**).
5. Samples can be stored in 70% EtOH.

3.2 Decalcification (Not Necessary for Bouin's Fixed Samples or Isolated Gonads)

1. After fixation, wash samples 3× 30 min in PBS.
2. Transfer to decalcification solution (0.5 M EDTA, pH 8) and incubate at 7–14 days at room temperature.
3. After decalcification wash 3× in PBS.
4. Dehydrate samples by washing 25%, 50%, then 70% EtOH (*see* **Note 5**).
5. Samples can be stored in 70% EtOH.

3.3 Infiltration and Embedding

1. If desired, isolate gonads from the trunk by dissecting in 50% EtOH then place isolated gonads into 70% EtOH in 2 ml microfuge tubes and incubate with agitation for 10 min (whole ovaries may need a larger tube or vial size).
2. Dehydrate samples completely with 2× 95% washes followed by 2× 100% EtOH washes.
3. Wash 1× in freshly prepared 50/50 EtOH/Citrisolv.
4. Wash 3× in Citrisolv.
5. Remove the Citrisolv and replace with approximately 1/3–1/2 the tube volume fresh Citrisolv.
6. Add an approximately equal volume of molten paraffin, mix gently and incubate for 2 h to overnight at 60 °C.
7. Over the next 1–2 days (minimum 3 changes), remove the paraffin and replace with fresh molten paraffin twice per day, once in the morning and once in the evening (*see* **Notes 6** and **7**). After changing solutions mix the tissue gently by swirling or inverting before placing back into the incubator.
8. To embed samples, place molds on a warm hotplate (~60 °C) and nearly fill with molten paraffin (*see* **Note 8**).
9. Gently pour the sample into the mold and fill as full as possible with additional molten paraffin.
10. Allow to stand for a minute or two to allow bubbles to rise to the surface.
11. Remove the mold from the heat block and place on a hard level surface. Use warmed utensils (dissection needles/forceps) to move the sample to the middle of the mold and orient as desired. Use an ethanol burner, or similar, for warming utensil during this process (*see* **Note 9**).
12. Allow the paraffin to cool and harden without disturbing (*see* **Note 10**).
13. Paraffin blocks can be stored at room temperature indefinitely.

3.4 Sectioning

1. Trim off excess paraffin using a razor blade (*see* **Note 11**).
2. Using a rotary microtome cut the desired thickness of sections (usually 5 μm) (*see* **Note 15**).

3. Float sections in a tissue floatation bath set to 42 °C for ~1 min to flatten out the sections (*see* **Note 12**).
4. Mount sections onto charged slides (*see* **Note 13**).
5. Label slides using a pencil and dry on a slide warmer at 42 °C overnight (*see* **Note 14**).

3.5 Histological Staining

3.5.1 Hematoxylin and Eosin (see Note 16)

1. Processes slides in staining jars or dishes.
2. To remove paraffin, incubate 3× 5 min in Citrisolv.
3. Wash 2× 5 min in 100% EtOH.
4. Wash 2× 5 min in 95% EtOH.
5. Wash 2× 5 min tap H_2O.
6. Incubate 8 min in Modified Harris hematoxylin.
7. Transfer slides to tap water.
8. Rinse with gentle stream tap water until water is clear (~1–5 min) (*see* **Note 17**).
9. Add slides briefly (10–30 s) to acid alcohol.
10. Wash briefly in tap water.
11. Put slides into bluing solution until they start to turn blue (~10 s) (*see* **Note 18**).
12. Rinse briefly in tap water.
13. Add slides to eosin for 1 min.
14. Wash 95% EtOH for 3 min.
15. Wash 3× 5 min in 100% EtOH.
16. 3× 5 min in Citrisolv.
17. Keep the slides in Citrisolv and coverslip one at a time.
18. Remove slide from Citrisolv and briefly drain excess Citrislov onto a paper towel, but do not allow the sections to dry.
19. Add 2–3 drops of mounting medium to the slide and slowly lower a coverslip using forceps (*see* **Note 19**).
20. Allow the slides to dry overnight.

3.5.2 Periodic Acid Schiff's With Hematoxylin (see Note 16)

1. Process slides in staining jars or dishes.
2. To remove paraffin, incubate 3× 5 min in Citrisolv.
3. Wash 2× 5 min in 100% EtOH.
4. Wash 2× 5 min in 95% EtOH.
5. Wash 2× 5 min in distilled water.
6. Oxidize in 0.5% periodic acid for 5 min.
7. Rinse briefly (10–30 s) in distilled water.
8. Incubate in Schiff's reagent for 15 min.

9. Wash in tap water 5 min.
10. Counterstain with Mayer's Hematoxylin for 1 min.
11. Transfer slides to tap water.
12. Rinse with gentle stream tap water until water is clear (~5 min) (*see* **Note 17**).
13. Wash 5 min each in 30% EtOH, 50% EtOH, 70% EtOH, then 3× 5 min in 100% EtOH and 3× 5 min in Citrisolv.
14. Keep the slides in Citrisolv and coverslip one at a time.
15. Remove slide from Citrisolv and briefly drain excess Citrisolv onto a paper towel, but do not allow the sections to dry.
16. Add a 2–3 drops of mounting medium to the slide and slowly lower a coverslip using forceps (*see* **Note 19**).
17. Allow the slides to dry overnight.

4 Notes

1. Bouin's fixative or Davidson's fixative is recommend for optimal histology. If antibody labeling may be carried out on the tissue, Davidson's fixative is recommended.
2. For histology of isolated gonads, euthanize fish and remove the head and tail as described above. Cut along the belly using scissors and remove the viscera using fine forceps, the testis/ovary should remain attached to the body wall. Using fine forceps, poke a hole in each lobe of the swim bladder, which will allow the sample to sink and facilitate processing. Immediately place the eviscerated trunk into fixative in a glass scintillation vial. Fix overnight up to 24 h at room temperature. Smaller samples can be fixed for fewer hours if desired. Dissect gonads from the trunk using fine forceps after washing out fixative, as described above. Alternatively, the whole trunk can be fixed and viscera removed when dissecting out the gonads.
3. If carrying out antibody labeling on the tissue in addition to histology be careful not to overfix the samples. Smaller samples can be fixed at 4 °C for up to 24 h to avoid overfixation.
4. Decalcification is necessary for larger whole-trunk samples fixed in Davidson's fixative or other fixative with low/no acid content. Smaller fish with thin bones or samples without bone do not require decalcification. The high acid content in Bouin's fixative sufficiently softens bones without decalcification.
5. Wash times for whole tissues depend on the sample size. Isolated gonads can be incubated for 15–30 min per wash (longer wash times are needed for ovaries than testes) while

whole trunks should be incubated for 30–60 min. All wash steps can be carried out for longer than the times listed.

6. For incubations in molten paraffin, we use a standard incubator for these steps and do not agitate the samples during incubations.
7. When processing samples in molten paraffin, if processing samples in microfuge tubes, removal of molten paraffin can be carried out with a glass pasture pipet with the tip broken off—warm the pipet using an ethanol burner to avoid paraffin solidifying during pipetting. If processing many samples in microfuge tubes, it can be helpful to keep samples in a metal block (such as those designed for dry baths). The entire block can be removed from the incubator and placed on a warm hotplate while changing paraffin.
8. Choose embedding molds that are suitable for the microtome that will be used and the sample size.
9. To orient whole trunks during embedding, stand the tissue on the desired surface using warmed forceps or dissecting needle (e.g., anterior surface for transverse sections or dorsal surface for longitudinal sections) as the paraffin cools, until it no longer falls over (may take up to a minute). An alcohol burner can be used to warm utensils.
10. Paraffin blocks can be labeled directly on the embedding mold or by fixing a paper label in one corner of the paraffin before it cools. If embedding many samples typing up labels ahead of time can save time.
11. Prior to sectioning, paraffin blocks can be cooled on ice to help attain smoother sections.
12. When sectioning, long ribbons can be cut and later separated to the desired size to fit on the slides. Separate sections by laying on a clean piece of paper and cutting using a chilled razor (chill by keeping in on ice). Transfer to the floatation bath using clean forceps. Alternatively, ribbons can be floated in a room temperature water bath and teased apart using forceps, then transferred to the tissue floatation bath using a plain uncharged microscope slide.
13. To identify sections containing the desired tissues, sections can be viewed briefly using a plain, non-charged, glass slide to gently lift from the water bath and view on a microscope—do not let the sections dry on the slide during this time. After viewing, place sections back into the tissue floatation bath then mount onto charged slides.
14. Labeling slides with a pencil is important as the label will be resistant to chemicals used for processing the slides.

15. Mature ovaries embedded in paraffin are sometimes difficult to section due to numerous large yolk-filled oocytes that can interfere with paraffin infiltration. Be sure to carry out all dehydration wash steps for the maximum time (or extend the time) and allow enough time for the tissue to be fully infiltrated. If large oocytes are "crumbling" during sectioning, it is often helpful to make slightly thicker sections. Resin-embedded ovaries do not have this problem.
16. Generally, solutions used for slide staining can be reused multiple times, except tap water. Store solutions in sealed containers between use to avoid evaporation. When staining becomes more faint than desired, replace the staining solutions with fresh working solutions. Replace solutions used after staining steps after too much stain accumulates.
17. The hematoxylin wash step should be carried out such that excess stain is washed out but adequate nuclear stain remains. When first optimizing this step, one can briefly check the slides on a microscope and either continue washing if not enough stain is washed from the tissue or add back to the stain if too much stain was washed from the tissue.
18. Watch slides in the bluing solution carefully and remove as soon as blue color is visible. This may take between 5 and 30 s.
19. If there is a large bubble in the mounting medium after adding the coverslip, remove the coverslip by immersing in Citrisolv to remove the coverslip and re-coverslip with mounting medium and a fresh coverslip.

References

1. Leal MC, Cardoso ER, Nóbrega RH, Batlouni SR, Bogerd J, França LR et al (2009) Histological and stereological evaluation of zebrafish (*Danio rerio*) spermatogenesis with an emphasis on spermatogonial generations. Biol Reprod 81:177–187. https://doi.org/10.1095/biolreprod.109.076299
2. Draper BW, Siegfried KR (2019) The reproductive system. In: Cartner SC, Eisen JS, Farmer SC, Guillemin KJ, Kent ML, Sanders GE (eds) The zebrafish in biomedical research: biology, husbandry, diseases, and research applications. Elsevier, Amsterdam
3. Selman K, Wallace RA, Sarka A, Qi X. Stages of oocyte development in the zebrafish, Brachydanio rerio,J Morphol 1993;218: 203–224. doi:https://doi.org/10.1002/jmor.1052180209
4. Houwing S, Kamminga LM, Berezikov E, Cronembold D, Girard A, van den Elst H et al (2007) A rRole for Piwi and piRNAs in germ cell maintenance and transposon silencing in zebrafish. Cell 129:69–82. https://doi.org/10.1016/j.cell.2007.03.026
5. Maack G, Segner H (2003) Morphological development of the gonads in zebrafish. J Fish Biol. https://doi.org/10.1046/j.1095-8649.2003.00074.x
6. Rodríguez-Marí A, Wilson C, Titus TA, Cañestro C, BreMiller RA, Yan Y-L et al (2011) Roles of brca2 (fancd1) in oocyte nuclear architecture, gametogenesis, gonad tumors, and genome stability in zebrafish. PLoS Genet 7:e1001357. https://doi.org/10.1371/journal.pgen.1001357
7. Webster KA, Schach U, Ordaz A, Steinfeld JS, Draper BW, Siegfried KR (2017) Dmrt1 is necessary for male sexual development in zebrafish. Dev Biol 422:33–46. https://doi.org/10.1016/j.ydbio.2016.12.008
8. Saito K, Siegfried KR, Nüsslein-Volhard C, Sakai N (2011) Isolation and cytogenetic

characterization of zebrafish meiotic prophase I mutants. Dev Dyn 240:1779–1792. https://doi.org/10.1002/dvdy.22661

9. Braat AK, Zandbergen T, Van De Water S, Goos HJT, Zivkovic D (1999) Characterization of zebrafish primordial germ cells: morphology and early distribution of vasa RNA. Dev Dyn 216:153–167. https://doi.org/10.1002/(SICI)1097-0177(199910)216:2<153::AID-DVDY6>3.0.CO;2-1
10. Gupta T, Marlow FL, Ferriola D, Mackiewicz K, Dapprich J, Monos D et al (2010) Microtubule actin crosslinking factor 1 regulates the balbiani body and animal-vegetal polarity of the zebrafish oocyte. PLoS Genet 6:e1001073. https://doi.org/10.1371/journal.pgen.1001073
11. Köhler A, Collymore C, Finger-Baier K, Geisler R, Kaufmann L, Pounder KC et al (2017) Report of workshop on euthanasia for zebrafish – a matter of welfare and science. Zebrafish. https://doi.org/10.1089/zeb.2017.1508

Chapter 21

Manipulating and Visualizing the Germline with Transgenic Lines

Ding Ye and Yonghua Sun

Abstract

Transgenic zebrafish in which the germline is specifically labeled with enhanced green fluorescent protein (eGFP) can be used for continuous observation of germline development during the lifetime, from the primordial germ cells (PGCs) in the early embryo to the gametes in the mature gonad. In this chapter, we describe a procedure for the generation of transgenic fish *Tg(piwil1:egfp-UTRnanos3)*, the sample preparation for live imaging of PGCs, for high-resolution imaging of germ cells in developing gonads, and quantifying PGC numbers. The methods described in this chapter are not only applicable to the study of germ cells, but also provide general advices for researchers who are willing to generate transgenic zebrafish and do observation on live embryos as well as on fixed tissues.

Key words Germline, Transgenic fish, Live imaging, Confocal imaging, Sample preparation

1 Introduction

Manipulating and visualizing the germline largely relies on transgenic zebrafish whose germline is specifically labeled with fluorescent proteins. There are several transgenic lines available for the visualization of germ cells, such as the transgenic line driven by promoters of *askop* [1, 2], *ddx4* [3], *piwil1* [4], *zp3b* [5], *buc* [6], etc. In addition, the gonadal somatic cell can also be labeled using promotors of *gsdf* [7] and *cyp19a1a* [8]. Here, in this chapter, we take *Tg(piwil1:egfp-UTRnanos3)* [9] as an example.

Tg(piwil1:egfp-UTRnanos3) labels the germline cells as early as a few hours post-fertilization till sexual maturation in adults. From the embryonic stage to larval stage (5 h post-fertilization (hpf) to 20 dpf) in zebrafish, it is possible to directly observe the germ cells under a fluorescent microscope in live fish because the pigment cells do not cover the gonadal area. However, in order to manipulate and visualize the gonads after 21 dpf, it is necessary to dissect fish to image the gonad. During the lifetime of the transgenic zebrafish, the germ cells are specifically labeled with eGFP, and different types

Roland Dosch (ed.), *Germline Development in the Zebrafish: Methods and Protocols*, Methods in Molecular Biology, vol. 2218, https://doi.org/10.1007/978-1-0716-0970-5_21, © Springer Science+Business Media, LLC, part of Springer Nature 2021

of germ cells are distinctly labeled with eGFP of different intensities. From 5 h to 2-week after fertilization, the eGFP labels the undifferentiated germ cells; in the larval gonads, the eGFP can label the oogonia and primary oocytes in the ovary, it labels the type A spermatogonia, primary spermatocytes, secondary spermatocytes, and all types of sperm cells but shows very weak expression in type B spermatogonia. This feature provides the opportunity to identify different germ cell types based on morphological criteria. Here, in this chapter, we will describe a step-by-step procedure on generation of transgenic fish, and the manipulation and visualization of embryonic germ cells and gonads of the *Tg(piwil1:egfp-UTRnanos3)* transgenic zebrafish.

2 Materials

1. Zebrafish:AB strain (*see* **Note 1**).
2. Tol2-plasmid containing transgene: 50 ng/μL pTol2(piwil1: egfp-UTRnanos3).
3. 100 ng/μL *transposase* mRNA.
4. Mating tanks.
5. Watch tweezers with ultra-fine tips.
6. Glass plate.
7. Dissecting scissors and dissecting needle.
8. 30× Danieau's buffer: 101.7 g NaCl, 1.56 g KCl, 2.96 g $MgSO_4 \cdot 7H_2O$, 2.65 g $CaCl_2 \cdot 2H_2O$, HEPES 35.75 g, fill to 1 L with distilled water, the pH value of the buffer should be 7.5, store at 4 °C.
9. Embryo medium: 0.3× Danieau's buffer. Add 100 mL 30× Danieau's buffer to 900 mL distilled water, agitate to mix it well. Freshly prepare it and store at room temperature (25 °C).
10. 1.5% agarose mold for injection (Fig. 1): Weigh 0.75 g agarose powder, transfer into a 250 mL conical flask, add 50 mL embryo medium, heat in a microwave oven using medium power until just starting boiling, take out and shake it gently, heat to boil it one more time, pour it in a 70 mm Petri dish, float a commercial injection plate mold (World Precision Instrument), wait for 30 min till it cools down, and take out the plate mold.
11. Blunt glass needle: burn used microinjection needles on an alcohol lamp.
12. Glass needles for microinjection: microinjection needles are made from capillaries with filament inside (TW100F-4,

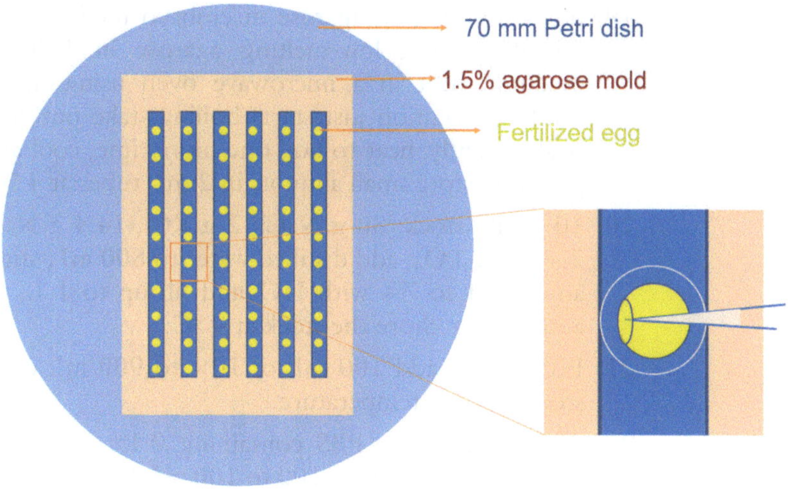

Fig. 1 Scheme of injection plate. Fertilized eggs are aligned in the groove of the injection plate with animal pole to the left. After the blastodisc forms at the animal pole, inject the transgenic plasmid directly into the cytoplasm of the blastodisc cell. Insert the injection needle through the yolk until it finally reaches into the cell as illustrated in the magnified scheme on the right

1/0.75 OD/ID (mm), World Precision Instrument) using a Micropipette Puller according to the manufacturer's manual.

13. 20× anesthetic stock solution: 4 mg/mL MS-222. Weigh 400 mg tricaine and put into 95 mL water, adjust pH with Tris-HCI (pH 9) to 7.0–7.5, make up to 100 mL. Store in small aliquots in 1.5 mL microcentrifuge tubes at −20 °C.
14. Anesthetic medium: 0.2 mg/mL MS-222 in embryo medium. Dilute the 20× stock solution into 1× working solution by adding 5 mL into 95 mL embryo medium.
15. 10× dechorionation solution: 20 mg/mL pronase. Dissolve 2 g pronase, in 100 mL embryo medium. Store small aliquots in 1.5 mL microcentrifuge tubes at −20 °C.
16. Fixation buffer: 4% paraformaldehyde (PFA) in 1× PBS. Dissolve 40 g PFA powder in 800 mL distilled water, add several drops of 1 M NaOH to help dissolve the PFA. Then add 100 mL 10× PBS, adjust pH to 7.5, make up to 1 L with distilled water. Store aliquots in 15- or 50-mL tube at −20 °C. Before use, thaw and pre-warm aliquot at 28 °C. After the first use, store the remaining solution at 4 °C not longer than 1 week.
17. 1000× PTU solution: dissolve 0.45 g 1-phenyl-2-thiourea in 10 mL of ethanol and store at −20 °C.
18. 10% Triton X-100:measure 45 mL distilled water in a 50 mL conical centrifuge tube with scale, add 5 mL TritonX-100, vortex well, wrap with aluminum foil and store at 4 °C.

19. 0.5% low melting agarose in embryo medium (*see* **Note 2**): Dissolve 0.5 g low-melting agarose in 100 mL embryo medium, heat in a microwave oven using medium power until the solution just starts boiling, take out and shake the beaker gently, heat to boil one more time, cool at room temperature, store small aliquots in 2 mL tubes at 4 °C.
20. 10× PBS stock: 80 g NaCl, 2 g KCl, 14.4 g Na_2HPO_4 and 2.4 g KH_2PO_4, add distilled water to 800 mL, stir to dissolve, adjust pH to 7.4 with HCl and fill up to 1 L, sterilized by autoclaving. Store the stock at 4 °C.
21. PBS buffer: add 100 mL of 10× to 900 mL distilled water, store at room temperature.
22. PBST buffer: 1× PBS containing 0.1% Triton X-100. Add 1 mL of 10% Triton X-100 to 100 mL PBS in a blue cap bottle and shake to mix well.
23. Permeabilization buffer: 1× PBS containing 0.5% Triton X-100. Add 200 μL of 10% Triton X-100 to 4 mL PBS in a 5 mL tube, vortex to mix well.
24. Mounting medium with anti-fade reagent: 50% glycerol-PBS. Add 50 mL 100% glycerol to 50 mL 2× PBS in a Reagent bottle, add 0.2 g propylgallate, shake to mix well, wrap with foil and store at 4 °C.
25. Staining/washing buffer (S/W buffer): 1× PBS with 1% DMSO, 1% BSA and 0.1% Triton X-100. Add 1 g BSA and transfer into 100 mL 1× PBS in a blue-cap reagent bottle, add 1 mL of DMSO and 1 mL of 10% Triton X-100, shake gently to mix well. Prepared freshly just before use, or store aliquots at −20 °C if necessary.
26. Staining solution with phalloidin and DAPI: add 0.2 μL of DAPI (5 mg/mL) and 5 μL of phalloidin (1 mg/mL) to 1 mL of S/W buffer, invert up and down to mix.
27. 3% methylcellulose in embryo medium: add about 100 mL embryo medium in a 100 mL blue-cap reagent bottle, stir the water at high speed with a magnetic bar on magnetic stirring apparatus, weigh 3 g methylcellulose and add carefully bit by bit into the stirring water avoiding the formation of any large lumps; incubate the bottle at 50 °C in a water bath for 30 min, and then cool in an ice bath for 30 min; shake it well; repeat the above steps of heat and cool down one more time; store overnight at 4 °C; prepare 5 mL aliquots on the next day.
28. Anesthetic methylcellulose: add 250 μL of 4 mg/mL MS-222 into 5 mL 3% methylcellulose in embryo medium, invert ten times to mix well.
29. 3-bridge slide: glue three layers of cover glasses (22 mm × 22 mm, 1#, 0.13–0.16 mm thickness) to a cover

glass (24 mm × 60 mm, 1#, 0.13–0.16 mm thickness) by a UV glue (Fig. 2a, b) and expose it under UV light for 5 min or until set.

30. Stereomicroscope: e.g., Olympus SZ61 or equivalent.
31. Macro Zoom Fluorescence microscope: e.g., Olympus MVX10 or equivalent.
32. Confocal microscope: e.g., Leica SP8 or equivalent.

3 Methods

3.1 Generation of Germ-Line Labeling Transgenic Zebrafish

3.1.1 Harvest Fertilized Eggs for Microinjection

1. At 5 pm the day before the spawning day, fill the mating tank with system water reaching about 2 cm to the cover.
2. Put two wildtype females and two wildtype males in the tank, separate the male and the female by a separator and keep them in the fish room at 28.5 °C.
3. At 8 am on the spawning day, replace the water in the mating tank with fresh fish system water. The water level should be 2 cm above the fish body. Remove the separator, wait 30 min and check whether there are eggs on the bottom.
4. Collect the eggs with a net and carefully transfer them to a 70 mm Petri dish containing embryo medium.

3.1.2 Injection of Transgenic Constructs (Fig. 1)

1. Align the embryos with chorion into the trenches of the agarose mold and cover them with embryo medium.
2. With a blunt glass needle orient the embryos with animal pole to the left if you use the right hand to control the injector.
3. Load *transposase* mRNA from the end of the microinjection needle, break the tip, and inject 1 nL of 100 ng/μL transposase mRNA into the yolk of the aligned embryos. Wait until the animal pole expands.
4. Load the transgene plasmid from the end of another microinjection needle, break the tip (*see* **Note 3**), and inject 1 nL of 50 ng/μL transgene plasmid into the cytoplasm of the zygotes (*see* **Note 4**).
5. Flush out the injected embryos with embryo medium from the mold by pipetting with a Pasteur pipette and raise them in embryo medium in a 28.5 °C incubator.
6. Sort out the dead embryos in the afternoon and change half of the embryo medium. On day 3, sort out the eGFP-positive embryos under the fluorescent microscope and raise them to adulthood.

3.1.3 Raising Transgenic Founders

1. Outcross potential founders to wildtype fish and obtain their fertilized embryos.
2. Screen the embryos at 3 pf for eGFP-positive PGCs.
3. Screen embryos again between 14 and 25 days pf to remove fish with non-specific eGFP in other regions than the germinal ridge, such as abdomen, trunk, head, etc.
4. Raise the positive adult founders in labeled fish tanks (*see* **Note 5**).

3.1.4 Maintenance of Tg(piwil1: egfp-UTRnanos3) Transgenic Line

1. At the juvenile stage between 14 and 25 dpf, sort the germline-eGFP-positive fish.
2. Anesthetize the juvenile fish in a 70 mm Petri dish containing anesthetic medium for 5 min.
3. Sort the positive fish under a fluorescent microscope.
4. Immediately transfer the eGFP-positive juvenile fish into a container filled with fresh fish system water (*see* **Note 6**).

3.2 Observation on Live Embryo

1. Transfer about 100 fertilized eggs with a Pasteur pipette into a 35 mm Petri dish with 2 mL embryo medium, add 200 μL of 10× dechorionation solution (Subheading 2, **item 10**), incubate 5 min at room temperature, gently agitate the plate, and watch under the stereomicroscope.
2. As soon as there are a few embryos out of the chorion after agitation, pour all the embryos into a 250 mL beaker containing 200 mL fish system water.
3. Remove most of the water, and add fresh fish system water along the wall of the beaker to wash the embryos.
4. After repeating this washing step for six times, most of the embryos should be out of chorion.
5. Carefully pick embryos with a fire-polished Pasteur pipette and transfer them into a 70 mm Petri dish containing embryo medium (*see* **Note** 7).
6. Keep the embryos in an incubator at 28.5 °C.
7. At 5 pm on the spawning day, remove the dead embryos and unfertilized eggs under the stereomicroscope with a fire-polished Pasteur pipette, and change half of the embryo medium.
8. To image embryos beyond 35 hpf, add 10 μL of PTU stock solution to a dish (70 mm) of embryos before 24 hpf to inhibit the formation of pigment.

3.3 Mounting Embryos for PGCs Imaging (Fig. 2)

1. Add 2 mL of anesthetic medium to a 35 mm Petri dish.
2. Select embryos with visible PGCs under fluorescent microscope.

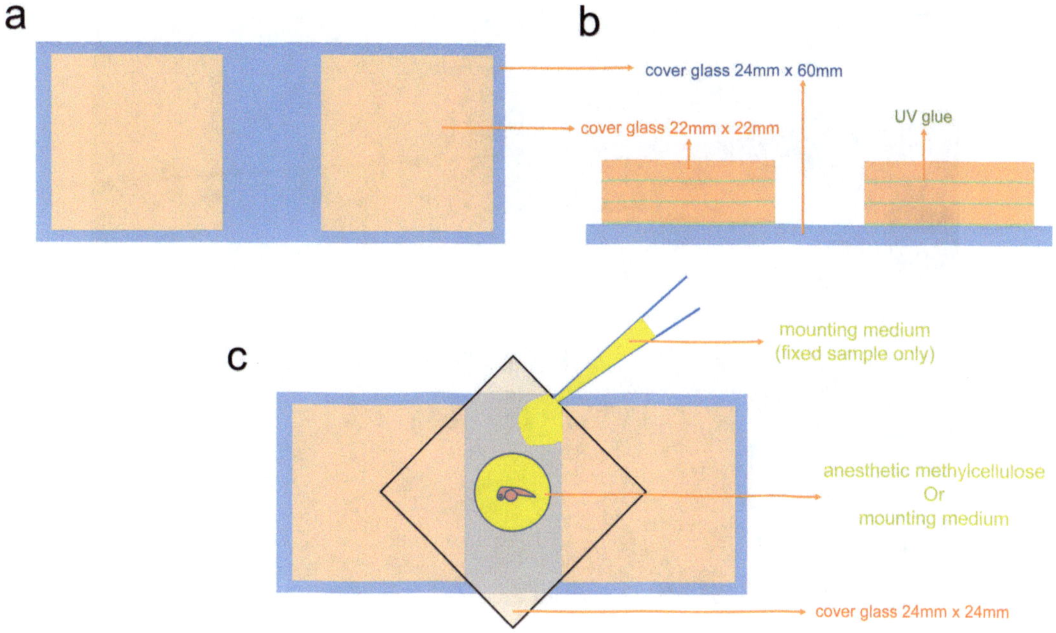

Fig. 2 Preparation of a 3-bridge slide for mounting of the embryo. (a, b) Scheme showing the top view (a) and lateral view (b) of a 3-bridge slide. (c) Scheme illustrating mounting of an embryo on a 3-bridge slide and how mounting medium is added from the edge of the slide

3. Wait for about 5 min to anesthetize the embryos.
4. Put a drop of anesthetic methylcellulose in the center of 3-bridge glass slide.
5. Transfer one embryo with plastic pipette into a tube containing the anesthetic methylcellulose.
6. Then transfer this embryo from the tube to the center of a 3-bridge glass slide and cover it with a coverslip (24 mm × 24 mm) (Fig. 2c).
7. Watch and position the embryos by moving the cover glass under the stereomicroscope. Finally, this embryo can be used for imaging on a fluorescent microscope (Fig. 3a, b).

3.4 Imaging Fixed PGCs

1. Thaw one tube of frozen fixation buffer from −20 °C at 28.5 °C for about 30 min.
2. Transfer no more than 20 dechorionated embryos into 1 mL of pre-warmed fixation buffer in a 1.5 mL microcentrifuge tube.
3. Add 1 μL of 10% Triton X-100 to prevent embryos sticking to each other and invert the tube several times.
4. Wrap the tube with aluminum foil.
5. Agitate the tube gently on a rotator at room temperature for 1 h.

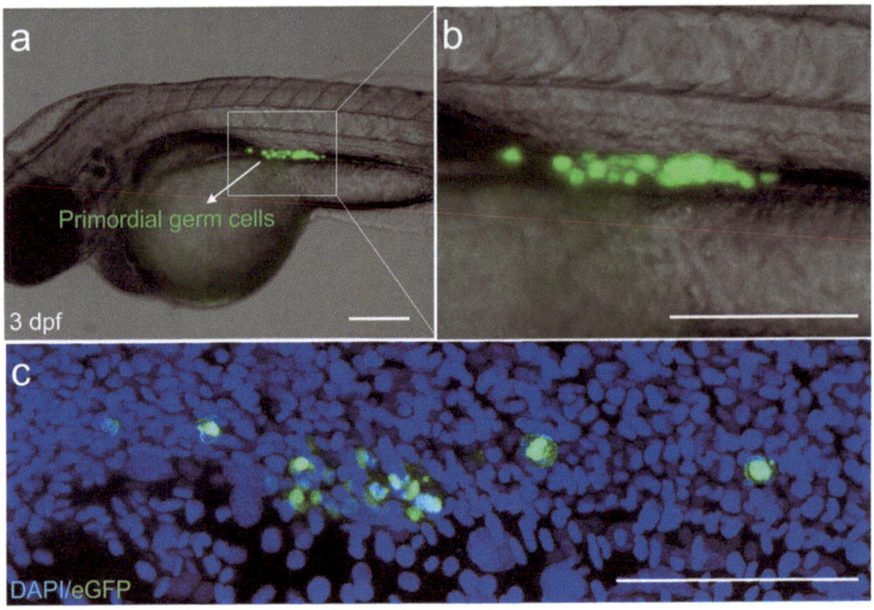

Fig. 3 Representative images of embryonic PGCs. (**a**) eGFP-positive PGCs at 3 dpf in a live embryo of *Tg(piwil1: egfp-UTRnanos3)*. The embryo is at a lateral view with dorsal to the top. (**b**) Magnification of the white box in (**a**) showing the PGCs. (**c**) Fixed eGFP-positive PGCs (green) and cell nuclei (DAPI, blue). Scale bars in all of the images: 100 μm

6. Remove the fixation buffer with a Pasture pipette and wash the embryos with PBST buffer for 3 × 5 min with gentle agitation.
7. Transfer the embryos into distilled water and incubate on a rotator for 30 min.
8. Change the water with permeabilization buffer and incubate on a rotator for 20 min.
9. Wash with PBST buffer for 3 × 5 min with gentle agitation on a rotator.
10. Change the PBST buffer with staining solution containing phalloidin and DAPI and incubate at room temperature for 2 h.
11. Wash with S/W buffer 4 × 15 min with gentle agitation on a rotator.
12. Transfer the stained embryos into the mounting medium with anti-fade reagent in a 1.5 microcentrifuge tube wrapped with foil (*see* **Note 8**) and incubate at 4 °C for at least 5 h (*see* **Note 9**).

3.5 Imaging Stained Larval Gonads

1. Transfer transgenic zebrafish at 30–60 dpf to a 70 mm Petri dish containing 1× PBS with 0.2 mg/mL of MS-222.
2. Wait for about 5 min to anesthetize the animals.

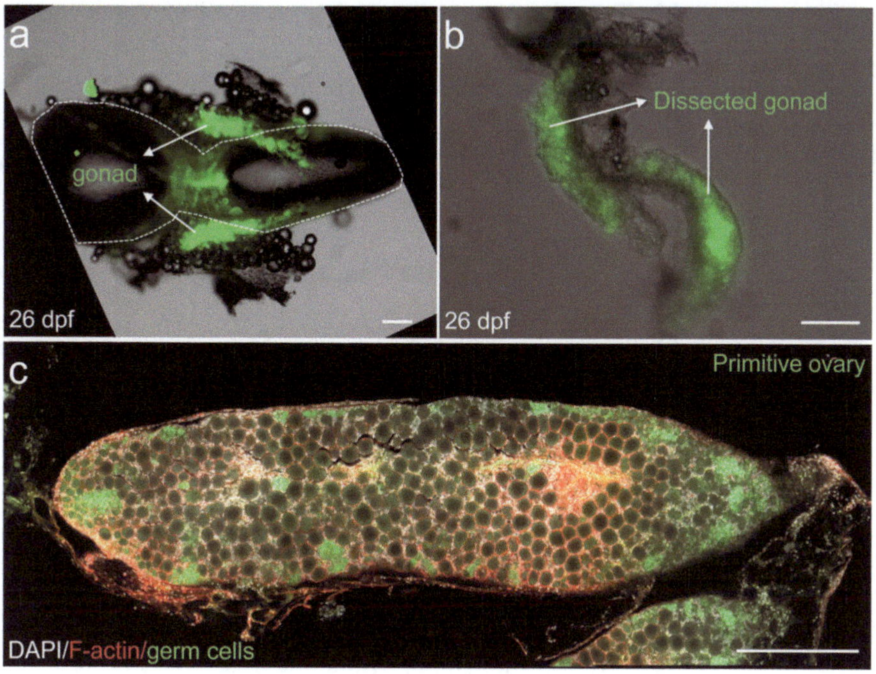

Fig. 4 Gonad in juvenile fish at 26 dpf. (**a**) Fluorescent gonads locate on both sides of the swimming bladder (outlined with stippled line). This image only shows the ventral view of gonads with a neighboring swimming bladder, while the rest of the fish body is removed. (**b**) Dissected intact gonads. (**c**) A confocal image of the fixed primitive ovary from a 26-dpf *Tg(piwil1:egfp-UTRnanos3)* transgenic zebrafish with labeled germ cells (eGFP, green), cortical actin (phalloidin, red), and cell nuclei (DAPI, white). Scale bars in all of the images: 100 μm

3. Under the stereoscope, use a pair of microscissors to open the abdomen from the cloacal pore toward the head until you reach the pectoral fins.
4. With watch tweezers grip the viscera (mostly the gut and liver) to expose the swim bladder. Cut off the viscera with microscissors. Two stripes of gonads are located at both sides of the swim bladder (Fig. 4a).
5. With watch tweezers, grip the front end of the swimming bladder and carefully remove it.
6. Put the remaining fish with gonads into a 5 mL centrifuge tube containing 3 mL of fixation buffer.
7. Add 3 μL of 10% Triton X-100 and incubate at room temperature for 2 h with agitation on a rotator.
8. Wash the fixed fish with PBST buffer for 3 × 10 min.
9. Transfer the animal into a 35 mm Petri dish containing PBST buffer.
10. Take out the gonad with watch tweezers and remove fat tissue as much as possible (Fig. 4b).

11. Transfer the fixed gonads into distilled water and rotate for 30 min.
12. Exchange the water with permeabilization buffer and rotate for 20 min.
13. Wash with PBST buffer for 3 × 5 min with gentle agitation on a rotator.
14. Change the PBST buffer with staining solution containing phalloidin and DAPI and incubate at room temperature for 2 h.
15. Wash with S/W buffer for 4 × 15 min with gentle agitation on a rotator.
16. Transfer the stained gonads into the mounting medium in a 1.5 microcentrifuge tube wrapped with foil (*see* **Note 8**) and incubate at 4 °C for at least 5 h (*see* **Note 9**).

3.6 Confocal Imaging of Gonads and Counting PGCs

1. Transfer an embryo or a gonad with a Pasteur pipette along with the mounting medium to the center of a 3-bridge slide for confocal imaging.
2. Position the sample with a fine dissecting needle.
3. Cover the sample with a #1 coverslip and fill the chamber with the mounting medium (*see* **Note 9**).
4. Take confocal images for counting embryonic PGCs using a 20× objective (Fig. 3c) (*see* **Note 10**). A set of reference scanning parameters is: 2048 × 2048 for frame size; bi-direction mode; 400 MHz for scanning speed; 1 AU for pinhole.
5. Take confocal images of the gonads using either 20× objective (Fig. 4c) or 40× objective (*see* **Note 11**). A set of reference scanning parameters is: 2048 × 2048 for frame size; bi-direction mode; 400 MHz for scanning speed; 1 AU for pinhole.
6. Import image stacks into Fiji software.
7. Count the PGCs numbers with the cell counter plugin of Fiji (*see* **Note 12**).
8. Export the data to Excel for calculation and statistical analysis.

4 Notes

1. The zebrafish of AB strain are raised under the condition of a 14-h light/10-h dark cycle. For good reproductive performance, the zebrafish should be fed twice with live fairy shrimp and once with dry food per day.
2. Wear baker's glove to handle the hot agarose solution.

3. The open end of the needle should be ultra-fine to break the chorion and plasma membrane without hurting it.
4. The injection can be performed in embryos from 1-cell stage to 2-cell stage.
5. They could be used for other rounds of crosses to generate more transgenic fish.
6. To prevent over-anesthesia, finish the whole screening process within 30 min.
7. Make sure there are no more than 50 embryos per dish.
8. The purpose of wrapping foil is to protect fluorescence from bleaching by light. Alternatively, the embryos containing tubes can be stored in a dark box.
9. The samples can be stored at 4 °C in the mounting medium for several weeks.
10. For each embryo, take images for each gonad arm separately. One image is for the PGCs on one side of an embryo, and the other image is for the PGCs on the other side of the same embryo. After taking one image, turn the slide upside down and shoot the other one. The image stacks were taken by setting a scanning range and a HC PL APO 20×/0.75 CS2 objective is used for getting good resolution, adequate working distance, and enough scanning field coving the whole germinal ridge.
11. The HC PL APO 20×/0.75 CS2 objective was used to give consideration to the field of scanning and resolution. If higher resolution is required, the HC PL APO 40×/1.10 water objective may be a good choice for both high resolution and long working distance.
12. The confocal images of gonads can be used for counting different types of cells within a gonad using the cell counter plugin of Fiji software.

Acknowledgments

This work was supported by the National Key R&D Program of China (grant nos: 2018YFD0901205 and 2018YFA0801000), National Natural Science Foundation of China (grant nos. 31721005 and 31872550), Strategic Priority Research Program of Chinese Academy of Sciences (grant no. XDA08010106), and Projects from State Key Laboratory of Freshwater Ecology and Biotechnology (grant no. 2019FBZ05).

References

1. Blaser H, Eisenbeiss S, Neumann M et al (2005) Transition from non-motile behaviour to directed migration during early pgc development in zebrafish. J Cell Sci 118 (17):4027–4038
2. Kardash E, Reichman-Fried M, Maitre JL et al (2010) A role for rho gtpases and cell-cell adhesion in single-cell motility in vivo. Nat Cell Biol 12(1):47–53
3. Krovel AV, Olsen LC (2002) Expression of a vas::Egfp transgene in primordial germ cells of the zebrafish. Mech Dev 116(1–2):141–150
4. Leu DH, Draper BW (2010) The ziwi promoter drives germ line-specific gene expression in zebrafish. Dev Dyn 239(10):2714–2721
5. Onichtchouk D, Aduroja K, Belting HG et al (2003) Transgene driving gfp expression from the promoter of the zona pellucida gene zpc is expressed in oocytes and provides an early marker for gonad differentiation in zebrafish. Dev Dyn 228(3):393–404
6. Riemer S, Bontems F, Krishnakumar P et al (2015) A functional bucky ball-gfp transgene visualizes germ plasm in living zebrafish. Gene Expr Patterns 18(1–2):44–52
7. Gautier A, Sohm F, Joly JS et al (2011) The proximal promoter region of the zebrafish gsdf gene is sufficient to mimic the spatio-temporal expression pattern of the endogenous gene in sertoli and granulosa cells. Biol Reprod 85 (6):1240–1251
8. Dranow DB, Hu K, Bird AM et al (2016) Bmp15 is an oocyte-produced signal required for maintenance of the adult female sexual phenotype in zebrafish. PLoS Genet 12(9): e1006323
9. Ye D, Zhu L, Zhang Q et al (2019) Abundance of early embryonic primordial germ cells promotes zebrafish female differentiation as revealed by lifetime labeling of germline. Mar Biotechnol (NY) 21(2):217–228

Chapter 22

Liquid Chromatography and Tandem Mass Spectrometry in Label-Free Protein Quantification of Zebrafish (*Danio rerio*) Eggs

Ozlem Yilmaz, Emmanuelle Com, Regis Lavigne, Charles Pineau, and Julien Bobe

Abstract

Rapid innovations in core proteomic technologies and proteome-based bioinformatics fortified by recent genome sequencing allow the characterization and quantification of proteins on a global scale. These capabilities empower research to develop a more comprehensive understanding of how changes in protein expression and modification can affect complex signaling and regulatory networks. The consequences of these studies have significant implications for understanding how myriad activities are regulated in biological systems.

Proteomic approaches have been applied to investigate the physiology, developmental biology, and impact of contaminants in fishes as model organisms. Here, we describe the use of label-free protein quantification and global proteome profiling to characterize eggs of different quality grades in the zebrafish.

Key words LC-MS/MS, Protein quantification, Normalized spectral counts, Global proteome, Zebrafish, Eggs

1 Introduction

The present field of proteomics employs rapidly evolving cuttin-g-edge methodologies that enable thousands of proteins to be studied concomitantly in a single experiment. This development provides an opportunity to scrutinize the abundance and dynamics of a vast amount of cellular proteins and their relationship to critically important biological processes. Proteomics is generally a discovery-driven approach where vast quantities of newly acquired data provide details about the system under study, largely without inclination of predictable results [1]. With these large datasets in hand, we seek to better understand how cells and tissues perform under functional and dysfunctional conditions.

Roland Dosch (ed.), *Germline Development in the Zebrafish: Methods and Protocols*, Methods in Molecular Biology, vol. 2218, https://doi.org/10.1007/978-1-0716-0970-5_22, © Springer Science+Business Media, LLC, part of Springer Nature 2021

In most common proteomic studies employing tandem mass spectrometry (MS/MS), proteins are digested by trypsin and generated peptides are separated by liquid chromatography before their ionization and detection by a first analyzer of the mass spectrometer. Some peptide ions are then selected and fragmented by collision-induced dissociation (CID) using an inert gas before their detection in a second analyzer. The MS/MS spectrum provides a measure of the various fragment ions produced by CID, from which partial or complete amino acid sequence information can be obtained. All these ions are separated and detected in the mass spectrometer according to their mass-to-charge ratio (m/z). A mass spectrum of the molecule is thus produced as a plot of ion intensity versus mass-to-charge ratio. Since peptides typically fragment across the amide bond, ensembles of ion series are produced that contain varying lengths of residues originating from the parent peptide. The measured tandem mass spectrum is then compared with theoretical tandem mass spectra calculated from the in silico trypsin digestion of protein sequences from protein databases to obtain the identity of the intact peptide(s) [1, 2]. Both label-based and label-free strategies have been developed in mass spectrometry for the quantification of protein abundances in complex biological systems. Label-based quantitation methods utilize stable isotope labels, which are incorporated into the peptides, introducing an expectable mass difference between two or more experimental conditions. In contrast, label-free methods assess both relative and absolute peptide quantities based on signal intensity and spectral counting procedures [3]. Both labeling and label-free approaches have their own sets of strengths and limitations. The method of choice depends to a large extent on the biological question being addressed, the capabilities of the researcher, and the cost involved, as well as the quality of available MS/MS instrumentation [3].

We used liquid chromatography-tandem mass spectrometry (LC-MS/MS) and label-free protein quantification to study the molecular mechanisms underlying egg quality determination in a study of zebrafish [4], which we will use here as a case study to describe the relevant methodology. For this purpose, fertilized egg samples with different quality grades were selected based on embryonic developmental success from 136 total spawns and differences in their proteomic profiles were studied.

2 Materials

2.1 Sample Collection and Embryo Culture for Egg Quality Assessments

1. Embryo culture medium (modified Yamamoto's embryo solution): 17.1 mM NaCl, 0.4 mM KCl, 0.65 mM $MgSO_4$, 0.27 mM $CaCl_2$, and 0.01 mg/l methylene blue.
2. Liquid nitrogen.

2.2 Tissue Homogenization, Total Protein Extraction and Measurement

1. HEPES buffer (*see* **Note 1**): 20 mM HEPES, 1 mM EDTA, 0.5 mM DTT, 1 mM AEBSF (4-(2-Aminoethyl)benzensulfonylfluoride) (*see* **Note 2**) and 10 μM E-64 (*see* **Note 2**). Adjust pH to 7.4.
2. UreaTrisChaps Buffer: 8 M Urea, 30 mM Tris, 4% CHAPS (pH adjustment is not needed).
3. 200 mM Na_2EDTA.
4. 100 mM DTT.
5. Antiprotease stock solutions: 200 mM AEBSF and 2 mM E-64.
6. Refrigerated centrifuge (e.g., Eppendorf 5415R).
7. Ultracentrifuge (e.g., Beckman Coulter Optima™ L-80 XP).

2.3 Determination of Protein Concentration (Bradford Assay)

1. Coomassie (Bradford) Protein Assay Kit (e.g., Pierce™).
2. Protein standard: Bovine serum albumin stock solution at 2 mg/ml and serial dilutions at a range of 0–1.5 mg/ml.
3. Microtiter plate reader (e.g., Agilent Technologies Multiscan Ascent).

2.4 SDS-PAGE

1. Lower (outer) chamber buffer: 50 ml NuPAGE™ MOPS SDS Running Buffer, 20× MOPS completed to 1 l with MilliQ water (MQ).
2. Higher (inner) chamber buffer: 200 ml of lower chamber buffer mixed with 500 μl NuPAGE™ antioxidant.
3. Sample buffer and stabilizers: (4×) NuPAGE® LDS Sample Buffer and 5 M DTT stock solutions.
4. Protein marker: protein molecular weight standards, unlabeled (e.g., Amersham; multiple ranges).
5. Gel: NuPAGE™ 4–12% Bis-Tris Protein Gels, 1.0 mm, 12-well.
6. Fixation solution: 30% ethanol, 10% acetic acid, 60% MQ.
7. Gel staining solution (e.g., EZBlue™ Gel Staining Reagent, Sigma-Aldrich).
8. Orbital shaker (e.g., VWR® Standard Orbital Shaker).

2.5 In Gel Tryptic Digestion

1. 100 mM Ammonium Bicarbonate (NH_4HCO_3, Ambic).
2. 50 mM Ambic.
3. 100% Acetonitrile (ACN).
4. 100 mM Ambic/100% ACN.
5. 1 M DTT.
6. 135 mM Iodoacetamide.
7. 1 μg/10 μl Porcine trypsin.

2.6 Peptide Extraction

1. 70% ACN (v/v)/0.1% Formic acid (v/v).
2. 100% ACN.
3. 95% MQ (v/v)/5% Formic acid.
4. 95% MQ (v/v)/5% ACN (v/v)/5% Formic acid (v/v).
5. Vacuum concentrator (e.g., Speed-vac Vacuum Concentrator, Thermo Fisher Scientific).

2.7 LC-MS/MS Instrumentation and Software

1. High-performance liquid chromatography (HPLC) system (Dionex, LC Packings Ultimate 3000) connected to a hybrid LTQ-OrbiTrap XL (Thermo Fisher Scientific) equipped with a nanoelectrospray ion source (New Objective).
2. Analytical column (15 cm × 300 μm i.d., 300 Å pore size, Pepmap C18, 5 μm, Dionex, LC Packings).
3. Proteome Discoverer 1.2 software supported by Mascot (Mascot server v2.2.07) (Thermo Fisher Scientific).
4. ProteoIQ 2.8 (Premier Biosoft, Palo Alto, CA, USA).

3 Methods

3.1 Sample Collection and Egg Quality Assessments

1. Egg samples utilized in this case study were selected from among 136 total spawns from paired males and females bred from different parents. Approximately sixty eggs per female were collected at the 1-cell stage immediately after fertilization and incubated in 100 mm Petri dishes filled with embryo culture medium to assess egg quality based on embryo development as indicated in Fig. 1.
2. Prior to assessing developmental competence, groups of 40 eggs per spawn were removed, frozen in liquid nitrogen, and stored at −80 °C until being used for proteomic analyses.
3. Incubated eggs/embryos were serially sampled for observation at different stages of embryonic development following standard developmental staging made by Kimmel et al. [5] (Fig. 1).
4. At the time of sampling, the number of surviving eggs/embryos was recorded, those not surviving were removed and, for samples taken at 8 hpf (hours post fertilization) and beyond, the number of abnormal embryos was recorded.
5. Based on this evaluation, spawns for which survival of normal embryos to 24 hpf was >90% were considered to contain good quality eggs, with those having a survival rate of <30% considered to contain poor quality eggs.
6. By these criteria, 55 of the 136 spawns (40.4%) contained good quality eggs and 15 spawns (11.0%) contained poor quality eggs.

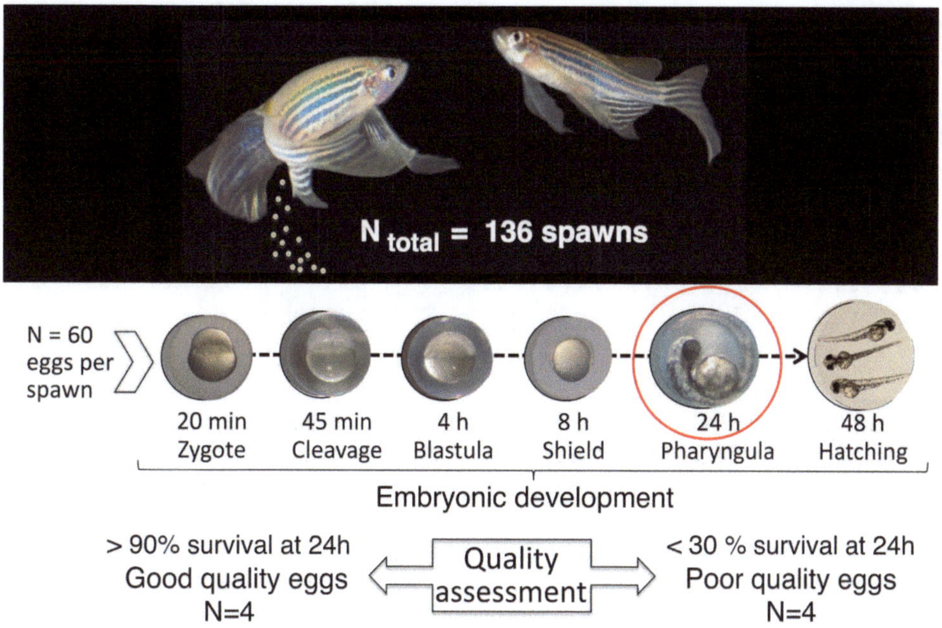

Fig. 1 Egg quality assessment based on embryonic developmental success. Egg samples utilized in this study were selected from among $N = 136$ spawns total. Approximately sixty eggs per batch were collected immediately after fertilization and incubated in embryo culture medium to assess egg quality based on embryo development. Incubated eggs/embryos were serially sampled for observation at the early blastula (~256 cells) stage (~4 hpf), at the shield to 75% epiboly stages (~8 hpf), at the early pharyngula stage (~24 hpf), and during the hatching period at 48 and 72 hpf (long-pec to protruding-mouth stages) following standard developmental staging [5]. At the time of sampling, the number of surviving eggs/embryos was recorded, those not surviving were removed and, for samples taken at 8 hpf and beyond, the number of abnormal embryos was recorded. Based on this evaluation spawns for which survival of normal embryos to 24 hpf was >90% were considered to contain good quality eggs, with those having a survival rate of <30% considered to contain poor quality eggs

7. From this sample set, good and poor quality spawns were selected for our experiments at random, with the exception that the contribution of individual male and female breeders was restricted to a single spawn in this study.
8. These egg batches of good and poor quality were used to create pooled (Fig. 2a; Pooled Samples Experiment) or replicated (Fig. 2b; Multiple Samples Experiment) sample sets whose protein extracts were subjected to different levels of SDS-PAGE fractionation during preparation for submission to LC-MS/MS.

3.2 Tissue Homogenization and Total Protein Extraction

1. Add 100 μl of 20 mM HEPES buffer pH 7.4 (including EDTA, DTT, and antiproteases) to egg samples ($N = 40$).

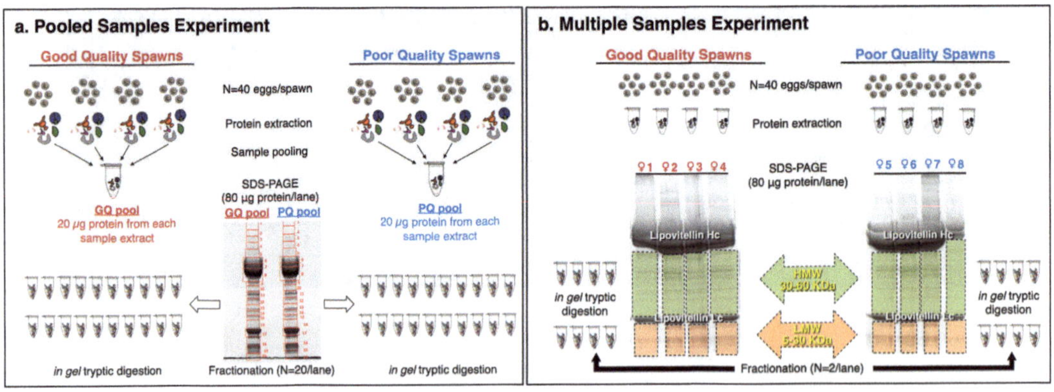

Fig. 2 Sample fractionation prior to LC-MS/MS in the egg quality experiments (from Yilmaz et al. [4]). (**a**) Sample fractionation after SDS-PAGE in the Pooled Samples Experiment. Egg samples from $N = 4$ females spawning good quality eggs and egg samples from $N = 4$ females spawning poor quality eggs were collected. Total protein from those samples was extracted and a volume equal to 20 μg of protein from each sample was pooled according to the quality grade as Good Quality Pool (*GQ Pool*) and as Poor Quality Pool (*PQ Pool*). This pool of protein extracts (80 μg protein total) was loaded on SDS-PAGE in two lanes and each lane was excised into $N = 20$ fractions in total (indicated in red numbered frames). Each gel fraction was subsequently subjected to *in gel* tryptic digestion separately before submission to LC-MS/MS. (**b**) Sample fractionation after SDS-PAGE in the Multiple Samples Experiment. Egg samples from $N = 4$ females spawning good quality eggs and egg samples from $N = 4$ females spawning poor quality eggs were collected. Total protein from those samples was extracted and a volume equal to 80 μg of protein from each sample was loaded on SDS-PAGE separately. Each lane of gel was excised at regions labeled as HMW (containing proteins with 30–60 kDa shown in dash lined frames and dash lined double-headed arrows colored in green) and LMW (containing proteins with 5–30 kDa shown in dash lined frames and dash lined double-headed arrows colored in orange). Protein bands at positions corresponding to the molecular weight of intact lipovitellin heavy chain (Lipovitellin Hc) and light chain (Lipovitellin Lc) were excluded in this experiment to better resolve other proteins. Each gel fraction ($N = 2$ fractions/lane for each sample) was subsequently subjected to *in gel* tryptic digestion separately before submission to LC-MS/MS. In total, from both Pooled Samples Experiment and from Multiple Samples Experiment, 56 egg protein fractions were individually subjected to LC-MS/MS and downstream proteomic analyses. HMW: high molecular weight, LMW: low molecular weight [4]

2. Sonicate on ice for 6 pulses of 10 s at 40% amplitude with 30 s pause between each pulse to avoid overheating of the samples (*see* **Note 3**).
3. Incubate homogenates on ice for 30 min and then centrifuge at 15,000 × *g* and +4 °C for 30 min.
4. Transfer the supernatant into new Eppendorf tubes and place supernatants on ice while completing the second-round of extraction with UreaTrisChaps buffer.
5. Add 20 μl UreaTrisChaps buffer (*see* **Note 4**) to the remnant pellet and mix thoroughly by pipetting or vortexing.
6. Sonicate on ice for 6 pulses of 10 s at 40% amplitude with 30 s pause between each pulse to avoid overheating of the samples.
7. Incubate homogenates on ice for 30 min and then centrifuge at 15,000 × *g* and +4 °C for 30 min.

8. Pool supernatant with the previously collected extract.
9. Ultracentrifuge at 105,000 × *g* and +4 °C for 1 h.
10. Recover the supernatant and proceed to protein concentration determination by Bradford assay.
11. Aliquot and store at −80 °C.

3.3 Bradford Assay

1. Bring Bradford solution to room temperature (RT).
2. Add 5 μl of BSA standard (*see* **Note 5**) and samples to be measured in duplicate or triplicate in microplate wells.
3. Add 250 μl of Bradford solution.
4. Read plate on a standard plate reader at 595 nm.
5. Use mean protein concentrations for each sample to calculate the amount of total protein to load on your gel.

3.4 SDS-PAGE

1. Calculate protein amounts to be loaded on each well of the gel (80 μg/well).
2. Mix sample volume corresponding to 80 μg protein with 5 μl NuPAGE® LDS sample buffer (4× stock solution) and 2 μl of DTT (5 M stock solution) and complete with water to 20 μl final volume.
3. Apply the same procedure for protein marker (e.g., Amersham protein molecular weight standards unlabeled).
4. Heat samples and protein marker at 70 °C for 10 min.
5. Prepare SDS-PAGE running buffer and gel.
6. Dilute 50 ml of 20× MOPS buffer to 1 l with MQ. Add 500 μl antioxidant (*see* **Note 6**) into 200 ml of diluted MOPS buffer.
7. Cut envelope and remove gel, rinse with MQ.
8. Remove sealer tape from the bottom of the gel.
9. Remove comb and rinse wells with 1 ml of higher (inner) chamber buffer.
10. Repeat rinsing twice and leave the buffer in the wells at last rinsing.
11. Place gel into the tank (if a single gel is used, place a buffer dam to close up the inner chamber circle).
12. Make sure the chamber is well closed and gel(s) are placed properly to avoid buffer leakage during the run.
13. Fill in the higher (inner) chamber with the 200 ml MOPS buffer including the antioxidant.
14. Fill in the lower (outer) chamber with about 600 ml MOPS buffer.
15. Load 20 μl of samples (and marker) into each well.

16. Run gel at 200 V (with maximum current) for ~1 h (until front dye reaches the bottom line of the gel).
17. Remove gel and open the cassette using a spatula.
18. Place gel into fixative solution and incubate at RT for 15 min to fix proteins on gel.
19. Wash gel three times at RT for 5 min each in MQ.
20. Stain gel in EZBlue™ Gel Staining Reagent at RT for 2 h on agitation.
21. Rinse gel with MQ and leave at RT in MQ without agitation overnight.
22. Cut bands at predetermined locations (*see* **Note 7**) according to your fragmentation strategy (*see* **Note 8**) (*see* Fig. 2 as example) and proceed to the *in gel* tryptic digestion process.

3.5 In Gel Tryptic Digestion

3.5.1 Destaining of Gel Pieces

1. Wash gel pieces twice in MQ.
2. Add 1 volume 100 mM Ambic and 1 volume 100% ACN (*see* **Note 9**).
3. Incubate at RT with agitation until decoloration is complete (*see* **Note 10**).
4. Remove liquid and add 300 μl of 100% ACN.
5. Continue agitation until the gel piece is dehydrated.
6. Replace liquid with 150 μl of 100 mM Ambic.
7. Continue agitation for 5 min.
8. Remove liquid and add the same volume of 100% ACN.
9. Continue agitation for 5 min.
10. Remove liquid and add the same volume of 100% ACN.
11. Continue agitation for 5 min.
12. Remove liquid and dry gel pieces at 37 °C for maximum 20 min (*see* **Note 11**).

3.5.2 Alkylation and Reduction

1. Mix 100 mM Ambic and 1 M DTT to a final DTT concentration of 65 mM.
2. Add 150 μl of this mix and incubate at 37 °C for 15 min.
3. Remove liquid.
4. Add 300 μl of 135 mM Iodoacetamide (*see* **Note 12**) and incubate at RT in dark for 15 min and remove liquid.
5. Add 150 μl of 100 mM Ambic/100% ACN 1:1 (v/v) and incubate at RT with agitation for 5 min.
6. Remove liquid and add 150 μl of 100% ACN.
7. Incubate at RT with agitation until gel pieces are milky white.
8. Replace liquid with 100 mM Ambic.

9. Incubate at RT with agitation for 5 min.
10. Remove liquid and add 3 volumes of 100% ACN.
11. Incubate at RT with agitation for 10 min.
12. Remove liquid and dry gel pieces at 37 °C for maximum 20 min (*see* **Note 13**).

3.5.3 Tryptic Digestion

1. Thaw previously aliquoted and frozen trypsin (1 μg/10 μl) on ice.
2. Add 70 μl of 50 mM Ambic mix pipetting and place on ice.
3. Add 75 μl trypsin solution.
4. Incubate on ice for 15 min.
5. Add 75 μl of 50 mM Ambic.
6. Incubate at 37 °C overnight.

3.5.4 Extraction of Digested Peptides

1. Transfer liquid including peptide digests into a new Eppendorf tube.
2. Add 150 μl of 70% ACN: 0.1% Formic acid (*see* **Note 14**).
3. Incubate at RT with agitation for 20 min.
4. Recover liquid and combine with the previously recovered peptide digests.
5. Add 75 μl of 100% of ACN to gel pieces.
6. Incubate at RT with agitation for 5 min.
7. Add 75 μl of 70% ACN: 0.1% Formic acid to gel pieces.
8. Incubate at RT with agitation for 15 min.
9. Recover liquid and pool with previously combined peptide digests.
10. Evaporate ACN using a Speed-vac device.
11. Resuspend pellet in 20 μl of 95% H_2O: 5% Formic acid.
12. Submit samples to LC-MS/MS.

3.6 LC-MS/MS, Mascot Search and Data Retrieval and Analysis

3.6.1 LC-MS/MS

In our case study, the MS measurements were performed with a nanoflow high-performance liquid chromatography (HPLC) system (Dionex, LC Packings Ultimate 3000) connected to a hybrid LTQ-OrbiTrap XL (Thermo Fisher Scientific) equipped with a nanoelectrospray ion source (New Objective, as previously described, [6]). The HPLC system consists of a solvent degasser nanoflow pump, a thermostatic column oven kept at 30 °C, and a thermostatic autosampler kept at 8 °C to reduce sample evaporation. Mobile A (99.9% MilliQ water and 0.1% formic acid (v:v)) and B (99.9% acetonitrile and 0.1% formic acid (v:v)) phases for HPLC were delivered by the Ultimate 3000 nanoflow LC system (Dionex, LC Packings).

1. Ten microliters of prepared peptide mixture were loaded on a trapping precolumn (5 mm × 300 μm i.d., 300 Å pore size, Pepmap C18, 5 μm) for 3 min in 2% buffer B at a flow rate of 25 μl/min.
2. This step was followed by reverse-phase separations at a flow rate of 0.250 μl/min using an analytical column (15 cm × 300 μm i.d., 300 Å pore size, Pepmap C18, 5 μm, Dionex, LC Packings). We ran a gradient ranging from 2% to 35% buffer B for the first 60 min, 35–60% buffer B from min 60–85, and 60–90% buffer B from min 85–105.
3. Finally, the column was washed with 90% buffer B for 16 min, and with 2% buffer B for 19 min prior to loading of the next sample.
4. The peptides were detected by directly eluting them from the HPLC column into the electrospray ion source of the mass spectrometer. An electrospray ionization voltage of 1.5 kV was applied to the HPLC buffer using the liquid junction provided by the nanoelectrospray ion source and the ion transfer tube temperature was set to 200 °C.
5. The MS instrument was operated in its data-dependent mode by automatically switching between full survey scan MS and consecutive MS/MS acquisition. Survey full-scan MS spectra (mass range 400–2000) were acquired in the OrbiTrap section of the instrument with a resolution of $r = 60,000$ at m/z 400; ion injection times are calculated for each spectrum to allow for the accumulation of 10^6 ions in the OrbiTrap.
6. The ten most intense peptide ions in each survey scan with an intensity above 2000 counts (to avoid triggering too early fragmentation during the peptide elution profile) and a charge state ≥ 2 were sequentially isolated at a target value of 10,000 and fragmented in the linear ion trap by collision-induced dissociation. The normalized collision energy was set to 35% with an activation time of 30 ms.
7. Peaks selected for fragmentation were automatically put on a dynamic exclusion list for 120 s with a mass tolerance of ±10 ppm to avoid selecting the same ion for fragmentation more than once.
8. The following parameters were used:
   (a) The repeat count: 1.
   (b) The exclusion list size limit: 500.
   (c) Singly charged precursors: rejected.

   (d) A maximum injection time: 500 ms for full MS scans and 300 ms for MS/MS scans.

9. For an optimal duty cycle, the fragment ion spectra were recorded in the LTQ mass spectrometer in parallel with the OrbiTrap full scan detection.
10. For OrbiTrap measurements, an external calibration was used before each injection series ensuring an overall error mass accuracy below 5 ppm for the detected peptides. MS data were saved in RAW file format (Thermo Fisher Scientific) using XCalibur 2.0.7 with tune 2.4.

3.6.2 Mascot Search and Data Retrieval

1. The spectra search was performed with the Proteome Discoverer 1.2 software supported by Mascot (Mascot server v2.2.07; http://www.matrixscience.com).
2. Obtained MS/MS spectra were searched against a target-decoy concatenated database created from the zebrafish Ensembl proteome database (*Danio rerio*_Zv9, March 2015) using Mascot (Matrix Science) with the following settings:
   (a) Mass tolerance: 10 ppm and 0.5 Da for MS and MS/MS, respectively.
   (b) Enzyme selectivity: full trypsin, with one miscleavage allowed.
   (c) The allowed protein modifications: carbamidomethylation of cysteines and variable oxidation of methionine.
3. MS data from multiple gel pieces were eventually combined for each biological sample during the Mascot search.

3.6.3 Data Analysis

1. Attributed spectra were then analyzed using ProteoIQ 2.8 (Premier Biosoft, Palo Alto, CA, USA) at <1% FDR, 0.5% minimum protein group probability, and six amino acids minimum peptide length, in order to identify egg quality group-specific and common proteins and quantify them based on their normalized spectral count (N-SC) values.
2. For each protein, obtained spectral counts were normalized in three sequential steps: (a) normalization by apportion of shared peptides based on the number of unique peptides each protein group possessed, (b) normalization by the total spectral counts between replicates and biological samples, and (c) normalization by the size of each protein (amino acid residues).
3. Protein annotations were performed using the GO, KEGG, and DAVID (Database for Annotation, Visualization, and Integrated Discovery) [7, 8] functional annotation tools.
4. Only proteins exhibiting a ≥2-fold difference (*see* **Note 15**) in N-SC between egg quality groups, or proteins unique to an

egg quality group, were considered to be differentially expressed between groups and subjected to further analyses for classifications by functional category, for enrichment analyses, and statistical analyses to detect significant differences in mean N-SC values between egg quality groups.

5. Differentially expressed proteins were classified into thirteen arbitrarily chosen functional categories that would account for >90% of the proteins. These functional categories are:
   (a) Protein synthesis (PS).
   (b) Energy metabolism (EM).
   (c) Lipid metabolism (LM).
   (d) Cell cycle, division, growth, and fate (CC).
   (e) Protein degradation and synthesis inhibition (PD).
   (f) Apoptosis-related (AP).
   (g) Oncogene-related (OG).
   (h) Endosome/lysosome-related (EL).
   (i) Immune function-related (IF).
   (j) Oxidoreductase (REDOX)- and detoxification (Detox)-related (RD).
   (k) Lectins.
   (l) Zona pellucida (egg envelope) proteins (ZPs).
   (m) Vitellogenins (Vtgs) including their derived yolk proteins.

   Upregulated proteins that could not be attributed to any of these categories and were placed in the category "Other." For simplicity, proteins were attributed to only one category considered as the "best" fit. Annotations were made by two independent observers before any other analyses categorizing the proteins (i.e., observations made "blind") (*see* Yilmaz et al. [4] for details).

6. Enrichment analyses were conducted separately for each experiment using the PANTHER-GO Slim enrichment tool from GO Consortium [9] available online at http://geneontology.org/ for Biological Process, Molecular Function, Protein Class, and Cellular Component.

7. Upregulated proteins were submitted to PANTHER Pathway [10] analysis and proteins attributed to enriched pathways were submitted to analysis using the STRING search tool for retrieval of protein-protein interaction networks [11] available from the STRING Consortium online at http://string-db.org/, with the data settings Confidence: High (0.70), Max Number of Interactions to Show: None/query proteins only. For the PANTHER and STRING analyses, only statistically significant enrichment results ($p < 0.05$) were reported.

8. To detect significant differences between egg quality groups in mean N-SC values for differentially expressed proteins (when a sufficient number of replicates exist) an independent *t*-test ($p < 0.05$) followed by Benjamini Hochberg correction for multiple tests ($p < 0.1$) was used (IBM SPSS Statistics Version 19.0.0, Armonk, NY).

4 Notes

1. HEPES buffer (w/o DTT, EDTA, and antiproteases) can be stored in aliquots at −20 °C.
2. Antiproteases should be added freshly into buffers since they are highly sensitive to freeze/thaw cycles and they also quickly degrade even under optimal storage conditions.
3. Sonication cycles and duration may need optimization for different types of tissues.
4. UreaTrisChaps buffer can be stored in aliquots at −20 °C.
5. Standard dilutions can be stored in aliquots at −20 °C. For standard concentration ranges see directions in Pierce™ Coomassie (Bradford) Protein Assay Kit.
6. Antioxidants should be added freshly before initiating the SDS-PAGE process.
7. Make sure to excise bands completely when defining the fragment frontiers.
8. Fractionation intensity will define the depth of your study. The higher number of fractions you have, the more coverage of the protein repertoire of your sample you will achieve. The fractionation intensity level can be adjusted according to the general purposes of the study (*see* Fig. 2a, b as an example).
9. Acetonitrile is toxic to skin and also flammable. Take the necessary precautions while working with this material.
10. For some fragments, destaining might take longer than others. Duration of this step should be adjusted accordingly until all fragments are completely decolorated.
11. If gel pieces are not contracted and white repeat process again from **step 10**.
12. Iodoacetamide is toxic when inhaled. Appropriate precautions should be taken while working with this material.
13. If gel pieces are not contracted and white, repeat process again from **step 10**.
14. Inhalation of formic acid vapors may cause acute toxicity. Wear protective clothing/eye protection and work under fume hood during processes that involve the use of formic acid.

15. Fold difference in N-SC as well as other parameters such as N-SC Log2 Relative Expression can be taken as reference and cut-off for these parameters can be set at different levels according to the purpose and the biological hypothesis of the study.

References

1. Hood BL, Veenstra TD, Conrads TP (2004) Mass spectrometry-based proteomics. Int Congr Ser 1266:375–380
2. McDonald WH, Yates JR III (2002) Shotgun proteomics and biomarker discovery. Dis Markers 18:99–105
3. Anand S, Samuel M, Ang CS, Keerthikumar S, Mathivanan S (2017) Label-based and label-free strategies for protein quantitation. In: Keerthikumar S, Mathivanan S (eds) Proteome bioinformatics. Methods in molecular biology, vol 1549. Humana, New York, NY
4. Yilmaz O, Patinote A, Nguyen TV, Com E, Lavigne R, Pineau C, Sullivan CV (2017) Scrambled eggs: proteomic portraits and novel biomarkers of egg quality in zebrafish (*Danio rerio*). PLoS One 12(11):e0188084. https://doi.org/10.1371/journal.pone.0188084
5. Kimmel CB, Ballard WW, Kimmel SR, Ullmann B, Schilling TF (1995) Stages of embryonic development of the zebrafish. Dev Dyn 203:253–310
6. Lavigne R, Becker E, Liu Y, Evrard B, Lardenois A, Primig M et al (2012) Direct iterative protein profiling (DIPP) – an innovative method for large-scale protein detection applied to budding yeast mitosis. Mol Cell Proteomics 11:M111-012682
7. Huang DW, Sherman BT, Lempicki RA (2009a) Systematic and integrative analysis of large gene lists using DAVID Bioinformatics Resources. Nat Protoc 4:44–57
8. Huang DW, Sherman BT, Lempicki RA (2009b) Bioinformatics enrichment tools: paths toward the comprehensive functional analysis of large gene lists. Nucleic Acids Res 37:1–13. https://doi.org/10.1093/nar/gkn923
9. Mi H, Poudel S, Muruganujan A, Casagrande JT, Thomas PD (2016) PANTHER version 10: expanded protein families and functions, and analysis tools. Nucleic Acids Res 44: D336–D342. https://doi.org/10.1093/nar/gkv1194
10. Huaiyu MH, Thomas P (2009) PANTHER pathway: an ontology-based pathway database coupled with data analysis tools. In: Nikolsky Y, Bryant J (eds) Methods in molecular biology, vol 563, Part 2. Humana, New York, NY, pp 123–140
11. Szklarczyk D, Franceschini A, Wyder S, Forslund K, Heller D, Huerta-Cepas J et al (2015) STRING v10: protein-protein interaction networks, integrated over the tree of life. Nucleic Acids Res 43:D447–D452. https://doi.org/10.1093/nar/gku1003

Chapter 23

In Vivo Protein Lifetime Measurements Across Multiple Organs in the Zebrafish

Sunit Mandad, Gudrun Kracht, and Eugenio F. Fornasiero

Abstract

Protein production and degradation are tightly regulated to prevent cellular structures from accumulating damage and to allow their correct functioning. A key aspect of this regulation is the protein half-life, corresponding to the time in which half of a specific protein population is exchanged with respect to its initial state. Proteome-wide techniques to investigate protein half-lives in vivo are emerging. Recently, we have established and thoroughly tested a metabolic labeling approach using ^{13}C lysine (Lys(6)) for measuring protein lifetimes in mice. The approach is based on the fact that different proteins will incorporate a metabolic label at a rate that is dependent on their half-life. Using amino acid pool modeling and mass spectrometry, it is possible to measure the fraction of newly synthesized proteins and determine protein half-lives. In this chapter, we show how to extend this approach to zebrafish (*Danio rerio*), using a commercially available fish diet based on the stable isotope labeling by amino acids in cell culture (SILAC) technology. We describe the methods for labeling animals and subsequently use mass spectrometry to determine the lifetimes of a large number of proteins. In the mass spectrometry workflow proposed here, we have implemented the BoxCar data acquisition approach for increasing sample coverage and optimize machine use. To establish the proteome library used in the BoxCar approach, we recommend performing an in-solution digestion followed by peptide fractionation through basic reversed-phase chromatography. Overall, this chapter extends the use of current proteome technologies for the quantification of protein turnover to zebrafish and similar organisms and permits the study of germline changes following specific manipulations.

Key words Protein lifetime, Protein turnover, Metabolic labeling, Protein mass spectrometry, BoxCar, SILAC

1 Introduction

Proteins are essential macromolecules at the basis of life. To ensure cellular efficiency and health, protein synthesis and degradation are tightly regulated in a process known as protein homeostasis or proteostasis [1]. Aberrant proteostasis can disturb several biochemical pathways and lead to diseases such as premature aging, neurodegeneration, and infertility [2]. Protein turnover studies are relevant for several aspects of biology, including cell differentiation,

Roland Dosch (ed.), *Germline Development in the Zebrafish: Methods and Protocols*, Methods in Molecular Biology, vol. 2218, https://doi.org/10.1007/978-1-0716-0970-5_23,

© Springer Science+Business Media, LLC, part of Springer Nature 2021

organism development, and can be applied to the analysis of several cell types including germlines [1, 3–6]. Mass spectrometry applied to proteomics allows to study changes at the level of the entire proteome and can be used for protein lifetime determinations [3, 5, 7, 8].

Traditionally, protein turnover studies were performed through the metabolic labeling of radioactive isotopes [9–11]. Less hazardous methods have been developed more recently, including the use of deuterated water (^2H_2O) [6], of specific diets containing the stable non-radioactive isotope of nitrogen ^{15}N [12] and other more precise methods based on essential amino acids labeled with a variety of stable non-radioactive isotopes [3, 5, 13–15]. By definition, essential amino acids cannot be synthetized by organisms and need to be absorbed through the diet [11, 16]. Cells and animals can be thus labeled with an approach called stable isotope labeling by amino acids in cell culture (SILAC) [17, 18]. In this approach, the essential amino acid lysine is synthetized substituting all the six carbon atoms (^{12}C) by a stable isotope of carbon (^{13}C), giving rise to a label (Lys(6)) that is incorporated in peptides with the same efficiency of conventional lysine (Lys(0); Fig. 1a). While being functionally identical, the two lysine forms can be distinguished with mass spectrometry, since they differ by 6 Da in molecular weight (Fig. 1b). A typical SILAC pulse labeling approach in vivo requires an animal model that can be fed with a diet where all Lys(0) is substituted with Lys(6). In an optimized workflow, a group of animals is fed for different time periods (Fig. 1c). Following labeling, tissue dissection, and sample preparation, it is possible to reveal the incorporation of Lys(6) by mass spectrometry (Fig. 1d). Proteins that incorporate Lys(6) faster are turned over with shorter times than proteins that incorporate the label slower [3, 5, 13]. The precise determination of the incorporation of Lys(6) allows to study the dynamic of the proteome and define protein lifetimes with precision [5, 13]. Our group has shown that, despite being based on expensive food sources, a well-designed pulse SILAC approaches outperform classical techniques of studying protein turnover in several aspects [5, 13].

Originally zebrafish became popular as a model organism for embryonic development studies, thanks to its handling simplicity and to the possibility of manipulating and observing changes in its phenotype [19, 20]. Over the past decades, a number of genetic studies indicated that ~70% of zebrafish genes are shared with humans [19]. Moreover, large scale forward genetic screenings have shown that ~84% of human disease-associated genes show similar pathological phenotypes in zebrafish [20]. This has rendered the small fish one of the most popular organisms for modeling pathology development and study disease to date. Metabolically labeled diets for protein turnover in zebrafish have been previously introduced and tested, confirming that the metabolic pulsing of this fish is feasible and effective [21, 22].

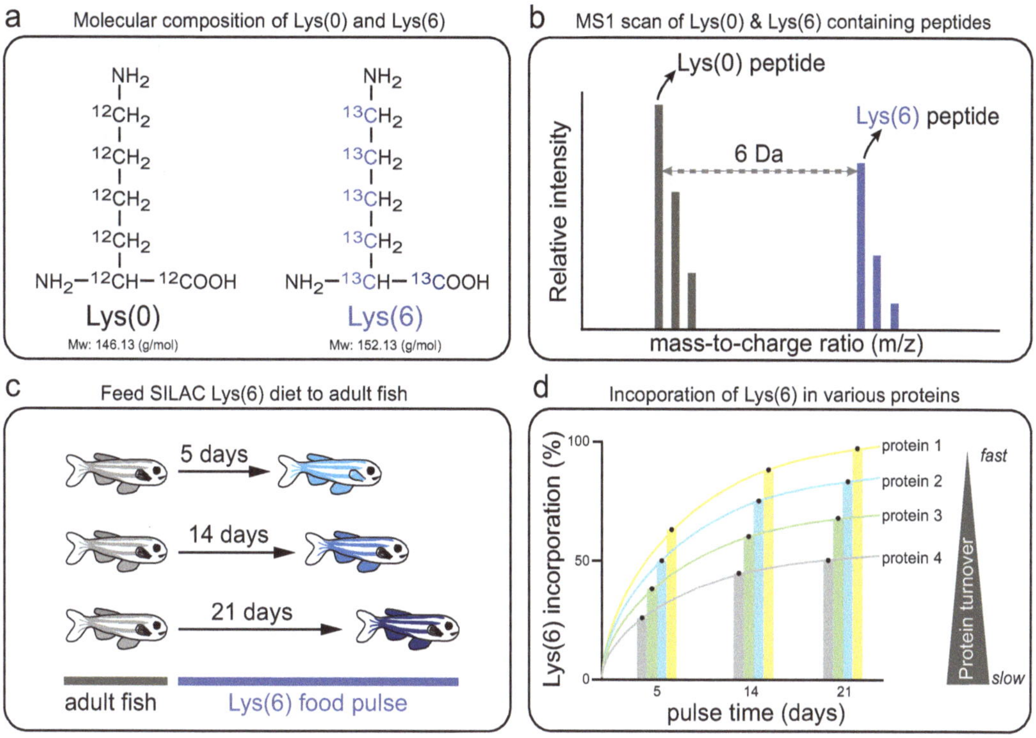

Fig. 1 Overview of the labeling approach. (**a**) Schematic representation of the molecular composition of light lysine Lys(0) containing all the naturally abundant elements (left) and stable isotopic SILAC Lysine Lys (6) where the 6 naturally occurring ^{12}C atoms were replaced with ^{13}C. (**b**) Mass spectrometric scan of peptides containing Lys(0) (left, grey) and Lys(6) (right, blue). The *x*-axis represents the mass-to-charge ratio (m/z). Peptides containing either one Lys(0) or one Lys(6) of a given charge will differ by a m/z interval of 6 Da in the MS scan. (**c**) Timeline for feeding the SILAC diet to animals. (**d**) Incorporation of Lys(6) will proceed faster in proteins that have shorter turnover times and would allow to determine the lifetime of a given protein

Recently, new methods have been developed to increase proteome coverage, including several data acquisition (DA) workflows [23, 24]. BoxCar is one of these methods, which improves proteome coverage with a small amount of starting material and optimizes instrument usage [24]. The most time-consuming aspect of BoxCar is that it requires a large peptide library, which can be created by fractionating some of the samples with a gel-free approach such as high pH chromatography [25]. Once the library is defined, BoxCar can be used to analyze a large number of samples in parallel and is compatible with SILAC measurements.

In this chapter, we detail a method to metabolically label zebrafish in combination with mass spectrometry which enables to determine lifetimes at the proteome level in several tissues (*see also* **Notes 1** and **2**). The comparative study of protein turnover in several tissues of zebrafish and eventually of various disease models will be instrumental for the understanding of basic biological processes

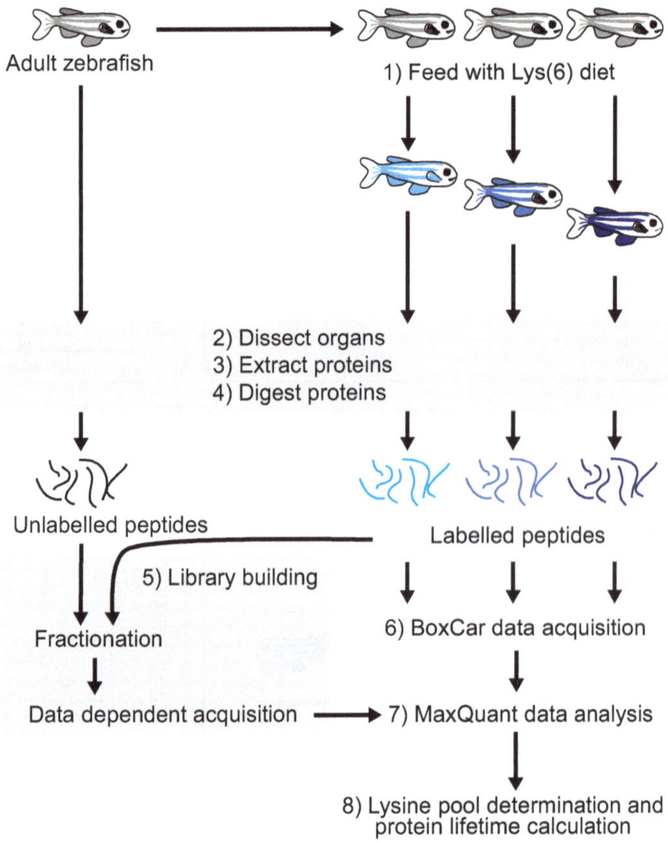

Fig. 2 Schematic representation of the workflow used for performing protein turnover measurements in zebrafish. For details see main text

and for the development of new therapies. The major steps of the workflow presented here are summarized in Fig. 2 and include: (1) feeding zebrafish with the labeled diet, (2) sacrificing the animals and dissecting the organ of interest, (3) lysing and solubilizing proteins from tissues, (4) digesting proteins, (5) fractionating the peptides for library preparation, (6) acquiring the data with the BoxCar method, (7) analyzing the data, and (8) determining the protein lifetimes.

2 Materials

To ensure high-quality results, prepare all solutions right before use and follow the indications for the storage of samples and materials.

2.1 Feeding Zebrafish

1. Zebrafish (*D. rerio*) strain AB maintained under standard conditions.

2. Diet: SILAC diet (Silantes, adult fish feed; ZFAd), granules size 4 (Aqua Schwarz).
3. Water parameters: pH 7.2–8.0, conductivity 500–600 μS, housing temperature 24–25 °C.
4. Weighing: analytical balance.
5. Dissection: forceps Inox Dumont No. 5 (biological tip), scissors, 10 cm Petri dish, 1.5 ml vial, liquid Nitrogen.

2.2 Tissue Preparation and Protein Extraction

1. 1% Rapigest (Waters).
2. Lysis buffer: 100 mM HEPES, 0.1% Rapigest, 0.1% NP-40, and protease inhibitor cocktail, pH 8.
3. Acetonitrile (ACN): (LiChrosolv grade; Merck).
4. Water (LiChrosolv grade; Merck).
5. Bioruptor (Diagenode).
6. BCA Protein Assay Kit.

2.3 Protein Digestion and Desalting

1. Digestion buffer: 100 mM ammonium bicarbonate, pH 8. Prepare freshly prior to use.
2. 10 mM dithiothreitol in digestion buffer. Prepare freshly prior to use.
3. 55 mM chloracetamide in digestion buffer. Prepare freshly prior to use.
4. Sequence Grade Modified Trypsin (Promega): 0.1 μg/μl in digestion buffer. Always prepare freshly prior to use.
5. C18 tip column (Empore, Fisher Scientific).
6. Trifluoroacetic acid.
7. SpeedVac Vacuum Concentrator.
8. Tabletop centrifuge.
9. Waterbath sonicator.
10. Thermomixer.

2.4 Basic Reversed-Phase Chromatography

1. Fractionation mobile phase A: 10 mM ammonia, pH 10.
2. Fractionation mobile phase B: 10 mM ammonia in 80% ACN (LiChrosolv grade) pH 10.
3. HPLC system coupled with fraction collector.
4. Basic reverse C18 XBridge column (3.5 μm; 1 × 150 mm; Waters).

2.5 Mass Spectrometry

1. Glass LC vials.
2. Sample suspension buffer: 5% ACN containing 0.1% TFA.
3. Self-packed fused silica capillary (Polymicro Technologies) columns filled with 75 μm inner diameter, 30 cm length packed

with Reprosil-Pur Basic C18-AQ 1.9 μm pore size beads (Dr. Maisch GmbH). Alternatively, use the commercially available Acclaim PepMap 100 C18 LC columns (Thermo Fisher Scientific), following the instrument setup provided by the supplier.

4. Trapping C18 column AcclaimPepMap100 C18–300 μm i.d. × 5 mm, 5 μm, 100 Å.
5. Buffer A: Water (LiChrosolv Grade) containing 0.01% TFA.
6. Buffer B: 95% ACN (LiChrosolv Grade) containing 0.01% TFA.
7. Q Exactive HF- Mass spectrometer.

2.6 Online Tools and Tutorials

1. MaxQuant: https://www.maxquant.org/.
2. MaxQuantLive: http://www.maxquant.live/.

3 Methods

3.1 Feeding and Dissecting Zebrafish

1. For simplicity, aliquot and store the tubes containing the amount of food for one meal (10 mg of fish diet, either SILAC or regular).
2. Select four AB of the same age and size.
3. Determine their weight in accordance with animal handling regulations.
4. Separate each fish in a single container.
5. Feed the fish for either 7, 14, or 21 days with 10 mg of food daily.
6. Also include one fish that will be housed separately, kept on a conventional diet, and used as an unlabeled control.
7. At the end of pulse determine and record the weight again.
8. Sacrifice the fish in accordance with regulations.
9. Dissect the various organs of fish:
   (a) Cut out a piece of muscle with scissors beginning at the anal fin along the belly to the gills along the gills to the spine, along the side and down to the anal fin again. Store the sample in 1.5 ml vial, label correctly and freeze immediately in liquid Nitrogen.
   (b) Now the organs are visible. With forceps, take out organs of interest such as testis, stomach, liver, kidney, heart, and gills. Store the samples in 1.5 ml vial, label correctly and freeze immediately in liquid Nitrogen.
   (c) Decapitate the fish, open the head with forceps and take out the brain. Store the sample in 1.5 ml vial, label correctly and freeze immediately in liquid Nitrogen.

(d) Open the spine with forceps and take out the central nervous system. Store the sample in 1.5 ml vial, label correctly and freeze immediately in liquid Nitrogen.

10. Transfer samples from liquid Nitrogen into a − 80 °C freezer. Store them until ready for sample preparation.

3.2 Tissue Preparation and Protein Extraction

1. Add 100 μl of lysis buffer. Sonicate in a bioruptor using the following parameters: total time 30 min, 10 duty cycle, 30 s ON/OFF. Repeat until the particulates solubilize completely making the sample a homogeneous mixture (for denser tissues, such as muscle, *see* **Note 3**).
2. Determine the protein concentration using a BCA assay kit following the instructions provided by the company.

3.3 Protein Digestion and Desalting

1. Aliquot 10 μg of SILAC-labeled protein lysate. For library building, aliquot 100 μg of lysate in a separate vial.
2. Add 10 μl of 1% Rapigest and heat at 70 °C for 10 min to solubilize the proteins. Spin the vial.
3. Add 10 μl of digestion buffer. Mix well by vortexing and spin.
4. Add 10 μl of 10 mM dithiothreitol solution. Incubate at 37 °C for 30 min on the thermomixer at 750 rpm to reduce the disulfide bonds.
5. Spin the vials. Add 10 μl of 55 mM chloracetamide solution to the vial (mix well).
6. Spin and incubate the vials on a thermomixer at 750 rpm in dark vial to alkylate the reduced cysteines. Cover the thermomixer lid with aluminum foil to avoid light.
7. Add 200 μl of digestion buffer to dilute the Rapigest detergent concentration from 1% to 0.1%. Vortex and spin.
8. Add 1:50 protein-to-trypsin amount to the vial to trypsinize the proteins into peptides. Briefly add 2 μl and 20 μl of trypsin (0.1 μg/μl) to 10 μg labeled and 100 μg unlabeled protein sample respectively.
9. Mix well and spin. Incubate the vials on thermomixer at 37 °C, 750 rpm overnight. About 16–18 h of lysate incubation are sufficient for efficient trypsinization of proteins.
10. Next day, spin the vials. Add 20 μl of 10% TFA to each vial. Mix well and spin. Check the pH using litmus paper. The acidified peptide solution has a pH of about 2. Incubate the vials on thermomixer at 37 °C, 750 rpm for 2 h. This (a) stops the trypsin activity and (b) degrades and precipitates the Rapigest detergent.
11. Spin the vials at maximum speed on tabletop centrifuge for 5 min. Carefully pipette out the supernatant in fresh vials. Discard the pellet.

12. Wash C18 columns tips using 50 μl of methanol by centrifuging at 1000 × *g* for 5 min. Subsequently, equilibrate the column tips by passing 50 μl of water containing 0.1% TFA. Note: for 100 μg of lysate use two C18 column tips.
13. Load the peptide solution from **step 11** on the washed, equilibrated C18 column tips.
14. Wash four times with 50 μl of water containing 0.1% TFA.
15. Finally, elute the peptides in fresh vials by passing 20 μl of 80% of ACN containing 0.1% TFA through the column twice.
16. Dry the eluted peptides in SpeedVac.

3.4 Fractionation of Unlabeled Lysate Peptides for Setting the Library

1. Add 100 μl of fractionation mobile phase A to the dried 100 μg of digested, desalted lysate (Subheading 3.3, **step 16**).
2. Sonicate for 3 min in a waterbath to resuspend the peptide in the buffer properly.
3. Spin for 2 min on the maximum speed of tabletop centrifuge to pellet any particulates.
4. Inject the solubilized ~50 μg peptides on a basic reversed-phase XBridge C18 column (Waters) connected to the Agilent LC system.
5. To fractionate the loaded peptides, use a 2-step linear gradient from 0% to 50% and 50% in 45 min and to 100% in next 10 min. Continue the isocratic separation at 100% for 10 min and switch back to 100% of buffer A for 5 min in the end.
6. Monitor the peptides at wavelength 214, 260, and 280 nm to observe peptides, RNA/DNA and aromatic amino acid peptides respectively.
7. Collect the fractions at every 1 min.
8. Dry the eluted peptides in SpeedVac.

3.5 DDA Runs of Unlabeled to Set Up the Library

1. Resuspend the dried peptides of fractions from basic reversed-phase chromatography in 20 μl of sample suspension buffer *see also* **Notes 4–6** for suggestions).
2. Sonicate in a waterbath for 3 min. Centrifuge to max speed. Discard pellet.
3. Transfer the supernatant in clean glass LC vials.
4. Inject 8 μl of resuspended peptide solution on a trap column followed by 30 cm C18 column coupled to Q-Exactive HF-X instrument for 88 min for Top30 method. Briefly, load the peptides in 5% buffer B. Fractionate the peptides using a 2-step linear gradient, i.e., for 3 min to 10% of buffer B and then until 75.9 min to 42% of buffer B. Finally, wash the column isocratically in 90% of buffer B for 5 min and switch back 5% of buffer B. For MS1, set the resolution at 60,000 with AGC target of 1e6 and max. IT of 50 ms. Set the scan range

from 350 to 1600 m/z. For MS2, set the resolution at 15,000 with AGC target as 1e5 and max. IT of 54 ms. Set the loop count parameter to 30.

3.6 BoxCar Mass Spectrometry Runs for the Labeled Lysate Digest

1. Resuspend 10 μg digested lysate in 50 μl of sample suspension buffer.
2. Sonicate for 3 min to resuspend peptides.
3. Centrifuge at maximum speed on tabletop centrifuge for 2 min to pellet the particulates. Pipette the supernatant in clean LC glass vials.
4. Inject 5 μl of resuspended peptide solution on trap column followed by 30 cm C18 column coupled to Q-Exactive HF-X with BoxCar method. For MS1 and MS2, use BoxCar settings mentioned in MQLive tutorial available. Briefly, set the loop count to 1. Set two Full MS-SIM scans for Scan Group tab at 120,000 resolution. Set the first Full MS SIM scan from 909 to 1001 m/z range, while the second Full MS SIM scan from 908 to 1088 m/z. For MQLive, save 1001 as BoxCar "magic number."

3.7 Data Analysis and Protein Turnover Interpretation

1. The MaxQuant software coupled to Andromeda search engine can be used to determine the protein turnover. Submit library and BoxCar RAW files as separate groups. Follow rest of the instruction as given in the BoxCar publication [24].
2. For protein lifetime determinations refer to our published set of tools [26]. Upload the H/L ratios of proteins. As a result, the scripts will give an output file containing the lifetimes of the input proteins (*see* **Notes** 7 and **8**).

4 Notes

1. The described protocol for studying protein turnover in zebrafish is also applicable to other model organisms. The detailed step-wise protocol for performing protein turnover in mouse has been tested and is available online [26].
2. Our original mass spectrometric approach for pulsed SILAC quantification was based on the fractionation of the tissue lysate with sodium dodecyl sulfate polyacrylamide gel electrophoresis (SDS-PAGE). This approach required cutting each gel lane into 23 slices followed by in-gel digestion and mass spectrometric analysis of each individual gel slice. In general, while being reliable and precise, this approach was tedious and time-consuming. If the protein lysate is directly processed for in-solution digestion (without gel fractionation), the number of proteins identified is limited, due to the complexity of

Table 1
Estimate of protein amounts obtained by the dissection of one fish for several organs/tissues

Organ	Protein amount (μg)
Brain	400
Gills	40
Heart	70
Kidney	10
Liver	10
Muscle (~2 mm^3)	300
Stomach	300
Testis	70
Ovary	200

peptide spectra which overlap and render the quantification inefficient [7, 8, 27, 28].

3. Dense tissues like muscles are difficult to lyse, therefore, it is recommended to sonicate the tissue in a bioruptor for longer times (up to 1 h). It is recommended to perform the sonication at 4 °C so that the samples do not heat up during long sonication cycles of bioruptor. If the sample is not solubilized, we recommend to add 10 μl of 8 M urea to the 100 μl of lysis buffer ad sonicate for an additional 10 min.
4. For setting up the library, it is recommended to start with relatively high sample amounts (~100 μg). Table 1 reports a rough estimate of protein amounts obtained by the dissection of one fish for several organs/tissues. Depending on the final protein amount required, a higher number of fishes can be sacrificed and pooled. Note that the exact protein amounts may vary from lab to lab due to dissection practices.
5. Gel-based fractionation techniques through SDS-PAGE or gel-free fractionation approaches such as strong cation exchange and size-exclusion chromatography can be applied as alternative methods to basic reversed-phase chromatography. Nevertheless, in our hands, basic reversed-phase chromatography outperforms the aforementioned fractionation methods.
6. For setting up the library, it is recommended to load at least 50 μg of digested protein lysate. Lowering the loading amounts might decrease the number of identifications in the experiment. Increasing excessively the amounts loaded might block the column or saturate the MS.

7. Protein turnover studies on various genetically modified zebrafish lines could be aimed at testing the causes of alterations at the basis of pathologies such as aging or germline exhaustion. For instance, the swimming ability of zebrafish slows down over aging, thus protein turnover study at various ages of zebrafish may pinpoint the molecular causes of this impairment [29].
8. Zebrafish show high dimorphism between males and females, and the time course of protein production/stabilization during gonad development could be analyzed in detail with metabolic labeling. The technique would be also able to pinpoint differences between models of infertility, clarifying which of the essential proteins are either stabilized or turned over more rapidly.

Acknowledgments

This work was in part supported by an EMBO Long Term Fellowship and a HFSP Fellowship to E.F.F. (EMBO_LT_797_2012 and HFSP_LT000830/2013).

References

1. Sala AJ, Bott LC, Morimoto RI (2017) Shaping proteostasis at the cellular, tissue, and organismal level. J Cell Biol 216:1231–1241. https://doi.org/10.1083/jcb.201612111
2. Basisty N, Meyer JG, Schilling B (2018) Protein turnover in aging and longevity. Proteomics 18. https://doi.org/10.1002/pmic.201700108
3. Schwanhäusser B, Gossen M, Dittmar G, Selbach M (2009) Global analysis of cellular protein translation by pulsed SILAC. Proteomics 9:205–209. https://doi.org/10.1002/pmic.200800275
4. Selbach M, Schwanhäusser B, Thierfelder N, Fang Z, Khanin R, Rajewsky N (2008) Widespread changes in protein synthesis induced by microRNAs. Nature 455:58–63. https://doi.org/10.1038/nature07228
5. Fornasiero EF, Mandad S, Wildhagen H, Alevra M, Rammner B, Keihani S, Opazo F, Urban I, Ischebeck T, Sakib MS, Fard MK, Kirli K, Centeno TP, Vidal RO, Rahman R-U, Benito E, Fischer A, Dennerlein S, Rehling P, Feussner I, Bonn S, Simons M, Urlaub H, Rizzoli SO (2018) Precisely measured protein lifetimes in the mouse brain reveal differences across tissues and subcellular fractions. Nat Commun 9:4230. https://doi.org/10.1038/s41467-018-06519-0
6. Lau E, Cao Q, Ng DCM, Bleakley BJ, Dincer TU, Bot BM, Wang D, Liem DA, Lam MPY, Ge J (2016) A large dataset of protein dynamics in the mammalian heart proteome. Sci Data 3:160015. https://doi.org/10.1038/sdata.2016.15
7. Mann M, Kulak NA, Nagaraj N, Cox J (2013) The coming age of complete, accurate, and ubiquitous proteomes. Mol Cell 49:583–590. https://doi.org/10.1016/j.molcel.2013.01.029
8. Aebersold R, Mann M (2003) Mass spectrometry-based proteomics. Nature 422:198–207. https://doi.org/10.1038/nature01511
9. Humphrey TJ, Davies DD (1975) A new method for the measurement of protein turnover. Biochem J 148:119–127. https://doi.org/10.1042/bj1480119
10. Dietz WH, Wolfe MH, Wolfe RR (1982) A method for the rapid determination of protein turnover. Metabolism 31:749–754. https://doi.org/10.1016/0026-0495(82)90070-1
11. Schoenheimer R, Rittenberg D (1935) Deuterium as an indicator in the study of intermediary metabolism. Science 82:156–157. https://doi.org/10.1126/science.82.2120.156
12. Toyama BH, Savas JN, Park SK, Harris MS, Ingolia NT, Yates JR, Hetzer MW (2013)

Identification of long-lived proteins reveals exceptional stability of essential cellular structures. Cell 154:971–982. https://doi.org/10.1016/j.cell.2013.07.037
13. Mandad S, Rahman R-U, Centeno TP, Vidal RO, Wildhagen H, Rammner B, Keihani S, Opazo F, Urban I, Ischebeck T, Kirli K, Benito E, Fischer A, Yousefi RY, Dennerlein S, Rehling P, Feussner I, Urlaub H, Bonn S, Rizzoli SO, Fornasiero EF (2018) The codon sequences predict protein lifetimes and other parameters of the protein life cycle in the mouse brain. Sci Rep 8:16913. https://doi.org/10.1038/s41598-018-35277-8
14. Zhang Y, Reckow S, Webhofer C, Boehme M, Gormanns P, Egge-Jacobsen WM, Turck CW (2011) Proteome scale turnover analysis in live animals using stable isotope metabolic labeling. Anal Chem 83:1665–1672. https://doi.org/10.1021/ac102755n
15. Price JC, Guan S, Burlingame A, Prusiner SB, Ghaemmaghami S (2010) Analysis of proteome dynamics in the mouse brain. Proc Natl Acad Sci U S A 107:14508–14513. https://doi.org/10.1073/pnas.1006551107
16. Schoenheimer R, Ratner S, Rittenberg D (2009) Studies in protein metabolism. Nutr Rev 40:23–26. https://doi.org/10.1111/j.1753-4887.1982.tb06822.x
17. Ong S-E, Mann M (2006) A practical recipe for stable isotope labeling by amino acids in cell culture (SILAC). Nat Protoc 1:2650–2660. https://doi.org/10.1038/nprot.2006.427
18. Krüger M, Moser M, Ussar S, Thievessen I, Luber CA, Forner F, Schmidt S, Zanivan S, Fässler R, Mann M (2008) SILAC mouse for quantitative proteomics uncovers kindlin-3 as an essential factor for red blood cell function. Cell 134:353–364. https://doi.org/10.1016/j.cell.2008.05.033
19. MacRae CA, Peterson RT (2015) Zebrafish as tools for drug discovery. Nat Rev Drug Discov 14:721–731. https://doi.org/10.1038/nrd4627
20. Lieschke GJ, Currie PD (2007) Animal models of human disease: zebrafish swim into view. Nat Rev Genet 8:353–367. https://doi.org/10.1038/nrg2091
21. Nolte H, Konzer A, Ruhs A, Jungblut B, Braun T, Krüger M (2014) Global protein expression profiling of zebrafish organs based on in vivo incorporation of stable isotopes. J Proteome Res 13:2162–2174. https://doi.org/10.1021/pr5000335
22. Geary B, Magee K, Cash P, Young IS, Whitfield PD, Doherty MK (2016) Determining synthesis rates of individual proteins in zebrafish (*Danio rerio*) with low levels of a stable isotope labelled amino acid. Proteomics 16:1398–1406. https://doi.org/10.1002/pmic.201500357
23. Gillet LC, Navarro P, Tate S, Röst H, Selevsek N, Reiter L, Bonner R, Aebersold R (2012) Targeted data extraction of the MS/MS spectra generated by data-independent acquisition: a new concept for consistent and accurate proteome analysis. Mol Cell Proteomics 11:O111.016717. https://doi.org/10.1074/mcp.O111.016717
24. Meier F, Geyer PE, Virreira Winter S, Cox J, Mann M (2018) BoxCar acquisition method enables single-shot proteomics at a depth of 10,000 proteins in 100 minutes. Nat Methods 15:440–448. https://doi.org/10.1038/s41592-018-0003-5
25. Batth TS, Francavilla C, Olsen JV (2014) Off-line high-pH reversed-phase fractionation for in-depth phosphoproteomics. J Proteome Res 13:6176–6186. https://doi.org/10.1021/pr500893m
26. Alevra M, Mandad S, Ischebeck T, Urlaub H, Rizzoli SO, Fornasiero EF (2019) A mass spectrometry workflow for measuring protein turnover rates in vivo. Nat Protoc 14 (12):3333–3365
27. Shevchenko A, Tomas H, Havlis J, Olsen JV, Mann M (2006) In-gel digestion for mass spectrometric characterization of proteins and proteomes. Nat Protoc 1:2856–2860. https://doi.org/10.1038/nprot.2006.468
28. Nakamura K, Aebersold R, Bairoch A, Dunn M, Celis J, Hanash S, Hochstrasser D, Humphrey-Smith I, James P, Klose J, LaBaer J, Langen H, Mann M, Parekh R, Patterson S, Pearce C, Poepstorff P, Simpson RJ, Tomlinson I, Tsugita A, Yates J (2004) From genome to proteome--aim of human proteomics. Seikagaku 76:1271–1274
29. Gilbert MJH, Zerulla TC, Tierney KB (2014) Zebrafish (*Danio rerio*) as a model for the study of aging and exercise: physical ability and trainability decrease with age. Exp Gerontol 50:106–113. https://doi.org/10.1016/j.exger.2013.11.013

Chapter 24

In Vivo Imaging of Protein Interactions in the Germplasm with Bimolecular Fluorescent Complementation

Roshan Priyarangana Perera and Roland Dosch

Abstract

Protein–protein interactions (PPIs) play a central role in all cellular processes. The discovery of green fluorescent protein (GFP) and split varieties, which are functionally reconstituted by complementation, led to the development of the bimolecular fluorescence complementation (BiFC) assay for the investigation of PPI in vivo. BiFC became a popular tool, as it is a convenient and quick technology to directly visualize PPI in a wide variety of living cells. In combination with the transparency of the early zebrafish embryo, it also permits detection of PPI in the context of an entire living organism, which performs all spatial and temporal regulations missing in in vitro systems like tissue culture. However, the application of BiFC in some research areas including the study of zebrafish is limited due to the lack of efficient and convenient BiFC expression vectors. Here, we describe the engineering of a novel set of Gateway®-adapted BiFC destination vectors to investigate PPI with all possible permutations for BiFC experiments. Moreover, we demonstrate the versatility of these destination vectors by confirming the interaction between zebrafish Bucky ball and RNA helicase Vasa in living embryos.

Key words Protein–protein interaction, Bimolecular fluorescence complementation (BiFC) assay, Zebrafish, Gateway vectors, Bucky ball, Vasa

1 Introduction

Many biological processes in living cells are mediated by protein–protein interactions (PPIs). Therefore, characterization of PPI is crucial to understand protein activity and to correlate their biological functions in the cellular context [1]. In vivo techniques like co-immunoprecipitation (co-IP) or pull-down assay followed by western blotting are considered the gold standard to investigate PPI because proteins interact in their physiological conditions and are post-translationally modified [2, 3]. However, unstable or instantaneous PPIs cannot be easily detected using these approaches. Further, these techniques do not reveal whether the interaction is direct or indirect.

Roland Dosch (ed.), *Germline Development in the Zebrafish: Methods and Protocols*, Methods in Molecular Biology, vol. 2218, https://doi.org/10.1007/978-1-0716-0970-5_24, © Springer Science+Business Media, LLC, part of Springer Nature 2021

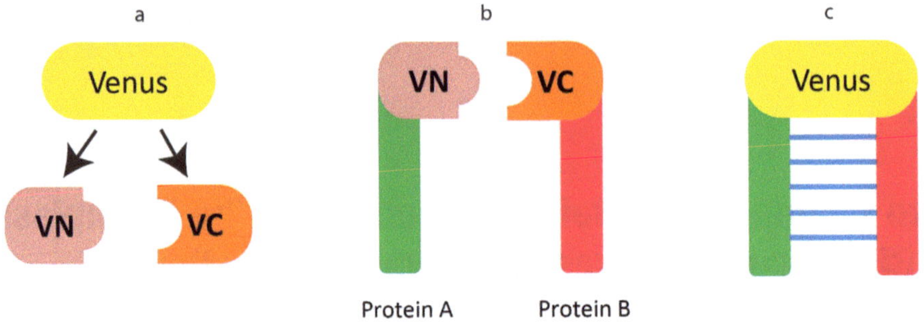

Fig. 1 Schematic of BiFC assay. (**a**) Venus protein (yellow) is divided into two non-fluorescent parts, Venus N-terminus (VN; rose) and Venus C-terminus (VC, orange). (**b**) Fusion of VN- and VC-fragments to the target proteins (protein A, green and protein B, pink). (**c**) Interaction of protein A and B reconstitutes a functional Venus protein forming a bimolecular fluorescent complex

Therefore, *in vitro* techniques such as affinity chromatography or tandem affinity purification determine direct PPI, albeit the target proteins do not interact in their native physiological environment [1, 3, 4]. Among PPI approaches, the bimolecular fluorescence complementation (BiFC) assay is popularly employed, as it provides the potential for direct visualization in a wide variety of living cells. In addition, analyzing PPI in cells reflects the properties of endogenous proteins including post-translational modifications [2, 4, 5].

The BiFC assay is based on the reconstitution of an intact fluorescent protein complex when two complementary non-fluorescent fragments are brought in close proximity by two interacting proteins upon their expression in living cells (Fig. 1) [2, 4]. So far, more than 15 different fluorescent proteins have been studied in the BiFC assay [4]. Among those, the Venus protein-based BiFC (Venus-BiFC) assay is more prominently applied due to significantly higher fluorescent signal-to-noise ratio while decreasing false-positive results. Furthermore, it is technically simple, expedient, and can be investigated using conventional fluorescence microscopy [6–9].

Besides PPI, the BiFC assay has been applied to other experimental setups. For example, split-GFP-based BiFC experiments have been applied to investigate the proximity of contact sites of organelles, e.g., between the endoplasmic reticulum and mitochondria in tissue culture and in the sensory neurons of zebrafish [10]. A split red fluorescent protein, sfChery3, has also been used to visualize contact sites between two cells. For instance, specific subsets of synapses between posterior neuron, PHB and anterior neuron AVA neurites in Caenorhabditis elegans were labeled, when the split red fluorescent protein was tagged to the transmembrane synaptic protein neuroligin, which localizes to both pre- and postsynaptic sites [11]. The far-red mNeptune-based trimolecular fluorescence complementation (TriFC) assay has been used to study protein–

RNA interaction in vivo. This technology is based on the bacteriophage MS2-MCP system, in which MCP protein is tagged with one-half of mNeptune, while the other half is fused to an RNA-binding protein (RBP). Split mNeptune protein is reconstituted and emits fluorescence when RBP bind to its RNA target while MCP binds to MS2-sites incorporated into the RNA. Further, the interaction between human polypyrimidine-tract-binding protein and HIV-1 mRNA elements was discovered using TriFC [12]. Interestingly, multicolored BiFC systems have been developed to visualize individual interactions between multiple protein complexes simultaneously [13]. Recently, for the first time, a split, orange-colored BiFC system using the Kusabira-Orange protein from the stony coral *Fungia concinna* was developed [14].

A major obstacle in Venus-BiFC experiments is finding the appropriate combination of split fluorescent fragments, Venus N-terminus (VN), and Venus C-terminus (VC) to produce a detectable signal in vivo, since the topology of the protein–protein complex plays a critical role during PPI [15]. To find the best combination of fusion fragments, each protein is fused with VN- and VC-fragments and each split fluorescent fragment is fused to both the N- and C-terminus of the candidate proteins. This creates eight different permutations of fusion fragment for two target proteins and time-consuming cloning work.

Previous BiFC expression vectors were generated with traditional restriction enzyme-based insertion sites. Generating the eight possible combinations with this method becomes troublesome, if we need to determine the best combination of fusion fragments. In recent years, recombination-based Gateway® cloning vectors have been developed to analyze robust and efficient gene functions and to create transgenic zebrafish [16–18].

To reduce this initial workload, we developed four Venus-BiFC Gateway-adapted destination vectors (Fig. 2). These vectors facilitate all possible permutations of BiFC expression vectors, which helps to investigate simultaneous multiple PPIs efficiently and conveniently. As a proof of principle, using the generated vectors, we confirmed a recently published PPI between zebrafish Bucky ball and the RNA helicase Vasa in vivo [19].

2 Materials

2.1 Common Materials

1. Nuclease-free water or HPLC grade water.
2. PCR tubes.
3. PCR machine.
4. Agarose powder.
5. 0.05% ethidium bromide.

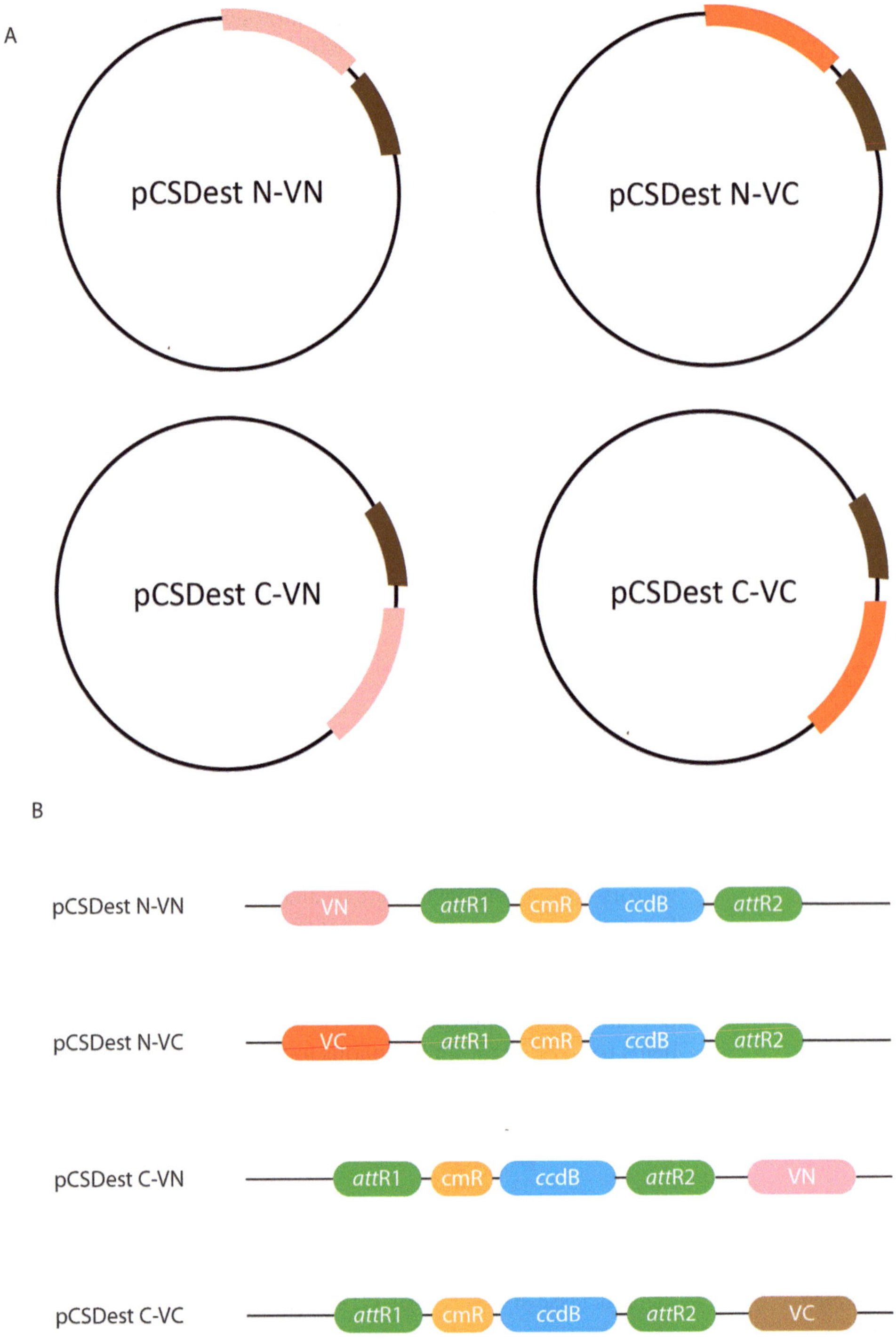

Fig. 2 Vector maps and key core elements of BiFC gateway destination vectors. (**a**) Vector maps of four different BiFC gateway destination vectors, pCSDest N-VN, pCSDest N-VC, pCSDest C-VN, and pCSDest C-VC. Black line represents the vector backbone and colored (rose) regions represent the core features of the vector (detailed in **b**) and the insert is denoted in brown. (**b**) The vectors possess all features of the core cassette of Gateway pCSDest destination vector for recombination reaction [17]. The core cassette contains two *att*R sites (green) flanking the bacterial toxin gene *ccd*B (light blue) for selection and the chloramphenicol resistance

6. 1× TBE buffer: 90 mM Tris (pH 8.0), 90 mM boric acid, 2 mM EDTA.
7. DNA ladder.
8. 10× loading dye: 50% glycerol, 0.4% bromophenol blue.
9. RNA gel loading buffer.
10. Gel documentation system.
11. UV transilluminator.
12. 1.5% Luria-Bertani (LB)-agar plate containing kanamycin as a selection antibiotic: (0.5% (w/v) NaCl, 0.5% (w/v) yeast extract, 0.1% (w/v) glucose; pH 7.0), kanamycin (final concentration 50 μg/ml).
13. 1.5% LB-agar plate containing ampicillin: 1.5% LB-agar plate containing kanamycin: (0.5% (w/v) NaCl, 0.5% (w/v) yeast extract, 0.1% (w/v) glucose; pH 7.0), ampicillin (final concentration 100 μg/ml).
14. Gel and PCR clean-up kit.
15. NanoDrop 2000c spectrophotometer.
16. Microcentrifuge tubes: 2 and 1.5 ml.
17. Ampicillin (100 mg/ml stock).
18. Kanamycin (50 mg/ml stock).
19. LB medium (pH 7.0) or Super Optimal broth with Catabolite repression (SOC) medium.
20. Plasmid miniprep kit.
21. Heating block.

2.2 Gateway Cloning BP Cloning Reaction

1. Template plasmid, pCSDest Vasa full-length.
2. *att*B-PCR primers (Table 1).
3. 5× BP Clonase II enzyme mix (Thermo Scientific).
4. pEXP7-tet positive control (Thermo Scientific).
5. DH5α competent cells.
6. 1.5% LB-agar plate containing kanamycin.

2.3 Gateway Cloning LR Cloning Reaction

1. pDONR vectors with target insert.
2. Gateway-adapted BiFC destination vectors.

Fig. 2 (continued) gene, cmR (light orange). In addition to the core cassette, Venus VN (rose) and VC (orange) are fused to the N-terminus of the core cassette to generate pCSDest N-VN or pCSDest N-VC, whereas VN and VC are fused to the C-terminus of the core cassette to generate pCSDest C-VN or pCSDest C-VC BiFC destination vectors

Table 1
Primers for gateway cloning

Primer name	Primer sequence
pDONR Vasa FL FW	GGGGACAAGTTTGTACAAAAAAGCAGGCTTA**ATGGATGAC TGGGAGGAAGATCAG**
pDONR Vasa FL STOP RV	GGGGACCACTTTGTACAAGAAAGCTGGGTT**TTATTCCCATTCCTCATCG TCTGCAG**
pDONR Vasa FL NO STOP RV	GGGGACCACTTTGTACAAGAAAGCTGGGTT**TTCCCATTCCTCATCGTC TGCAG**

Gene-specific sequence is in bold

3. 5× LR Clonase II enzyme mix (Thermo Scientific).
4. pENTR-gus positive control (Thermo Scientific).
5. DH5α competent cells.
6. 1.5% LB-agar plate containing ampicillin.

2.4 Restriction Enzyme Digestion

1. 10× FastDigest buffer.
2. FastDigest Restriction enzyme NotI.
3. Plasmid DNA.
4. Nuclease-free water.

2.5 *In Vitro* Transcription

1. 2× NTP/CAP, 10× reaction buffer (e.g., mMESSAGE mM ACHINE™ SP6 Transcription Kit, Thermo Scientific).
2. Linearized plasmid DNA.
3. Nuclease-free water.
4. Illustra ProbeQuant G-50 Micro Columns RNA purification kit.

2.6 Microinjection

1. Glass capillary (0.58 × 1.00 × 80 mm, GB100F-8P).
2. P-97 Flaming/Brown Micropipette Puller.
3. Eppendorf microloader pipette tips.
4. Microinjector (e.g., PV 820; WPI, Sarasota, USA).
5. Stereomicroscope.
6. Micrometer calibrating slide.
7. Mineral oil.
8. 1 M KCl.
9. Phenol red.
10. 1× E3 medium: 5 mM NaCl, 0.17 mM KCl, 0.33 mM $CaCl_2$, 0.33 mM $MgSO_4$, 20 μl of 0.03 M methylene blue per 1 l of 1× E3 medium.

11. Incubator 28.5 °C.
12. Dumont #5 watchmaker forceps.
13. Mold with wedge-shaped troughs.
14. 10 cm plastic petri dishes.
15. Fire-polished pasteur pipette.
16. Plastic transfer pipette.

3 Method

In this chapter, we describe in vivo imaging of zebrafish Bucky ball and RNA helicase Vasa interaction using BiFC assay. The experimental workflow is summarized in Fig. 3.

3.1 Construct Gateway DONR Vectors

1. Design *att*B-PCR primers to amplify full-length zebrafish Vasa with and without a stop codon (Table 1).
2. Set up a PCR reaction (Table 2) and run the PCR with cycling conditions listed in Table 3.
3. Resolve the PCR products on agarose gel.
4. Purify PCR products using Gel and PCR clean-up kit according to the manufacturer's instructions (*see* **Note 1**).
5. Measure the DNA concentration with the NanoDrop.
6. Set up a gateway cloning BP reaction in a 1.5-ml microcentrifuge tube (Table 4).

3.2 Transformation of Gateway Cloning BP Reaction

1. Add 2 μl of proteinase K enzyme into the BP reaction mixture and incubate at 37 °C for 10 min to inactivate the BP clonase II enzyme.
2. Thaw and gently mix DH5α competent cells to ensure even distribution and aliquot 50 μl of the competent cells into a 1.5-ml microcentrifuge tube.
3. Add 2.5 μl of the BP reaction mixture to the cells and gently pipette the cells for even distribution of the reaction mixture (*see* **Note 2**).
4. Incubate on ice for 30 min.
5. Heat-shock the competent cells for 45–60 s at 42 °C.
6. Immediately transfer the microcentrifuge tube on ice and keep for 2–3 min.
7. Add 450 μl SOC medium (*see* **Note 3**).
8. Tighten the tube cap and incubate for 1 h at 37 °C with horizontal shaking at 220 rpm.

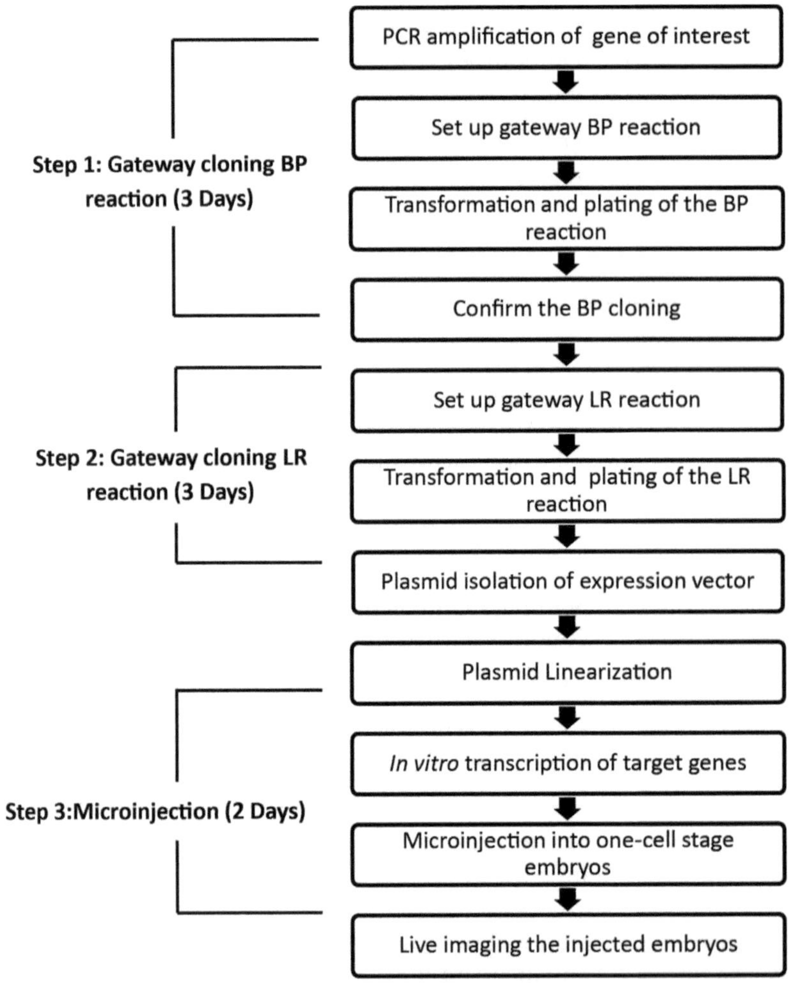

Fig. 3 Flowchart summarizing the experimental procedure for in vivo BiFC. The workflow is divided into three major steps. Step 1 and step 2 include the generation of the DONR vector (BP reaction) of the interaction candidates and fusing them into the set of BiFC destination vectors (LR reaction). Step 3 describes the microinjection

3.3 Plating and Making Mini-Cultures for Sequencing

1. Spread 250 μl of transformation reaction on a 1.5% LB-agar plate containing kanamycin as a selection antibiotic.
2. Incubate the plates overnight at 37 °C.
3. Prepare miniprep cultures adding a single colony into 2 ml of LB medium containing kanamycin as a selection antibiotic.
4. Incubate the miniprep cultures overnight at 37 °C with shaking at 220 rpm.

3.4 Plasmid Isolation

1. Isolate the plasmid DNA from mini-culture with the GeneJet Plasmid Miniprep Kit (Thermo Scientific) according to the manufacturer's instructions.

Table 2
PCR reaction for gateway cloning

Reagent	Volume	Final concentration
5× Phusion HF buffer	10.0 μl	1×
Forward primer (10 μM)	2.5 μl	0.5 μM
Reverse primer (10 μM)	2.5 μl	0.5 μM
dNTPs (5 mM)	2.0 μl	200 μM
Template DNA	Variable	<250 ng
Phusion DNA polymerase	1.0 μl	
Nuclease-free water	To 50 μl	

Table 3
PCR cycling condition for gateway cloning

Step	Temperature, °C	Time
Initial denaturation	98	3.00 min
35–40 cycles	98 Depending on primer T_m 72	0.15 s 0.15 s 15–30 s/kb
Final extension	72	10 min
Hold	20	

Table 4
Gateway cloning BP reaction

Reaction component	Sample, μl	Negative control	Positive control, μl
*att*B-PCR products (50–150 ng/μl)	1–7	1–7 μl	
pDONR 221 vector (150 ng/μl)	1.0	1.0 μl	2.0
pEXP7-tet positive control (50 ng/μl)	–	–	2.0
5× BP Clonase II reaction buffer	2.0	*2.0 μl* H_2O	2.0
Nuclease-free water	To 10.0	to 10.0 μl	4.0
Gently mix and spin the reaction mixture, then incubate overnight at 25 °C			

2. Measure the DNA concentration using NanoDrop.
3. Confirm the gateway DONR vector constructs by DNA sequencing.

3.5 Construct Gateway BiFC Destination Vectors

1. Set up gateway cloning LR reaction according to Table 5 (*see* **Note 4**).

3.6 Transformation Gateway Cloning LR Reaction

1. Follow the steps described in Subheading 3.2 using ampicillin as a selection antibiotic.

3.7 Plating and Making Mini-Cultures for Sequencing

1. Follow the steps described in Subheading 3.3 using ampicillin as a selection antibiotic.

3.8 Plasmid Isolation

1. Follow the steps described in Subheading 3.4 and validate the destination vectors.

3.9 Plasmid Linearization

1. Set up the plasmid linearization reaction in a 1.5-ml microcentrifuge tube using NotI restriction enzyme (Table 6).
2. Inactivate NotI by incubation at 80 °C for 20 min.
3. Resolve the linearized plasmid in a 0.5% agarose gel along with a non-linearized plasmid to confirm plasmid linearization.
4. Purify the linearized plasmid DNA using Gel and PCR clean-up kit according to the manufacturer's instructions (*see* **Note 5**).
5. Measure the purified DNA with the NanoDrop.

3.10 In Vitro Transcription of BiFC Destination Vectors

1. Set up an in vitro transcription reaction (Table 7).
2. Purify RNA using Illustra ProbeQuant G-50 Micro Columns according to the manufacturer's instructions.
3. Resolve RNA on 1% agarose gel to check the integrity of the purified RNA (*see* **Note 6**).

3.11 Microinjection

1. Pull 1.0-mm OD injection needles with a micropipette puller device (*see* **Note 7**).
2. Pour 1.5% agarose in 1× E3 medium into a petri dish and place a mold with wedge-shaped troughs to make grooves in the plates, which facilitates holding embryos during microinjection.
3. Prepare a working solution of 150 ng/μl RNA using nuclease-free water and a final concentration of 0.025–0.05% phenol red and 0.1 M KCl (*see* **Note 8**).
4. Load 4–5 μl of diluted RNA into the capillary tubes using a microloader tips and a pipetman.
5. Switch on the air source, microinjector, and stereomicroscope (*see* **Note 9**).

Table 5
Gateway cloning LR reaction

Reaction component	Sample, μl	Negative control	Positive control, μl
pDONR Vector (50–150 ng/μl)	1–7	1–7 μl	
Destination Vector (150 ng/μl)*	1.0	1.0 μl	2.0
pENTR-gus positive control (50 ng/μl)	–	–	2.0
5× LR Clonase II reaction buffer	2.0	2.0 μl H_2O	2.0
Nuclease-free water	To 10.0	To 10.0 μl	4.0
Gently mix and spin the reaction mixture, then incubate overnight at 25 °C			

*PCSDest N-VN , PCSDest N-VC, PCSDest C-VN, PCSDest C-VC

Table 6
Restriction enzyme digestion

Reaction component	Sample, μl
10× FastDigest Buffer	2.0
Plasmid DNA (<2 μg)	×
FastDigest Restriction Enzyme	2.0
Nuclease-free water	To 20
Gently mix and spin the reaction mixture, then incubate at 37 °C for 1–2 h	

Table 7
In vitro transcription

Reaction component	Volume
2× NTP/CAP	10.0 μl
10× reaction buffer	2.0 μl
Linearized plasmid DNA	100–1000 ng
SP6 RNA polymerase	2.0 μl
Nuclease-free water	To 20 μl
Gently mix the sample tube and spin the tube *Incubate the sample at 42 °C for 2–3 h*	
After 3 h, add 2 μl of TURBO DNase and incubate the sample for 15 min at 37 °C	

6. Mount the RNA-loaded capillary tube to the needle holder in the microinjector apparatus.
7. Break the tip of the capillary tube using a pair of sharp forceps under the microscope at 5× magnification.

8. Press the foot pedal to make sure that the tip has been broken and RNA can flow out of the needle.
9. Place one drop of mineral oil on a micrometer calibrating slide.
10. Inject a drop of working RNA solution onto the oil drop and check the diameter of the drop.
11. Adjust the injection duration and pressure to maintain the drop diameter to 100 μm, which injects approximately 0.5 nl of working solution (*see* **Note 10**).
12. Arrange one-cell stage embryos along the troughs in the microinjection plates under a stereomicroscope using a transfer pipette.
13. Remove the excess E3 medium from the microinjection plate.
14. Multiply calibrated injection time in **step 11** by 4 to inject 2 nl of RNA solution.
15. Carefully pierce the chorion and advance the tip of the needle into the yolk.
16. Press the foot pedal to inject the RNA into the yolk (*see* **Note 11**).
17. After injection, transfer the injected embryos into a new petri dish with 1× E3 medium.
18. Incubate the injected embryos at 28.5 °C for 3 h.

3.12 Visualization in Live Embryos

1. Manually dechorionate the embryos using a pair of sharp forceps.
2. Mount dechorionated embryos in 1.5% agarose-coated dishes filled with 1× E3 medium.
3. Visualize the embryos with the LSM780 confocal microscope using GFP fluorescence channel (*see* **Note 12**).
4. Analyze images using the ZEN 2011 software.

4 Notes

1. Directly purify the PCR products from the PCR reaction itself, if you observe a single band in the agarose gel. If there are multiple bands, cut the expected band after resolving PCR products from 0.5% agarose gel at a low wavelength (366 nm) UV light. To increase DNA yield, heat the gel piece with the NTI buffer (NucleoSpin Gel and PCR clean-up kit) between 50 and 70 °C in a shaking incubator at 1400 rpm. After washing with buffer NT3, keep the column in a new 1.5-ml microcentrifuge tube at 50–55 °C for 2–3 min to evaporate the remaining EtOH. Further, after adding elution buffer, keep the tubes at 50–55 °C for another 2–5 min to increase DNA yield.

2. If the transformation is not performed immediately, store the cloning reaction at −20 °C.
3. LB medium can also be used as a growth medium instead of SOC medium.
4. We used commercially synthesized DONR vectors for Bucky ball with and without stop codons. In the LR cloning, zebrafish Vasa fused to VN and Bucky ball fused to VC. Here we generated only four destination vectors (Fig. 4) as the above Bucky ball and Vasa expression vector combinations produce good fluorescent signal during their interaction.
5. Initially, check 1 μl from linearized reaction mixture on 0.5% agarose gel. If the plasmid is completely linearized compared to non-linearized plasmid, directly purify the linearized plasmid from the reaction mixture itself as it will increase the final DNA concentration. If the plasmid is not completely linearized, cut the linearized plasmid band after separation on a 0.5% agarose gel.
6. Aliquot the RNA in small quantities (5 μl) and store at −20 °C. This helps to avoid mRNA degradation and loss of activity due to frequent freeze and thaw cycles.
7. Needle puller settings: Heat = 515, Pull = 228, Vel = 80, Time = 250.

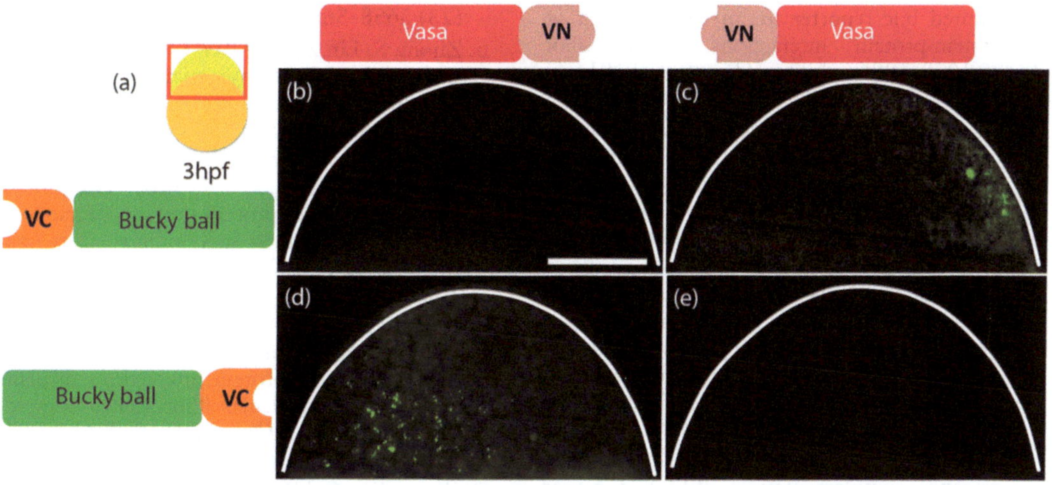

Fig. 4 BiFC assay shows Buc and Vasa interaction in vivo. (a) Cartoon representing the imaged area of embryos at 3 hpf focusing on the blastodisc as outlined with a white line in panels b–e (lateral view, animal to the top). (b) N-terminal Venus C-terminus tagged Bucky ball,VC-Buc and C-terminal Venus N-terminus tagged Vasa,Vasa-VN do not show fluorescent puncta in the blastodisc (0% ± 0, $n = 51$). (c) VC-Buc and N-terminal Venus N-terminus tagged Vasa,VN-Vasa show interaction (fluorescent embryos 47% ± 9.1, $n = 59$). (d) C-terminal Venus C-terminus tagged Bucky ball, Buc-VC and C-terminal Venus N-terminus tagged Vasa, Vasa-VN show interaction (fluorescent embryos 82.2% ± 3.9, $n = 58$). (e) Buc-VC and VN-Vasa do not show fluorescence in the blastodisc at 3 hpf (0% ± 0, $n = 72$). Scale bar 100 m

8. Titrate the in vitro transcribed RNAs, even if concentrations appear similar on 1% agarose gel or with NanoDrop measurement. In this BiFC experiment, 150 ng/μl of diluted RNA showed maximum expression with no toxicity to the embryos.
9. At the beginning, set the injection pressure between 10 and 20 psi. Further, set the back pressure between 0.2 and 0.4 psi to avoid backflow of the loaded injection solution.
10. The drop should hang in the oil to form a perfect sphere. However, if the drop falls and spreads out on the slide, use a larger volume of mineral oil to make a thicker oil layer.
11. Usually, injection volume is approximately 10% of the yolk volume. The red spot of phenol red in the yolk indicates successful injection. RNA is more actively expressed, if injected directly into the cell and not into the yolk of the embryo. However, piercing the cell membrane is tougher than piercing the yolk. To inject into the cell, first place the needle inside the yolk and slowly extend the tip of the needle into the cell. This can be achieved by arranging the animal pole of the embryo opposite to the direction of injection.
12. Signal for Venus protein can be visualized using the GFP channel since the emission spectra of both GFP and Venus proteins overlap.

References

1. Filonov GS, Verkhusha VV (2013) A near-infrared bifc reporter for in vivo imaging of protein-protein interactions. Chem Biol 20:1078–1086. https://doi.org/10.1016/j.chembiol.2013.06.009
2. Kerppola TK (2006) Design and implementation of bimolecular fluorescence complementation (BiFC) assays for the visualization of protein interactions in living cells. Nat Protoc 1:1278–1286. https://doi.org/10.1038/nprot.2006.201
3. Kerppola TK (2008) Bimolecular fluorescence complementation (BiFC) analysis as a probe of protein interactions in living cells. Annu Rev Biophys 37:465–487. https://doi.org/10.1146/annurev.biophys.37.032807.125842.BIMOLECULAR
4. Kodama Y, Hu CD (2012) Bimolecular fluorescence complementation (BiFC): a 5-year update and future perspectives. BioTechniques 53:285–298. https://doi.org/10.2144/000113943
5. Ghosh I, Hamilton AD, Regan L et al (2000) Antiparallel leucine zipper-directed protein reassembly: application to the green fluorescent protein. J Am Chem Soc 122 (23):5658–5659
6. Zhong S, Lin Z, Fray RG, Grierson D (2008) Improved plant transformation vectors for fluorescent protein tagging. Transgenic Res 17:985–989. https://doi.org/10.1007/s11248-008-9199-y
7. Free RB, Hazelwood LA, Sibley DR (2009) Identifying novel protein-protein interactions using co-IP and mass spectrometry. Curr Protoc Neurosci 2:1–19. https://doi.org/10.1002/0471142301.ns0528s46.Identifying
8. Kamigaki A, Nito K, Hikino K et al (2016) Gateway vectors for simultaneous detection of multiple protein-protein interactions in plant cells using bimolecular fluorescence complementation. PLoS One 11:1–15. https://doi.org/10.1371/journal.pone.0160717
9. Zhang XE, Cui Z, Wang D (2016) Sensing of biomolecular interactions using fluorescence complementing systems in living cells. Biosens Bioelectron 76:243–250. https://doi.org/10.1016/j.bios.2015.07.069
10. Cieri D, Vicario M, Giacomello M et al (2018) SPLICS: A split green fluorescent protein-based contact site sensor for narrow and wide

heterotypic organelle juxtaposition. Cell Death Differ 25:1131–1145. https://doi.org/10.1038/s41418-017-0033-z

11. Feng S, Varshney A, Villa DC et al (2018) Bright split red fluorescent proteins with enhanced complementation efficiency for the tagging of endogenous proteins and visualization of synapses. bioRxiv 2018:454041. https://doi.org/10.1101/454041
12. Han Y, Wang S, Zhang Z et al (2014) In vivo imaging of protein-protein and RNA-protein interactions using novel far-red fluorescence complementation systems. Nucleic Acids Res 2014:42. https://doi.org/10.1093/nar/gku408
13. Bischof J, Duffraisse M, Furger E et al (2018) Generation of a versatile bifc orfeome library for analyzing protein–protein interactions in live drosophila. elife 7:1–24. https://doi.org/10.7554/eLife.38853
14. Summary M (2018) Kusabira-Orange protein:64. https://doi.org/10.4155/btn-2017-0121
15. Verhoef LGGC, Wade M (2017) Visualization of protein interactions in living cells using bimolecular luminescence complementation (BiLC). Curr Protoc Protein Sci 90:30.5.1–30.5.14. https://doi.org/10.1002/cpps.42
16. Miles LB, Verkade H (2014) TA-cloning vectors for rapid and cheap cloning of zebrafish transgenesis constructs. Zebrafish 11:281–282. https://doi.org/10.1089/zeb.2013.0954
17. Villefranc JA, Amigo J, Lawson ND (2007) Gateway compatible vectors for analysis of gene function in the zebrafish. Dev Dyn 236:3077–3087. https://doi.org/10.1002/dvdy.21354
18. Kwan KM, Fujimoto E, Grabher C et al (2007) The Tol2kit: a multisite gateway-based construction Kit for Tol2 transposon transgenesis constructs. Dev Dyn 236:3088–3099. https://doi.org/10.1002/dvdy.21343
19. Krishnakumar P, Riemer S, Perera R et al (2018) Functional equivalence of germ plasm organizers. PLoS Genet 14:1–29. https://doi.org/10.1371/journal.pgen.1007696

Chapter 25

Zygotic Genome Activation: Critical Prelude to the Most Important Time of Your Life

Vladimir Korzh

Abstract

Activation of the embryonic genome during development represents a major developmental transition in all species. The history of its exploration began in the 1950s–1960s, when this idea was put forward and proven experimentally by Alexander Neyfakh. He observed the aberrant development of fish embryos upon X-ray irradiation and noted the different developmental outcomes depending on the stage when fertilized eggs were subjected to irradiation. Neyfakh also discriminated a regional difference of X-irradiation between the nucleus and the cytoplasm. By selecting the X-ray dose causing nuclear damage, he determined the beginning of zygotic transcription, which at that time became known as the morphogenetic function of nuclei. His team defined the link of zygotic transcription with the asynchronization of cell division and cell migration, the two other hallmarks, which along with the morphogenetic function (or the zygotic genome activation), are at the core of the mid-blastula transition during development. Within this framework, current studies using maternal mutants and application of modern methods of whole-embryo and single-cell transcriptomics begin to decipher the molecular mechanisms of the mid-blastula transition (or the maternal-zygotic transition).

Key words History of science, Morphogenetic function of nuclei, Mid-Blastula transition, Zygotic genome activation, Oocyte-to-embryo transition, Maternal-zygotic-transition, Loach *Misgurnus fossilis*

It is not birth, marriage or death, but gastrulation which is truly the most important time in your life. Lewis Wolpert.

1 Introduction

The history of the concept of zygotic genome activation (ZGA) depicts a chain of events, which led to an understanding of the nature of several biological phenomena, which are partially overlapping [1]. Upon fertilization, the mature oocyte produced by the mother begins to develop and eventually becomes an embryo. During development, the embryonic or zygotic genome is transcriptionally activated. ZGA manifests after the replacement of the parental factors governing embryogenesis (RNA and proteins) by

Roland Dosch (ed.), *Germline Development in the Zebrafish: Methods and Protocols*, Methods in Molecular Biology, vol. 2218, https://doi.org/10.1007/978-1-0716-0970-5_25, © Springer Science+Business Media, LLC, part of Springer Nature 2021

components encoded by the zygotic genome. Most of the parental factors were synthesized under control of the maternal genome and deposited into the egg. These processes led to the introduction of several three-letter words (OET, MFN, MBT, MZT, and ZGA, resolved below), which represent scientific terminology defining these phenomena in different species. This rather broad period and related mechanisms were defined recently as the oocyte-to-embryo transition (OET) [2]. The slightly narrower term "maternal-zygotic transition" (MZT) has been used largely to define transition from dependence of an embryo on maternal factors to dependence on zygotic ones in insects and mammals, i.e., in the species with relatively small eggs [3–9]. In contrast, the term "mid-blastula transition" (MBT) has been used in application to the relatively large embryos of aquatic vertebrate species (fish and amphibians), which develop outside the maternal organism [10].

Egg size varies in different species depending on how much cytoplasm and yolk is deposited during oogenesis. In all species studied, fertilization is followed by a variable period of transcriptional quiescence when the embryonic genome remains transcriptionally inactive. During this period, development is driven by maternal cytoplasmic factors such as mRNA, proteins, and even organelles. At some point, the zygotic genome becomes transcriptionally active, as illustrated by ZGA, i.e., initiation of transcription. From this point onward, the development starts to rely on newly made zygotic factors, whereas the maternal mRNAs are degraded and maternal proteins are slowly replaced by zygotic ones. ZGA represents only one of the interdependent elements of the MZT/MBT. ZGA triggers several closely linked processes such as desynchronization of the cell cycle and acquisition of cell motility. In the past, ZGA was known as the "morphogenetic function of nuclei" (MFN). Numerous studies of many scientists elucidated the cellular events occurring during ZGA. This includes an increase in the nucleo-cytoplasmic ratio, activation of zygotic transcription, and desynchronization of the cell cycle, among others. These days, this process looks more and more complex after the scenario looked relatively simple, when the "morphogenetic function of nuclei" was discovered. Today, ZGA consists of several additional partially overlapping stages, including polyadenylation of some maternal mRNA [10–13], protein phosphorylation, and activation of several "stemness" transcription factors (TFs) such as Pou5fl/3, SoxB1, and Nanog [14–16], chromatin modification, binding of RNA-binding proteins to mRNA 3′UTRs, and clearance of maternal mRNA [17–19].

2 Early Descriptive Period

The concept of MZT/MBT probably began with Theodore Boveri (1862–1915), who put forward the idea that cell nuclei start their specific developmental function just before the onset of gastrulation [20]. In parallel, Gunther Hertwig (1888–1970) noticed in anamniotes (fish and amphibians) of the same stage that a period of early fast synchronous cell divisions is replaced by a later period of slow and asynchronous cell divisions [21, 22]. After that, it was found in the study on lethal hybrids of different species (echinoderms, amphibians, and fish) that such hybrid embryos arrest their development during the blastula stage [23, 24].

In the mid-1950s, Alexander Neyfakh (1926–1997) studied the effect of high doses of X-rays on the development of the loach (*Misgurnus fossilis* L.). Neyfakh's romance with the loach started when he was looking for a useful species to make time-lapse education documentaries, and his mentor Prof. V.V. Popov suggested to use the loach, whose development has been studied already [25]. Upon returning from a film studio to a laboratory, Neyfakh found this species to be a convenient animal model to study development (Fig. 1a). Since X-ray irradiation destroyed the nuclear apparatus, Neyfakh discovered the developmental function of cell nuclei [26, 27]. The loach has been rather instrumental in these studies due to the ease of stimulating egg-laying using gonadotropin (similar to that in *Xenopus*). This usually results in hundreds to thousands of unfertilized eggs from a single female. Their simultaneous fertilization in vitro generates a large number of synchronously developing embryos sufficient for large-scale biochemical purifications [28, 29]. Neyfakh, along with other colleagues, who

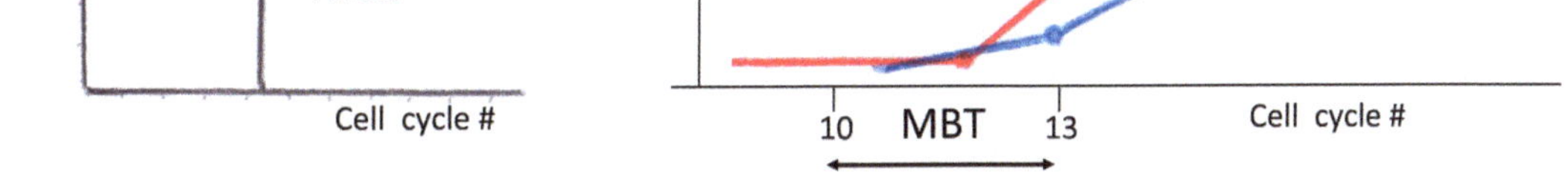

Fig. 1 X-Ray irradiation of fertilized loach eggs with a dose that inactivates nuclei (40,000 roentgen) prior to mid-blastula transition (MBT) causes an arrest in development in the late blastula, during MBT causes an arrest in development at stages between late blastula to organogenesis, irradiation after MBT causes an arrest in development during organogenesis

in parallel performed similar experiments using amphibians [30, 31], noted that X-ray irradiation of the anamniote embryo during early developmental stages leads to the cessation of development at the end of the blastula stage similar to that in the lethal interspecies hybrids. Based on his initial findings in 1956, Neyfakh developed a research program to characterize this phenomenon in more detail. During a period of over 30 years, he and his laboratory published many more papers to define various aspects (ZGA or MFN, cell cycle desynchronization, initiation of cell motility) of the biological process that he discovered in the mid-1950s. Some of these papers were published in as late as 1988 [32].

It all started when Neyfakh demonstrated that upon X-ray irradiation (6–20 kR) of the fertilized loach embryos shortly after they form a chorion, the damage was mostly to the nucleus. When he irradiated the embryos with such a dose, they developed only until the end of the blastula stage and then development was terminated. In contrast, a much higher dose of irradiation (60–100 kR) caused the developmental arrest almost immediately, indicating a cytoplasmic damage [27]. He also noted the periodic increase in radiosensitivity, which correlated with cell division [26]. Based on the synchronous cell divisions of the embryos, he linked the peak of radiosensitivity to the telophase [33], when chromosomes de-condense and repack into newly formed nuclei. Since the damage to the nuclear genetic apparatus did not block development immediately, Neyfakh thought that the nucleus generates a limited amount of substances necessary for the development of the embryo and that these substances move to the cytoplasm after synthesis in the nuclei. Thus, Neyfakh concluded that the "activity of the nucleus at the early and middle blastula stage supports gastrulation" [34]. Later on, Olga Stroeva, a colleague of Neyfakh coined the term "morphogenetic function of nuclei" for this phenomenon. In this way, a novel developmental process acquired its first three-lettered name [35] setting a trend to use such abbreviations.

Noteworthy, these findings were made prior to the discovery of mRNA by Brenner, Jacob, and Meselson in 1961 [36]. Later on, Neyfakh and his colleagues linked MFN with an important component of the developmental mechanism—the start of mRNA synthesis as demonstrated initially in loach (Fig. 1b) [37–39] and in amphibians [40–42]. In the meantime, Neyfakh decided to determine the timing of MFN in fish more accurately and demonstrated that this process begins during the mid-blastula stage [42, 43]. In 1974, Neyfakh completed an analysis of the development of lethal hybrids of the loach and other bony fish, including zebrafish, allowing a much more precise timing of MFN [44].

Soon, mainly by efforts of Neyfakh, his student Natalie Rott and several other colleagues, who followed in the footsteps of Gunther Hertwig, demonstrated that MFN correlates in time with the transition from a period of synchronous cell divisions to asynchronous ones in the cartilaginous (sturgeon) and teleost (loach) fish species [45–51]. These papers clearly defined the second essential component related to ZGA process, but only one became regularly cited [46]. During the early development of all anamniotes, MFN comes as a first manifestation of the cell nucleus activity and then, desynchronization of the cell cycles starts.

These findings were confirmed by experiments on the axolotl [52]. Importantly, French authors proposed the term "transition blastuleenne," which later on was transformed into the "mid-blastula transition" and abbreviated as "MBT" by John Gerhart [53]. The use of this term in modern literature has been validated further largely due to the efforts of John Newport and Marc Kirschner, who stated that they *"reinvestigated several phenomena associated with the midblastula to late blastula period in Xenopus embryos, including desynchronization of individual cell cycles, activation of transcription, and acquisition of cell motility. We have shown that all these events occur at the same time as the midblastula transition (MBT) and that none of them depends on the expression of the others. This concerted transition suggests a common mechanism capable of activating several new cellular processes at the MBT. Furthermore, we have shown that timing of the MBT does not depend on the activation of RNA synthesis, or on a mechanism for counting cleavages, rounds of DNA synthesis or elapsed time since fertilization. Rather, the MBT seems to respond to a critical ratio of nucleus to the cytoplasm. We suggest that a concerted class of cellular parameters responds to this ratio. The ratio may be sensed by a process in which cytoplasmic factors in the egg are titrated by the exponentially increasing nuclei in the embryo during this period of development."* [54]. Therefore, Newport and Kirschner clearly indicated their principal finding. It was neither the MBT itself nor related phenomena (desynchronization of individual cell cycles, activation of transcription, and acquisition of cell motility). They discovered that the timing of MBT is determined by an increase in the number of cell nuclei causing a change in nucleo-cytoplasmic ratio [54, 55].

As a European species, the loach was popular as a model for developmental biology only in the USSR and to some extent in its satellites. Its popularity greatly declined due to reduced funding and brain drain following the collapse of the USSR as well as other species that were more suitable for constant maintenance in the laboratory. Importantly, genetic analysis started to dominate developmental biology following identification of radiation mutants of zebrafish (*Danio rerio*) by George Streisinger (1927–1984) in 1980 [56], an isolation of the first genes regulating development in 1984 [57, 58], and development of gene knockouts in mice in 1987 [59]. In Japan, most studies of teleosts were done on medaka (*Oryzias latipes*), where in respect of these phenomena, first, an acquisition of cell motility has been shown at mid-blastula [60], followed by demonstration of ZGA during the same period [61]. By contrast, in the USA and Canada, other fish species were used at that time as model animals. John P. Trinkaus (1918–2003) described MBT for the seawater teleost mummichog (*Fundulus heteroclitus*), which he used to study how differentiation of deep cells results in the acquisition of cell motility after cleavages [62, 63]. This was paralleled by the efforts of Don Kane and Charles Kimmel, who studied MBT in the zebrafish [64]. Importantly, the studies of complex biological phenomena, which constitute MBT and, in particular ZGA, set the stage for transition to molecular studies of individual genes involved in the regulation of MBT. These efforts became much more focused when large-scale mutagenesis screens on zebrafish were completed in the mid-1990s, which isolated a number of the so-called "maternal" genes of vertebrates [65–68]. While now the studies of maternal genes in vertebrates became a routine and many new discoveries were made in this area, in the early 1990s, it was a matter of some controversy as the invertebrate species were thought to be superior as animal model to study maternal genes.

3 Molecular Studies of ZGA

This research area greatly benefited from the introduction of large-scale whole transcriptomics based on novel methodologies (microarrays, RNA sequencing, lineage tracing, etc.) [69–72]. This set the stage for molecular studies of MBT and ZGA. These experiments on the molecular mechanisms of ZGA have advanced significantly by the availability of maternal-zygotic and/or temperature-sensitive mutants in teleosts, in particular in zebrafish. Using these research tools in combination with morpholino-mediated transient gene knockdown, several important discoveries essential to understand the processes driving ZGA were made. For example, it was found that the maternal stem-loop binding protein 2 (Slbp2) plays a pivotal role in the storage of replication-dependent histone mRNAs and proteins during zebrafish oogenesis. In the absence of maternal Slbp2, early nuclear cleavage was disrupted, which resulted in a developmental arrest at the MBT stage [73]. The bulk of maternal mRNAs produced during oogenesis undergo translational repression necessary for the prevention of premature activation of the developmental program. This is achieved by the shortening of the poly(A) tail upon their export into the cytoplasm. Here the transcripts could be differentially localized into subcellular domains, which in turn determine their embryonic patterning. When mRNAs are differentially localized, they are differently inherited during cell cleavage by blastomeres and contribute to different cell lineages. The localization of dorsal determinants in oocytes and early embryos of anamniotes is determined by the microtubule-based transport of mRNAs in RNP complexes with RNA-binding proteins within the context of ooplasmic segregation. These proteins play a role in the regulation of mRNA processing and repression of their translation. For example, Ybx1 binds to the 3′-UTR of one of the Nodals (Squint), preventing its premature translation and ectopic Nodal signaling [74]. During oogenesis in *Xenopus*, maternal mRNAs are associated with the mitochondrial cloud, an electron-dense structure characterized by the presence of germinal granules and dense array of mitochondria. These mRNA are translationally repressed. During zebrafish oogenesis, the Balbiani body acts as a homolog of the mitochondrial cloud. However, it is known that ZGA results in the formation of distinct types of nuclear granules including those involved in the processing of histone mRNAs and assembly of splicing machinery [75]. To form the cytoplasmic bodies, mRNAs are required. In this context, ZGA could be an initiator of such granules.

Before the initiation of ZGA, many developmental processes, which constitute oogenesis and early embryogenesis, depend on maternal mRNAs only. During this species-dependent prolonged period characterized by an absence of mRNA synthesis, maternal RNAs could be regulated post-transcriptionally by various mechanisms. More than a third of zebrafish maternal mRNAs seems to be N6-methyladenosine (m6A) modification, which triggers the removal of these maternal mRNAs by an m6A-binding protein, Ythdf2. The loss of function of Ythdf2 in zebrafish embryos impedes the degradation of m6A-modified maternal mRNAs, which in turn delays ZGA and MBT and affects cell cycle progression, resulting in an overall developmental delay (Fig. 2) [19]. MBT is followed by the clearance of maternal mRNA by miR-430 [17, 76].

Another mode of post-transcriptional regulation is cytoplasmic polyadenylation of the multiple fractions of translationally inactive mRNAs with short polyA tails. Cytoplasmic polyadenylation triggers an increase in the polysome-bound mRNA fraction, which is also necessary for development past ZGA/MBT (Fig. 2) [12, 13, 69]. It has been shown that the regulation of ZGA in *Drosophila* involves TFs Zelda, Tramtrack, and chromatin remodeling factor Trithorax-like (Trl), which bind the regulatory sites of invertebrate genes activated during ZGA. The involvement of Trl in this process suggests that at least in invertebrates, ZGA probably depends on epigenetic regulation [77]. In the absence of evidence that essential components of this mechanism are evolutionarily conserved, it might be difficult at the moment to see the importance of these findings for vertebrate development.

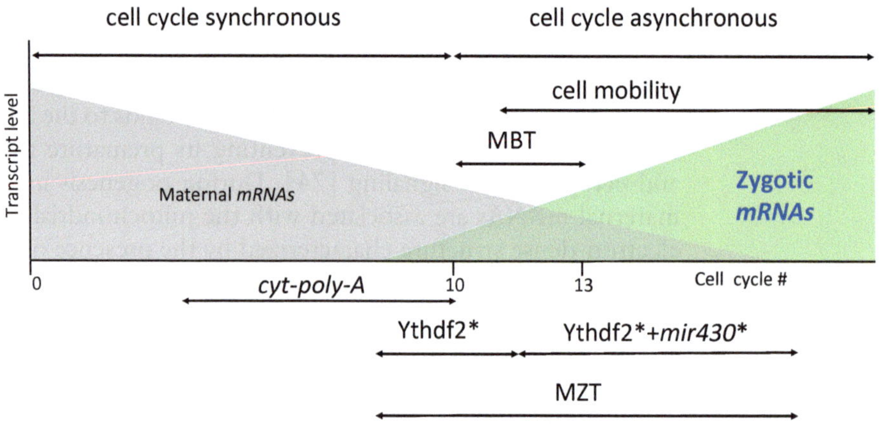

Fig. 2 MBT in the XXIth century (modified from Winata and Korzh, 2018). Prior to MBT, some maternal mRNAs are activated by polyadenylation. All maternal mRNAs are destroyed using the Ythdf2- and mir430-dependent mechanisms. The maternal mRNAs are replaced by newly synthesized zygotic mRNAs

4 Conclusion

In this review, I attempt to trace how development of research methodology led to a better understanding of fundamental biological processes such as MBT [27], which at least initially was a by-product of radiation biology. An introduction of a wide range of new genetic and molecular tools over a period of 60 years resulted in the blossoming of MBT research into a broad scientific field. Here novel and exciting discoveries, such as the role of polyadenylation in events prior to MBT or the role of microRNAs in degradation of maternal RNAs after MBT influenced our understanding of animal development as a whole [10, 78]. Oddly enough, it all began as an attempt to make a time-lapse education documentary. This is a nice illustration of how a principle of serendipity influences scientific discovery. This is also a lesson to those clocking long hours in the laboratories of the world, beginners and seasoned students of life who are on a long and winding road toward discovery.

Acknowledgments

The author was supported by an Opus grant (2016/21/B/NZ3/00354) from the NCN, Poland.

References

1. Jukam D, Shariati S, Skotheim J (2017) Zygotic genome activation in vertebrates. Dev Cell 42:316–332. https://doi.org/10.1016/j.devcel.2017.07.026
2. Svoboda P (2018) Mammalian zygotic genome activation. Semin Cell Dev Biol 84:118–126. https://doi.org/10.1016/j.semcdb.2017.12.006
3. Robbins L (1984) Developmental use of gene products in Drosophila: the maternal zygotic transition. Genetics 108:361–375
4. Robbins L (1980) Maternal zygotic lethal interactions in *Drosophila melanogaster*: the effects of deficiencies in the Zeste-White region of the X chromosome. Genetics 96:187–200
5. Barnes F, Eyestone W (1990) Early cleavage and maternal zygotic transition in bovine embryos. Theriogenology 33:141–152
6. Blythe S, Wieschaus E (2015) Coordinating cell cycle remodelling with transcriptional activation at the Drosophila MBT. Curr Top Dev Biol 113:113–148. https://doi.org/10.1016/bs.ctdb.2015.06.002
7. Harrison M, Eisen M (2015) Transcriptional activation of the zygotic genome in Drosophila. Curr Top Dev Biol 113:85–112. https://doi.org/10.1016/bs.ctdb.2015.07.028
8. Robertson S, Lin R (2015) The maternal-to-zygotic transition in *C. elegans*. Curr Top Dev Biol 113:1–42. https://doi.org/10.1016/bs.ctdb.2015.06.001
9. Tadros W, Lipshitz H (2009) The maternal-to-zygotic transition: a play in two acts. Development 136:3033–3042. https://doi.org/10.1242/dev.033183
10. Winata C, Korzh V (2018) The translational regulation of maternal mRNAs in time and space. FEBS Lett 592:3007–3023. https://doi.org/10.1002/1873-3468.13183
11. Winata C, Kondrychyn I, Kumar V et al (2013) Genome-wide analysis of Zic3 stage-specific interaction with distal regulatory elements of developmental genes. PLoS Genet 9: e1003852. https://doi.org/10.1371/journal.pgen.1003852
12. Winata C, Lapinski M, Pryszcz L et al (2018) Cytoplasmic polyadenylation mediated translational control of maternal mRNAs directs maternal-to-zygotic transition. Development 145:dev159566. https://doi.org/10.1242/dev.159566
13. Ulitsky I, Shkumatava A, Jan C et al (2012) Extensive alternative polyadenylation during zebrafish development. Genome Res 22:2054–2066. https://doi.org/10.1101/gr.139733.112
14. Lee M, Bonneau A, Takacs C et al (2013) Nanog, Pou5fl and SoxB1 activate zygotic gene expression during the maternal-to-zygotic transition. Nature 503:360–364. https://doi.org/10.1038/nature12632
15. Onichtchouk D, Driever W (2016) Zygotic genome activators, developmental timing, and pluripotency. Curr Top Dev Biol 116:273–297. https://doi.org/10.1016/bs.ctdb.2015.12.004
16. Veil M, Yampolsky LY, Grüning B et al (2019) Pou5f3, SoxB1, and Nanog remodel chromatin on high nucleosome affinity regions at zygotic genome activation. Genome Res 29:383–395. https://doi.org/10.1101/gr.240572.118
17. Giraldez A, Mishima Y, Rihel J et al (2006) Zebrafish MiR-430 promotes deadenylation and clearance of maternal mRNAs. Science 312:75–79. https://doi.org/10.1126/science.1122689
18. Schier A, Giraldez A (2006) MicroRNA function and mechanism: insights from zebra fish. Cold Spring Harb Symp Quant Biol 71:195–203. https://doi.org/10.1101/sqb.2006.71.055
19. Zhao B, Wang X, Beadell A et al (2017) m(6) A-dependent maternal mRNA clearance facilitates zebrafish maternal-to-zygotic transition. Nature 542:475–478. https://doi.org/10.1038/nature21355
20. Boveri T (1918) Zwei Fehlerquellen bei merogonien Versuchen und die Entwickslungfahigkeit merogonischer und partiellmerogonischer Seeigelbastarde. Arch Entw Mech Org 44:417–471
21. Hertwig G (1918) Kreuzungsversuche an amphibien. Arch Microsc Anat 91:reference in Schonmann, 1938
22. Schonmann W (1938) Der diploide Bastard Tritom palmatus x Salamandra. Roux Arch Entrwickslungbiol 138:345–375
23. Korovina V. (1939) Geterogennaya gibridizatsiya ryb: Tez. dokl. (Heterogeneous Hybridization of Fish), Leningrad: Leningr. Gos. Univ.
24. Moore J (1955) Abnormal combinations of nuclear and cytoplasmic systems in frogs and toads. Adv Genet 7:139–182
25. Roosen-Runge E (1938) On the early development-bipolar differentiation and cleavage of the zebrafish, *Brachidanio rerio*. Biol Bull 75:119–133

26. Neyfakh AA (1956) The changes of radiosensitivity in the course of fertilization in the loach, *Misgurnus fossilis*. Dokl Akad Nauk SSSR 109:943–946
27. Neyfakh AA (1956) Effect of ionizing radiation on gametes of the loach, *Misgurnus fossilis*. Dokl Akad Nauk SSSR 111:585–588
28. Kostomarova A, Neifakh AA (1964) Method of blastoderm separation in loach embryos and possibilities of its application. Zh Obshch Biol 25:386–388
29. Kostomarova AA (1974) Loach *Misgurnus fossilis* L. In: Dettlaff TA (ed) Ob'ekty biologii razvitiya (Objects of developmental biology). Moscow, Nauka, pp 308–323
30. Mangold O, Peters T (1956) Uber die Wirkung Gleicher Rontgendosen auf Verschiedene Stadien der Entwicklung von Triton alpestris. Beitr Path Anat 116:480–498
31. Sanides F (1956) Die Lethale und Teratogene Wirkung von Rontgenstrahlen auf Ungefurchte Keime und Gastrulen von Triton alpestris in Verschiedenem Milieu. Biol Zbl 75:149–177
32. Burakova T, Korzh V, Khainovskaya A et al (1988) Small nuclear RNAs increase the life span of enucleated loach embryos. Dokl Akad Nauk SSSR 303:733–735
33. Belyayeva VN, Pokrovskaya GL (1959) Change in radiosensitivity of the loach egg during early cleavage. Dokl Akad Nauk SSSR 125:632–635
34. Neyfakh AA (1959) X-ray inactivation of nuclei as a method for studying their function in the early development of fishes. J Embryol Exp Morphol 7:173–192
35. Neyfakh AA (1961) Comparative radiation investigation into morphogenetic functions of nuclei in development of animals. Zh Obshch Biol 22:42–57
36. Brenner S, Jacob F, Meselson M (1961) An unstable intermediate carrying information from genes to ribosomes for protein synthesis. Nature 190:576–581
37. Kafiani CA, Timofeeva MY (1965) The heterogeneity of messenger RNA synthetized at early stages of embryogeny. Dokl Akad Nauk SSSR 164:1183–1186
38. Kafiani CA, Timofeeva MY, Neufakh AA et al (1966) The effect of X-irradiation on the synthesis of messenger RNA at the early stages of loach embryogenesis. Biokhimiia 31:365–371
39. Kafiani CA, Timofeeva MY, Neufakh AA et al (1969) RNA synthesis in the early embryogenesis of a fish (*Misgurnus fossilis*). J Embryol Exp Morphol 21:295–308
40. Bachvarova R, Davidson EH (1966) Nuclear activation at the onset of amphibian gastrulation. J Exp Zool 163:285–295
41. Bachvarova R, Davidson EH, Allfrey RV et al (1966) Activation of RNA synthesis associated with gastrulation. Proc Natl Acad Sci U S A 55:358–365
42. Neyfakh AA (1971) Steps in realization of genetic information in early development. Curr Top Dev Biol 6:45–77
43. Neyfakh AA (1964) Radiation investigation of nucleo-cytoplasmic interrelations in morphogenesis and biochemical differentiation. Nature 201:880–884
44. Neyfakh AA (1974) Nuclear function during early development of remote fish hybrids. Ontogenez:614–622
45. Neyfakh AA, Rott NN (1958) A study of the ways of realization of injury caused by radiation in fishes at early developmental stages. Dokl Akad Nauk SSSR 119:261–264
46. Rott NN, Sheveleva GA (1968) Changes in the rate of cell divisions in the course of early development of diploid and haploid loach embryos. J Embryol Exp Morphol 20:141–150
47. Rott NN (1987) Cell cycles in early development of animals. Nauka, Moscow
48. Rott NN, Sheveleva GA (1967) Changes in the type of cell divisions in the early stages of development of diploid and haploid loach embryos. Tsitologiia 9:1265–1275
49. Kostomarova AA, Ignatieva GM (1968) Correlation of processes of kario- and cytotomy during period of synchronous cleavage divisions in loach (*Misgurnus fossilis* L.). Dokl Acad Nauk SSSR 183:490–492
50. Chulitskaya EV (1970) Desynchronization of cell divisions in the course of egg cleavage and an attempt at experimental shift of its onset. J Embryol Exp Morphol 23:359–374
51. Neyfakh AA, Rott NN (1968) A quantitative approach to the detection of nuclear activity after differential damage to nucleus and cytoplasm in early development. J Embryol Exp Morphol 20:129–140
52. Signoret J, Lefresne J (1971) Contribution a l'etude de la segmentation de l'oef d'axolotl. I. Definition de la transition blastuleenne. Ann Embryol Morphogen 4:113–123
53. Gerhart JC (1980) Mechanisms regulating pattern formation in the amphibian egg and early embryo. In: Goldberg RF (ed) Biological regulation and development, vol 2. Plenum, New York, NY, pp 133–315
54. Newport J, Kirschner (1982) A major developmental transition in early Xenopus embryos: I. Characterization and timing of cellular

changes at the midblastula stage. Cell 30:675–686
55. Newport J, Kirschner M (1982) A major developmental transition in early Xenopus embryos: II. Control of the onset of transcription. Cell 30:687–696
56. Walker C, Streisinger G (1983) Induction of mutations by gamma-rays in pregonial germ cells of zebrafish embryos. Genetics 103:125–136
57. McGinnis W, Garber RL, Wirz J et al (1984) A homologous protein-coding sequence in Drosophila homeotic genes and its conservation in other metazoans. Cell 37:403–408
58. Kuroiwa A, Hafen E, Gehring WJ (1984) Cloning and transcriptional analysis of the segmentation gene fushi tarazu of Drosophila. Cell 37:825–831
59. Thomas KR, Capecchi MR (1987) Site-directed mutagenesis by gene targeting in mouse embryo-derived stem cells. Cell 51:503–512. https://doi.org/10.1016/0092-8674(87)90646-5
60. Kageyama T (1977) Motility and locomotion of embryonic cells of the medaka, *Oryzias latipes*, during early development. Dev Growth Differ 19:103–110
61. Aizawa K, Shimada A, Naruse K et al (2003) The medaka midblastula transition as revealed by the expression of the paternal genome. Gene Expr Patterns 3:43–47
62. Trinkaus JP (1992) The midblastula transition, the YSL transition and the onset of gastrulation in Fundulus. Development 116(Suppl):75–80
63. Trinkaus JP, Erickson CA (1983) Locomotion of Fundulus deep cells during gastrulation. Am Zool 21:401–411
64. Kane DA, Kimmel CB (1993) The zebrafish midblastula transition. Development 119:447–456
65. Haffter P, Granato M, Brand M et al (1996) The identification of genes with unique and essential functions in the development of the zebrafish, *Danio rerio*. Development 123:1–36
66. Kane DA, Hammerschmidt M, Mullins MC (1996) The zebrafish epiboly mutants. Development 123:47–55
67. Kane DA, Maischein HM, Brand M (1996) The zebrafish early arrest mutants. Development 123:57–66
68. Dosch R, Wagner DS, Mintzer KA et al (2004) Maternal control of vertebrate development before the midblastula transition: mutants from the zebrafish I. Dev Cell 6:771–780
69. Aanes H, Winata CL, Lin CH et al (2011) Zebrafish mRNA sequencing reveals polyadenylation of maternal RNAs as a mechanism regulating gastrulation. Genome Res 21:1328–1338. https://doi.org/10.1101/gr.116012.110
70. Mathavan S, Lee S, Mak A et al (2005) Transcriptome analysis of zebrafish embryogenesis using microarrays. PLoS Genet 1:e29. https://doi.org/10.1371/journal.pgen.0010029
71. McKenna A, Findlay GM, Gagnon JA et al (2016) Whole-organism lineage tracing by combinatorial and cumulative genome editing. Science 353:aaf7907. https://doi.org/10.1126/science.aaf7907
72. Satija R, Farrell JA, Gennert D et al (2015) Spatial reconstruction of single-cell gene expression data. Nat Biotechnol 33:495–502. https://doi.org/10.1038/nbt.3192
73. He WX, Wu M, Liu Z et al (2018) Oocyte-specific maternal Slbp2 is required for replication-dependent histone storage and early nuclear cleavage in zebrafish oogenesis and embryogenesis. RNA 24:1738–1748. https://doi.org/10.1261/rna.067090.118
74. Kumari P, Gilligan PC, Lim S (2013) An essential role for maternal control of nodal signaling. eLife 2:e00683. https://doi.org/10.7554/eLife.00683
75. Heyn P, Salmonowicz H, Rodenfels J et al (2017) Activation of transcription enforces the formation of distinct nuclear bodies in zebrafish embryos. RNA Biol 14:752–760. https://doi.org/10.1080/15476286.2016.1255397
76. Bazzini AA, Lee MT, Giraldez AJ (2012) Ribosome profiling shows that miR-430 reduces translation before causing mRNA decay in zebrafish. Science 336:233–237. https://doi.org/10.1126/science.1215704
77. Darbo E, Herrmann C, Lecuit T et al (2013) Transcriptional and epigenetic signatures of zygotic genome activation during early Drosophila embryogenesis. BMC Genomics 14:226. https://doi.org/10.1186/1471-2164-14-226
78. Takacs CM, Giraldez AJ (2010) MicroRNAs as genetic sculptors: fishing for clues. Semin Cell Dev Biol 21:760–767

Chapter 26

Identification of RNA-Binding Protein Landscapes Across Zebrafish Embryonic Transcriptome via iCLIP Approach

Vladimir Despic

Abstract

In metazoans, fertilization initiates vast remodeling of the embryonic proteome and transcriptome. This is accomplished via complex post-transcriptional regulation of maternal and zygotic RNA. RNA-binding proteins (RBPs) are one of the major mediators of embryonic post-transcriptional RNA regulation. Thus, elucidation of the molecular mechanisms by which maternal and zygotic transcripts change their translational capacities and expression levels requires thorough and precise determination of the targets and binding sites of individual RBPs in embryonic transcriptomes. Here, I provide a detailed protocol for the UV crosslinking-based method, named iCLIP, to study RBP functions during early zebrafish embryogenesis.

Key words Zebrafish, Embryogenesis, RNA-binding proteins, UV crosslinking, iCLIP

1 Introduction

Several temporally orchestrated molecular events characterize and drive early embryogenesis in all metazoans. Following fertilization, the maternally inherited transcriptome undergoes extensive translational regulation to control the repertoire of the embryonic proteome [1–3]. This is followed by two tightly coupled processes, namely massive degradation of maternal mRNAs and their replacement by new zygotic mRNA complement as a consequence of the transcriptional awakening of the zygotic genome [3]. These tectonic shifts in the landscape of both the embryonic transcriptome and proteome are the result of the complex and dynamic RNA-protein interactions [3–5]. In this unique post-transcriptional regulatory network, RNA-binding proteins (RBPs) are important components that act to control various aspects of maternal and zygotic mRNA fates [3]. Therefore, understanding molecular principles that underlie embryonic transcriptome dynamics, and development itself, requires detailed identification

Roland Dosch (ed.), *Germline Development in the Zebrafish: Methods and Protocols*, Methods in Molecular Biology, vol. 2218,
https://doi.org/10.1007/978-1-0716-0970-5_26,

© Springer Science+Business Media, LLC, part of Springer Nature 2021

of the direct targets of individual RBPs, as well as their binding motifs across embryonic RNA.

Over the past two decades, a multitude of protein-centric UV crosslinking approaches (e.g., HITS-CLIP, iCLIP, PAR-CLIP, irCLIP, eCLIP) have been developed to catalog targets of RBPs and their binding sites in RNA at a single-nucleotide resolution [6–10]. UV crosslinking methods appear to be the most valuable tools for studying RNA-protein interactions in vivo, owing to the ability of UV light to capture RNA-protein contact sites at 0 Å resolution [11]. Recent adaptations of these methods to living organisms have started to elucidate the contribution of individual RBPs to the regulation of the embryonic transcriptome dynamics [5, 12–16]. Here, I provide a detailed individual, single-nucleotide resolution crosslinking and immunoprecipitation (iCLIP) protocol for identification of the RBP targets and binding motifs during early zebrafish development (Fig. 1).

2 Materials

2.1 Staging and Collection of Zebrafish Embryos

1. 10-cm petri dishes.
2. E3 medium: 5 mM NaCl, 0.17 mM KCl, 0.33 mM $CaCl_2$, 0.33 mM $MgSO_4$, 0.00005% methylene blue in water.
3. Incubator.
4. Stereomicroscope.
5. Plastic or glass Pasteur pipettes.
6. Sterile 1.5-ml tubes.
7. 1× phosphate-buffered saline (PBS).
8. UV crosslinker with UV 254-nm bulbs (e.g., UV stratalinker 2400, Stratagene).
9. Custom-made mRNA encoding an RBP fused to a protein tag.
10. Glass capillaries (1.0 mm O.D. × 0.58 mm I.D. capillaries).
11. Microinjection system.
12. Needle puller.

2.2 Preparation of the Embryonic Lysate

1. 4 °C centrifuge.
2. Lysis buffer: 50 mM Tris–HCl pH 7.4, 150 mM NaCl, 1% nonidet P-40 (NP-40), 0.1% sodium dodecylsulfate (SDS), 0.5% sodium deoxycholate, 5 μl/ml RNaseOUT, 10 μl/ml 100× Protease and Phosphatase Inhibitor Cocktail (use RNase-free water and filter the final buffer).
3. 22 G $1^{1/2}$ gauge needles.
4. 1-ml syringes.

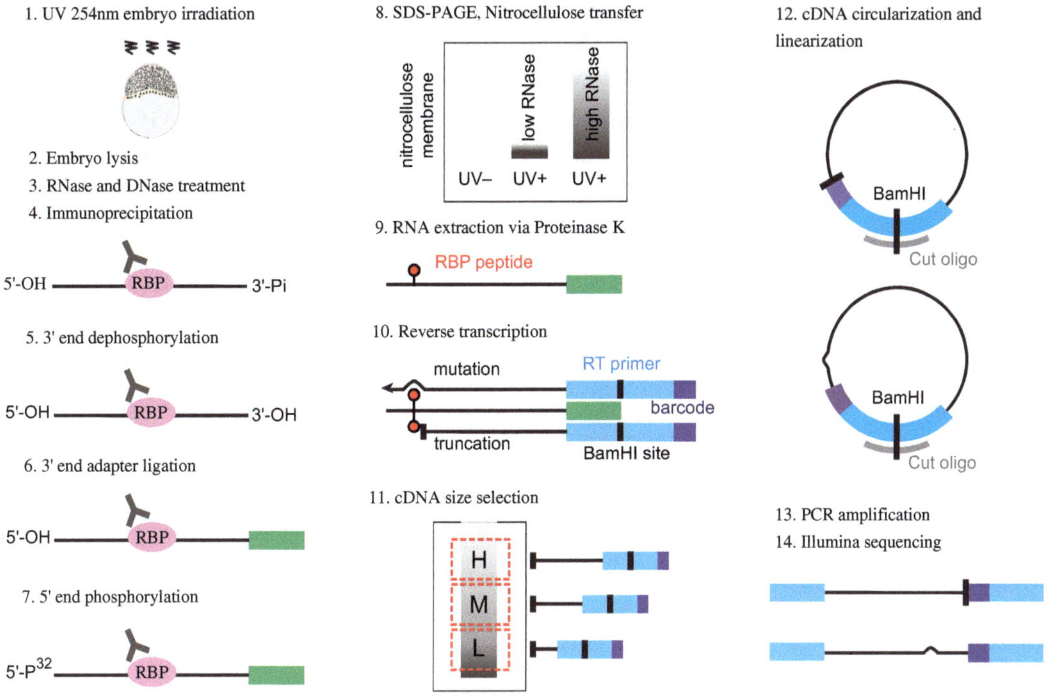

Fig. 1 iCLIP protocol overview. Irradiation of zebrafish embryos at the desired developmental stage (e.g., 1K-cell stage) with 254 nm UV light generates irreversible crosslinks between RNA and proteins in vivo (step 1). Subsequent embryo lysis (step 2) and RNase treatment (step 3) liberate small protein-RNA complexes that are immunopurified with RBP-specific antibody (step 4). RNA 3′ end phosphates are removed with PNK enzyme (step 5) to allow 3′ end L3 adapter ligation (step 6). Following radioactive labeling of RNA 5′ ends (step 7), RBP-RNA complexes are resolved on SDS-PAGE, transferred onto the nitrocellulose membrane, and visualized by autoradiography (step 8). Subsequent proteinase K treatment of the membrane pieces with RBP-RNA complexes elutes the RNA from the membrane, leaving a short RBP peptide attached to the RNA (step 9). Next, reverse transcription is conducted with primers containing barcode and sequences for PCR amplification (step 10). Reverse transcriptase prematurely terminates or creates specific mutational signature at the RBP peptide site. This results in the production of either truncated or mutated cDNA sequences. Three different sizes of cDNA are selected from the 6% TBE-urea gel (low, medium, and high) (step 11). cDNA fractions are further circularized. Next, the Cut oligo anneals to the portion of the 3′ end adapter and allows BamHI-induced linearization of cDNA (step 12). This linearization creates cDNA templates suitable for PCR amplification (step 13). Finally, PCR amplified libraries are sequenced on the Illumina sequencing platforms and resulting reads are analyzed for determination of RBP-binding sites (crosslink positions) (step 14)

5. RNaseOUT Recombinant Ribonuclease Inhibitor.
6. 100× Protease and Phosphatase Inhibitor Cocktail.

2.3 DNase and RNase Treatment

1. TURBO DNase.
2. RNase I.

2.4 Immunoprecipitation

1. RBP-specific antibody or alternatively protein tag-specific antibody (e.g., HA-, FLAG-tag).

2. High-salt buffer: 50 mM Tris–HCl pH 7.4, 1 M NaCl, 1 mM EDTA, 0.1% SDS, 0.5% sodium deoxycholate, 1% NP-40 (use RNase-free water and filter the final buffer).
3. Urea-containing lysis buffer: 50 mM Tris–HCl pH 7.4, 150 mM NaCl, 1 M urea, 1% NP-40, 0.1% SDS, 0.5% sodium deoxycholate (use RNase-free water and filter the final buffer).
4. PNK buffer: 20 mM Tris–HCl pH 7.4, 10 mM $MgCl_2$, 0.2% Tween-20 (use RNase-free water and filter the final buffer).
5. Dynabeads Protein A/G for Immunoprecipitation.
6. Magnetic rack for 1.5-ml tubes.

2.5 3′-End RNA Dephosphorylation

1. 1 M Tris–acetate, pH 6.0.
2. 0.1 M dithiothreitol (DTT).
3. 0.2 M $MgCl_2$.
4. T4 polynucleotide kinase (T4 PNK).

2.6 3′-End Adapter Ligation

1. T4 RNA Ligase 2, truncated KQ, (NEB).
2. 50% PEG8000.
3. Yeast 5′ deadenylase.
4. RecJf enzyme.
5. L3 adapter: /5rApp/ AGATCGGAAGAGCGGTTCAG / 3ddc/.

2.7 5′-End Radioactive Labeling of RNA

1. γ-P^{32}-ATP.
2. 4× NuPAGE LDS sample buffer.
3. 1 M DTT.

2.8 SDS-PAGE

1. NuPAGE 4–12% Bis-Tris protein gels, 1.0 mm, ten wells.
2. XCell SureLock Mini-Cell and XCell II Blot Module or alternatives.
3. Phase Lock Gel Heavy.
4. Amersham Hyperfilm MP.
5. Phenol:chloroform:isoamyl alcohol, 25:24:1, pH 6.6.
6. 20× NuPAGE MOPS SDS running buffer.
7. Saran wrap.
8. Proteinase K (PK) buffer: 100 mM Tris–HCl, pH 7.5, 50 mM NaCl, 10 mM EDTA, 0.5% SDS.
9. Proteinase K solution (20 mg/ml), RNA grade.
10. Ethanol, 100% and 75%.
11. Glycoblue.
12. 3 M sodium acetate, pH 5.5.

13. 20% SDS.
14. Sterile scalpels/razor blades.

2.9 Reverse Transcription

1. PCR machine.
2. SuperScript III reverse transcriptase kit.
3. 1× TE buffer: 10 mM Tris–HCl pH 7.5, 1 mM EDTA.
4. 10 mM dNTP mix (each).
5. RT primers:

   RT iCLIP1: /5phos/NNaaccNNNAGATCGGAAGAGCGTCGTGgatcCTGAACCGC.

   RT iCLIP2: /5phos/NNacaaNNNAGATCGGAAGAGCGTCGTGgatcCTGAACCGC.

   RT iCLIP3: /5phos/NNattgNNNAGATCGGAAGAGCGTCGTGgatcCTGAACCGC.

   RT iCLIP4: /5phos/NNaggtNNNAGATCGGAAGAGCGTCGTGgatcCTGAACCGC.

   RT iCLIP5: /5phos/NNcgccNNNAGATCGGAAGAGCGTCGTGgatcCTGAACCGC.

   RT iCLIP6: /5phos/NNccggNNNAGATCGGAAGAGCGTCGTGgatcCTGAACCGC.

   RT iCLIP7: /5phos/NNctaaNNNAGATCGGAAGAGCGTCGTGgatcCTGAACCGC.

   RT iCLIP8: /5phos/NNcattNNNAGATCGGAAGAGCGTCGTGgatcCTGAACCGC.

   RT iCLIP9: /5phos/NNgccaNNNAGATCGGAAGAGCGTCGTGgatcCTGAACCGC.

   RT iCLIP10: /5phos/NNgaccNNNAGATCGGAAGAGCGTCGTGgatcCTGAACCGC.

   RT iCLIP11: /5phos/NNggttNNNAGATCGGAAGAGCGTCGTGgatcCTGAACCGC.

   RT iCLIP12: /5phos/NNgtggNNNAGATCGGAAGAGCGTCGTGgatcCTGAACCGC.

   RT iCLIP13: /5phos/NNtccgNNNAGATCGGAAGAGCGTCGTGgatcCTGAACCGC.

   RT iCLIP14: /5phos/NNtgccNNNAGATCGGAAGAGCGTCGTGgatcCTGAACCGC.

   RT iCLIP15: /5phos/NNtattNNNAGATCGGAAGAGCGTCGTGgatcCTGAACCGC.

   RT iCLIP16: /5phos/NNttaaNNNAGATCGGAAGAGCGTCGTGgatcCTGAACCGC.

2.10 cDNA Size Selection

1. Novex 2× TBE-urea sample buffer.
2. Low-molecular-weight DNA ladder (25–766 bp from NEB or alternative).
3. Novex TBE-urea gels, 6%, ten wells.
4. 1× TBE buffer: 89 mM Tris base, 89 mM Boric acid, 2 mM EDTA.
5. Gel Doc system.
6. SYBR Gold Nucleic Acid stain, 1000×.
7. Gel breaker tubes.

2.11 cDNA Circularization and Linearization

1. CircLigase II ssDNA ligase.
2. 5 M Betaine.
3. 10 mM ATP.
4. FastDigest BamHI.
5. Cut oligo: 5'-GTTCAGGATCCACGACGCTCTTC/3ddC/.

2.12 PCR Amplification

1. Phusion High-Fidelity PCR Master Mix with HF buffer.
2. P3 Solexa primer: 5′- CAAGCAGAAGACGGCATACGA GATCGGTCTCGGCATTCCTGCTGAACCGCTCTTCCG ATCT-3′.
3. P5 Solexa primer: 5′-AATGATACGGCGACCACCGAGATC TACACTCTTTCCCTACACGACGCTCTTCCGATCT-3′.
4. AMPure XP, PCR purification.
5. Novex TBE Gels, 6%, ten wells.

3 Methods

3.1 Staging and Collection of Zebrafish Embryos

1. Prepare several breeding tanks with a single mating pair of fish the day before the experiment. The following day, allow the fish pairs to mate. The moment the eggs start appearing, start the timer and collect the eggs after 10 min (*see* **Note 1**).
2. In the absence of the specific antibody for an endogenous RBP of interest, collected fertilized eggs can be microinjected with an mRNA encoding the RBP fused to a protein tag for which commercially available antibodies exist. Successful immunoprecipitations from zebrafish lysates have been reported with the following antibodies: mouse monoclonal anti-HA.11 antibody, clone 16B12 from BioLegend, and rabbit polyclonal anti-FLAG antibody from Sigma [16, 17].
3. Grow (microinjected) embryos in 10-cm petri dishes with E3 medium at 28 °C. After 35–40 min, inspect the embryonic stage (2-cell stage) and place 200 synchronized embryos in a

new 10-cm petri dish filled with 10 ml E3 medium using a plastic Pasteur pipette. Continue growing embryos at 28 °C until they reach the desired developmental stage. In case of the mRNA-microinjected embryos, allow the embryos to develop at least until the 1 K-cell stage to ensure sufficient expression of the RBP.

4. Irradiate two groups of 200 embryos twice with 0.8 J/cm^2 UV 254 nm light. Perform irradiation on ice. Simultaneously prepare two batches of 200 non-UV-irradiated embryos at the same developmental stage. Non-irradiated embryos will serve as a negative control (*see* **Notes 2** and **3**).
5. Collect each embryo batch in a 1.5-ml tube, wash with 1× PBS and aspirate as much liquid as possible with a pipette without disturbing embryos. Snap freeze embryos in liquid nitrogen and keep the samples at −80 °C. Irradiated embryos can be kept at −80 °C for a long period of time.

3.2 Preparation of the Embryonic Lysate

1. Allow the embryos to thaw on ice and lyse each batch of embryos in 400 μl ice-cold lysis buffer using a 1-ml syringe and 22 G 1$^{1/2}$ gauge needle. Eight to ten passages through the syringe and needle will ensure complete embryo lysis. Simultaneously process UV-irradiated and non-irradiated lysates in the same manner.
2. Clear the embryonic lysates by centrifugation at 20,000 × *g* for 10 min at 4 °C. Combine supernatants from two 200-embryo lysates into a single, 1.5-ml tube. Keep the final lysate on ice (*see* **Note 4**).
3. Adjust the volume of the combined lysate to 1 ml with the appropriate volume of lysis buffer.
4. Centrifuge final lysate at 20,000 × *g* for 10 min at 4 °C, save the supernatant, and keep it on ice.

3.3 DNase and RNase Treatment

1. Add 5 μl TURBO DNase and 10 μl 1:500 dilution of RNase I to the lysate. Invert the tube twice and place it immediately in a thermo mixer and incubate for 10 min at 37 °C at 1,100 rpm. Transfer the tube on ice and incubate for at least 5 min (*see* **Note 5**).

3.4 Immunoprecipitation

1. Add 8–10 μg RBP-specific antibody to the RNase/DNase-digested lysate and incubate for 2 h to overnight at 4 °C on a rotating wheel. (*see* **Note 6**).
2. Wash 35 μl Protein A/G Dynabeads twice with 500 μl lysis buffer.
3. Reconstitute washed beads in 35 μl lysis buffer and add them to the antibody-lysate mixture and incubate for 1.5 h at 4 °C.

4. Collect the beads on the magnetic rack and discard the lysate.
5. Gently resuspend the beads in 850 μl high-salt buffer and incubate for 5 min at 4 °C.
6. Collect the beads on the magnetic rack and discard the supernatant.
7. Repeat **steps 5** and **6**.
8. Wash the beads twice with 850 μl urea-containing lysis buffer.
9. Wash the beads twice with 850 μl lysis buffer.
10. Wash the beads twice with 850 μl PNK buffer.
11. Collect the beads on the magnetic rack and discard the supernatant.

3.5 3′-End RNA Dephosphorylation

1. Resuspend the beads in the following mix: 14.5 μl nuclease-free water, 2 μl 1 M Tris–acetate, pH 6.0, 1 μl 0.1 M DTT, 1 μl 0.2 M $MgCl_2$.
2. Add 0.5 μl RNaseOUT and 1 μl PNK enzyme.
3. Mix the reaction and incubate in a thermo mixer for 30 min at 23 °C at 1,100 rpm (*see* **Note 7**).
4. Wash the beads with 850 μl urea-containing lysis buffer.
5. Wash the beads with 850 μl lysis buffer.
6. Wash the beads twice with 850 μl PNK buffer.
7. Collect the beads on the magnetic rack and discard the supernatant.

3.6 3′-End Adapter Ligation

1. Gently resuspend the beads in 11 μl ligation mix (6.5 μl nuclease-free water, 2 μl 10× RNA ligase buffer, 0.5 μl RNase-OUT, 0.5 μl 40 mM DTT, 1.5 μl 20 μM L3 adapter).
2. Add 1 μl T4 RNA ligase 2, truncated KQ, and 8 μl 50% PEG8000; briefly vortex and incubate in a thermo mixer for 3 h at 25 °C with the following cycle: 15 s shaking at 1,200 rpm followed by 30 s pause.
3. Add 0.5 μl yeast 5′ deadenylase and 0.5 μl RecJf enzyme. Incubate reaction at 30 °C for 45 min with shaking at 1,100 rpm (*see* **Note 8**).
4. Wash the beads with 850 μl urea-containing lysis buffer.
5. Wash the beads with 850 μl lysis buffer.
6. Wash the beads twice with 850 μl PNK buffer.
7. Collect the beads on the magnetic rack and discard the supernatant.

3.7 5′-End Radioactive Labeling of RNA

1. Gently resuspend the beads in the labeling buffer (16 μl nuclease-free water, 2 μl 10× PNK buffer, 1 μl T4 polynucleotide kinase).
2. Add 1 μl γ-P^{32}-ATP and incubate at 37 °C for 10 min at 1,100 rpm (*see* **Note 9**).
3. Wash the beads twice with 850 μl PNK buffer and discard the supernatant containing free radioactive ATP.
4. Resuspend the beads in 20 μl DTT-containing 1.5× NuPAGE loading buffer and incubate for 10 min at 75 °C.
5. Collect the beads on the magnetic rack and save the eluate in a new 1.5-ml tube. Place the eluted RNA-protein complex mixture on ice.

3.8 SDS-PAGE, Nitrocellulose Transfer, and RNA Isolation

1. Load UV-irradiated and non-irradiated samples on a 4–12% NuPAGE Bis-Tris gel. Leave one empty lane between samples.
2. Run the gel in 1× MOPS running buffer for 1 h at 180 V (*see* **Note 10**).
3. Carefully remove the gel out of the cassette and quickly rinse in nuclease-free water. Remove the bottom, free ATP-containing part of the gel with a clean razor blade.
4. Transfer RNA-protein complexes to a 0.45-μm nitrocellulose membrane for 2 h at 30 V using XCell SureLock™ wet transfer apparatus or equivalent according to the manufacturer's instructions.
5. After the transfer is completed, wash the membrane in 1× PBS, close it with a Saran wrap, place the film on top of the membrane, and expose for 1 h, 4 h, or overnight at −80 °C (*see* **Note 11**).
6. Analyze the exposed film and cut the membrane piece containing RNA-protein complexes of interest (Fig. 2).
7. Cut the excised membrane part into smaller pieces, and submerge them into 300 μl PK buffer.
8. Add 10 μl PK solution (20 mg/ml) and incubate for 45 min at 50 °C at 1,300 rpm.
9. Transfer the solution into a 2 ml Phase Lock Gel Heavy tube and mix it with 300 μl phenol:chloroform:isoamyl alcohol, 25:24:1, pH 6.6.
10. Incubate the mixture for 5 min at 30 °C and 1,000 rpm.
11. Centrifuge for 5 min at 16,000 × *g* at room temperature and save the aqueous phase in a fresh 1.5-ml tube.
12. Add 2 μl glycoblue, 1/10 volume 3 M sodium acetate (pH 5.5), and 2.5–3 volumes 100% ethanol; vortex, and precipitate RNA overnight at −20 °C.

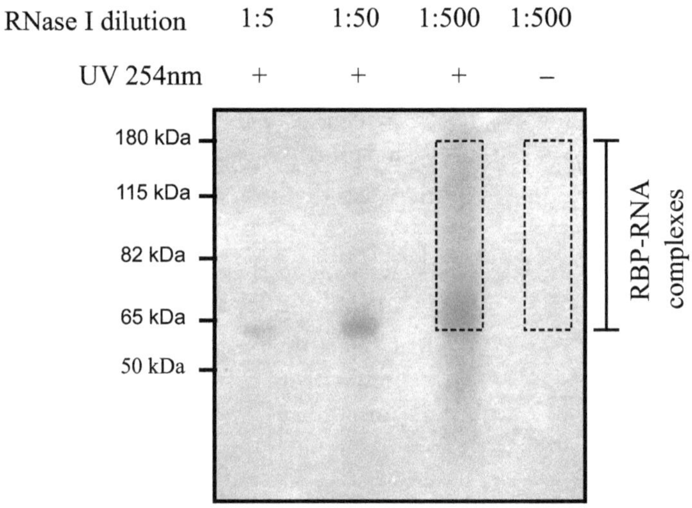

Fig. 2 hnRNP A1-protein complexes immunopurified from UV-irradiated (UV+) and non-irradiated (UV–) zebrafish embryonic lysates treated with a series of RNase I dilutions. 1:5 and 1:50 RNase I dilutions result in RNA overdigestion, as indicated by a single band that corresponds to the size of hnRNP A1 protein. 1:500 RNase I dilution generates larger hnRNP A1-RNA complexes as a result of longer RNA fragments attached to the protein (visible smear). Dashed rectangles show the part of the membrane that was subjected to proteinase K treatment and RNA extraction

13. Centrifuge at 4 °C for 30 min at 20,000 × *g*.
14. Wash RNA pellet with 500 μl 75% ethanol, and centrifuge at 20,000 × *g* for additional 15 min.
15. Take out the alcohol, air dry RNA pellet, and resuspend in 10 μl nuclease-free water.

3.9 Reverse Transcription

1. Transfer the RNA into PCR tubes.
2. Add 1 μl RT iCLIP primer (3 μM) and mix.
3. Heat the mixture for 2 min at 90 °C, and cool it down to 4 °C.
4. Add the following components into the reaction mixture: 4 μl 5× First strand buffer, 1 μl 0.1 M DTT, 1 μl dNTP mix (10 mM each), 0.5 μl RNaseOUT, and 1 μl SuperScript III.
5. Run the following program for reverse transcription: 5 min 25 °C, 15 min 42 °C, 45 min 56 °C, 10 min 80 °C, 4 °C on hold.
6. Add 330 μl 1× TE buffer, 2 μl glycoblue, 40 μl 3 M sodium acetate (pH 5.5), and 1 ml 100% ethanol; vortex and precipitate cDNA overnight at −20 °C.
7. Centrifuge at 20,000 × *g* for 30 min at 4 °C.

8. Wash cDNA pellet once with 75% ethanol and centrifuge for additional 15 min.
9. Take out the alcohol, air dry RNA pellet, and resuspend in 6 μl nuclease-free water.

3.10 cDNA Size Selection

1. Add 4 μl 2× TBE-urea sample buffer to the 6 μl cDNA, mix with a pipette, and heat at 75 °C for 3 min.
2. Allow the samples to cool on ice for 2 min.
3. Load low-molecular-weight DNA ladder and cDNA samples onto a 6% TBE-urea gel. Make sure to leave one empty lane between adjacent samples to avoid cross-contamination.
4. Run the gel at 180 V for 40 min in 1× TBE buffer.
5. Excise the entire lane with the DNA ladder and soak it in the staining buffer (30 ml 1× TBE and 3 μl SYBR Gold) for 5 min. UV-expose DNA ladder gel piece to visualize DNA marker. Take a picture and print it in a 100% scale. Use this print out as a reference to excise the cDNAs in the following size ranges: 66–80 nt (low), 80–150 nt (medium), and 150–300 nt (high) (*see* **Note 12**).
6. Crush the excised gel pieces by passing them through a gel breaker tube. Add 350 μl 1× TE buffer and incubate at 1,100 rpm for 2 h at 37 °C. Freeze the tubes for 2–3 min at −80 °C and continue incubation at 1,100 rpm overnight at 4 °C.
7. Purify the elution mixture using Costar SpinX columns by 2-min centrifugation at 16,000 × *g*.
8. Transfer the sample to a 2-ml Phase Lock Gel Heavy tube and mix it with 300 μl phenol:chloroform:isoamyl alcohol, 25:24:1, pH 6.6.
9. Incubate the mixture for 5 min at 30 °C at 1,000 rpm.
10. Centrifuge for 5 min at 16,000 × *g* at room temperature and save the aqueous phase in a fresh 1.5-ml tube.
11. Add 2 μl glycoblue, 1/10 volume 3 M sodium acetate (pH 5.5), and 2.5–3 volumes 100% ethanol; vortex and precipitate gel extracted cDNA overnight at −20 °C.
12. Centrifuge at 20,000 × *g* for 30 min at 4 °C.
13. Wash cDNA pellet once with 75% ethanol and centrifuge for additional 15 min.
14. Take out the alcohol, air dry RNA pellet, and resuspend in 6 μl nuclease-free water.

3.11 cDNA Circularization and Linearization

1. Resuspend the cDNA pellet in 11 μl nuclease-free water, add 8 μl of the following ligation mix (2 μl 10× Circligase buffer II, 1 μl 50 mM $MnCl_2$, 1 μl 1 mM ATP, 4 μl 5 M Betaine), and 1 μl

CircLigase II. Incubate the reaction for 2.5 h at 60 °C. Heat inactivate the enzyme for 10 min at 80 °C.

2. Add the annealing solution to the circularized cDNA: 23.75 μl nuclease-free water, 5 μl 10× FastDigest buffer, and 1.25 μl 10 μM Cut oligo.
3. Heat the samples to 95 °C for 2 min and start gradually decreasing temperature by 1 °C every 20 s until 25 °C is reached.
4. Add 2 μl FastDigest BamHI enzyme to the samples and incubate for 30 min at 37 °C.
5. Heat inactivate BamHI for 10 min at 80 °C and cool down the reaction to 4 °C.
6. Precipitate cDNA overnight at −20 °C by adding 300 μl 1× TE buffer, 1 μl glycoblue, 35 μl 3 M sodium acetate (pH 5.5), and 0.9 ml 100% ethanol.
7. Centrifuge at 20,000 × *g* for 30 min at 4 °C.
8. Wash the pellet once with 500 μl 75% ethanol and spin again for 10 min.
9. Resuspend air-dried pellet in 15 μl nuclease-free water.

3.12 PCR Amplification

1. For initial library amplification tests, mix 1 μl cDNA with 10 μl 2× Phusion High-Fidellity PCR master mix, 8.5 μl nuclease-free water, and 0.5 μl P3/P5 primer mix (10 μM each), and amplify with 15, 20, and 25 PCR cycles using the following program: 1× (98 °C, 2 min), 15× (98 °C, 15 s | 65 °C, 15 s, 72 °C, 20 s), 1× (72 °C, 3 min, 4 °C, hold).
2. Assess the quality and quantity of the amplified libraries by running them on a 6% TBE gel and staining with SYBR Gold for 5–10 min (Fig. 3, *see* **Note 13**).
3. For preparative library amplification, use 10 μl linearized cDNA and scale up the reaction mix from 20 to 40 μl.
4. Perform PCR with at least two-three cycles less than the optimal number of PCR cycles determined in the test PCR reaction at **step 2**.
5. Purify final libraries with AMPure XP beads according to the manufacturers' instructions.
6. Analyze iCLIP library quality and quantity on Agilent Bioanalyzer high sensitivity (HS) DNA chip.
7. Mix three different size-selected fractions of amplified cDNA libraries (high, medium, and low) in an equimolar ratio and sequence on Illumina sequencing platforms.

3.13 Bioinformatic Analysis of iCLIP Data

Consult the following literature for iCLIP read processing and determination of RBP-binding sites [7, 18–21].

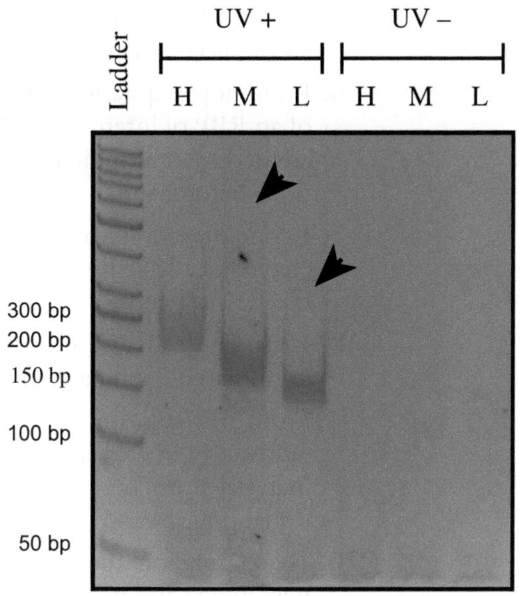

Fig. 3 PCR amplification of iCLIP libraries. SYBR Gold-stained 6% TBE gel with optimally amplified cDNA library is shown (18 PCR cycles). Black arrows indicate slight PCR overamplification in the low and medium cDNA fractions. Note the lack of PCR products in the negative, UV non-irradiated samples

4 Notes

1. It is possible to collect eggs over longer breeding periods, but it will result in an increased number of unsynchronized embryos.
2. Determine an appropriate UV energy dose before performing the entire experiment. UV crosslinking efficiency is expected to decrease as the embryo continues growing. Thus, an increase in the applied UV energy dose might be necessary for later developmental stages.
3. The number of embryos for an experiment will depend on several factors, including the expression level of the protein of interest, its RNA-binding capacity, and UV crosslinking efficiency. Thus, protocol users should determine the optimal number of embryos for their own iCLIP experiments.
4. Avoid pipetting small chorion pieces.
5. RNase treatment is an essential step of the protocol and must be optimized beforehand. The best iCLIP results are obtained from the RNase treatments that generate RNA fragments in 50–300 nt range. For example, 100-nt-long RNA fragment attached to the protein of interest will shift the molecular

weight of the protein for around 30 kDa on the SDS-PAGE autoradiograph (*see* Subheading 3.8, Fig. 2).

6. Immunoprecipitation efficiency will depend on the expression of an RBP of interest and the quality of the antibody. Users are advised to determine the optimal immunoprecipitation conditions beforehand.
7. RNase digestion leaves 3′ terminal phosphates that need to be enzymatically removed for efficient ligation of the adapter at the RNA 3′ ends.
8. Yeast 5′ deadenylase and RecJf treatment will remove the non-ligated 3′ end adapter species.
9. From this step onward, protocol users will be working with radioactive materials. Handling and disposal of radioactive material must be carried out in accordance with the institutional policies and regulations.
10. Depending on the size of an RBP of interest, different gel percentage and gel running time might be required.
11. Try all three recommended exposures and determine which one of them gives the clearest signal on the film.
12. Unless an RBP of interest is known to interact with small RNA species (e.g., mature miRNAs), I recommend avoiding the isolation of the low cDNA fraction. This fraction can be easily contaminated with cDNAs originated from unwashed, non-ligated adapter sequences.
13. The adequate number of PCR cycles is determined by the intensity of the SYBR Gold signal on the 6% TBE gel; iCLIP libraries should be faintly visible in the expected size range. Overamplification of iCLIP libraries usually results in the appearance of smears above the expected library size (Fig. 3).

Acknowledgments

I thank the entire Neugebauer lab at Yale University for their enormous help and support during the establishment of this protocol.

References

1. Eichhorn SW, Subtelny AO, Kronja I, Kwasnieski JC, Orr-Weaver TL, Bartel DP (2016) mRNA poly(A)-tail changes specified by deadenylation broadly reshape translation in drosophila oocytes and early embryos. Elife 5: e16955
2. Subtelny AO, Eichhorn SW, Chen GR, Sive H, Bartel DP (2014) Poly(A)-tail profiling reveals an embryonic switch in translational control. Nature 508:66–71
3. Despic V, Neugebauer KM (2018) RNA tales – how embryos read and discard messages from mom. J Cell Sci 131:pii: jcs201996
4. Sysoev VO, Fischer B, Frese CK, Gupta I, Krijgsveld J, Hentze MW, Castello A, Ephrussi

A (2016) Global changes of the RNA-bound proteome during the maternal-to-zygotic transition in Drosophila. Nat Commun 7:12128

5. Despic V, Dejung M, Gu M, Krishnan J, Zhang J, Herzel L, Straube K, Gerstein MB, Butter F, Neugebauer KM (2017) Dynamic RNA-protein interactions underlie the zebrafish maternal-to-zygotic transition. Genome Res 27(7):1184–1194
6. Licatalosi DD, Mele A, Fak JJ, Ule J, Kayikci M, Chi SW, Clark TA, Schweitzer AC, Blume JE, Wang X et al (2008) HITS-CLIP yields genome-wide insights into brain alternative RNA processing. Nature 456:464–469
7. Konig J, Zarnack K, Rot G, Curk T, Kayikci M, Zupan B, Turner DJ, Luscombe NM, Ule J (2010) iCLIP reveals the function of hnRNP particles in splicing at individual nucleotide resolution. Nat Struct Mol Biol 17:909–915
8. Hafner M, Landthaler M, Burger L, Khorshid M, Hausser J, Berninger P, Rothballer A, Ascano M Jr, Jungkamp AC, Munschauer M et al (2010) Transcriptome-wide identification of RNA-binding protein and microRNA target sites by PAR-CLIP. Cell 141:129–141
9. Zarnegar BJ, Flynn RA, Shen Y, Do BT, Chang HY, Khavari PA (2016) irCLIP platform for efficient characterization of protein-RNA interactions. Nat Methods 13:489–492
10. Van Nostrand EL, Pratt GA, Shishkin AA, Gelboin-Burkhart C, Fang MY, Sundararaman B, Blue SM, Nguyen TB, Surka C, Elkins K et al (2016) Robust transcriptome-wide discovery of RNA-binding protein binding sites with enhanced CLIP (eCLIP). Nat Methods 13:508–514
11. Brimacombe R, Stiege W, Kyriatsoulis A, Maly P (1988) Intra-RNA and RNA-protein cross-linking techniques in *Escherichia coli* ribosomes. Methods Enzymol 164:287–309
12. Barckmann B, Pierson S, Dufourt J, Papin C, Armenise C, Port F, Grentzinger T, Chambeyron S, Baronian G, Desvignes JP et al (2015) Aubergine iCLIP reveals piRNA-dependent decay of mRNAs involved in germ cell development in the early embryo. Cell Rep 12:1205–1216
13. Broughton JP, Pasquinelli AE (2013) Identifying argonaute binding sites in *Caenorhabditis elegans* using iCLIP. Methods 63:119–125
14. Zisoulis DG, Lovci MT, Wilbert ML, Hutt KR, Liang TY, Pasquinelli AE, Yeo GW (2010) Comprehensive discovery of endogenous argonaute binding sites in *Caenorhabditis elegans*. Nat Struct Mol Biol 17:173–179
15. Despic V, Dejung M, Butter F, Neugebauer KM (2017) Analysis of RNA-protein interactions in vertebrate embryos using UV crosslinking approaches. Methods 126:44–53
16. Vejnar CE, Abdel Messih M, Takacs CM, Yartseva V, Oikonomou P, Christiano R, Stoeckius M, Lau S, Lee MT, Beaudoin JD et al (2019) Genome wide analysis of 3′ UTR sequence elements and proteins regulating mRNA stability during maternal-to-zygotic transition in zebrafish. Genome Res 29:1100–1114
17. Matern MS, Beirl A, Ogawa Y, Song Y, Paladugu N, Kindt KS, Hertzano R (2018) Transcriptomic profiling of zebrafish hair cells using RiboTag. Front Cell Dev Biol 6:47
18. Sugimoto Y, Konig J, Hussain S, Zupan B, Curk T, Frye M, Ule J (2012) Analysis of CLIP and iCLIP methods for nucleotide-resolution studies of protein-RNA interactions. Genome Biol 13:R67
19. Wang Z, Kayikci M, Briese M, Zarnack K, Luscombe NM, Rot G, Zupan B, Curk T, Ule J (2010) iCLIP predicts the dual splicing effects of TIA-RNA interactions. PLoS Biol 8: e1000530
20. Chen B, Yun J, Kim MS, Mendell JT, Xie Y (2014) PIPE-CLIP: a comprehensive online tool for CLIP-seq data analysis. Genome Biol 15:R18
21. Haberman N, Huppertz I, Attig J, Konig J, Wang Z, Hauer C, Hentze MW, Kulozik AE, Le Hir H, Curk T et al (2017) Insights into the design and interpretation of iCLIP experiments. Genome Biol 18:7

Chapter 27

Tethered Function Assay to Study RNA-Regulatory Proteins in Zebrafish Embryos

Yuichiro Mishima and Kunio Inoue

Abstract

Many proteins are assumed to mediate post-transcriptional regulation of mRNAs. However, the lack of information about their target mRNAs and functional domains hampers the detailed analysis of their molecular function. Here we describe a method to analyze the post-transcriptional effects of proteins of interest by artificially tethering the protein to a reporter mRNA in zebrafish embryos.

Key words Zebrafish, Post-transcriptional regulation, mRNA, Translation, Tethering

1 Introduction

The control of mRNA translation and stability is fundamental to regulate gene expression during animal development [1, 2]. Numerous proteins associate with mRNAs and mediate post-transcriptional regulation. To dissect the molecular function of each RNA-associating protein, we need to recapitulate a situation in which a protein of interest specifically associates with a reporter mRNA and analyze the output from, or the status of, the reporter mRNA. However, reporter assay is often not applicable to the candidate protein for several reasons. For example, target mRNAs or binding sequence of the candidate protein might be uncharacterized. The candidate protein might lack an RNA-binding domain and therefore associate with mRNAs indirectly. Its functional domains might be uncharacterized or need to be analyzed apart from their RNA-binding domains.

To circumvent these limitations, one can artificially recruit a protein to a reporter mRNA by utilizing RNA-binding domains and their target sequences discovered from bacteriophages. These include MS2 coat protein [3], PP7 coat protein (PCP) [4], and λN22 peptide (λN) [5] that specifically bind to short stem-loop sequences. Hence, co-expression of a protein fused to one of these

Roland Dosch (ed.), *Germline Development in the Zebrafish: Methods and Protocols*, Methods in Molecular Biology, vol. 2218, https://doi.org/10.1007/978-1-0716-0970-5_27, © Springer Science+Business Media, LLC, part of Springer Nature 2021

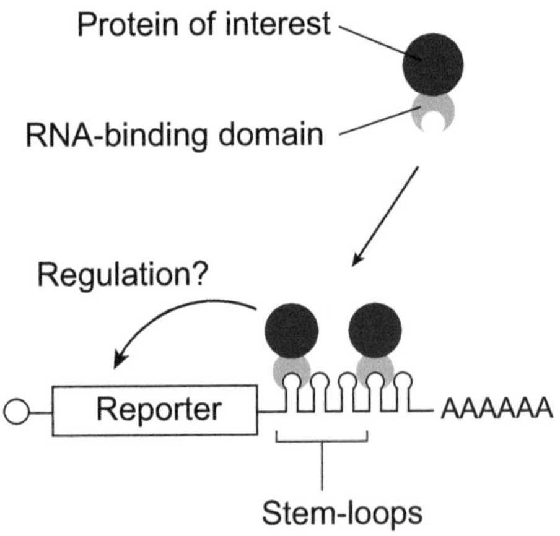

Fig. 1 Scheme of the tethered function assay. A protein of interest is fused to the RNA-binding domain from bacteriophages and expressed together with the reporter mRNA containing target stem-loops. Specific association of the fusion protein to the reporter mRNA may cause post-transcriptional regulation

domains and a reporter mRNA containing target stem-loops results in the specific association of the protein and reporter mRNA in vivo (Fig. 1). This technique is called tethered function assay or tethering and has revealed the roles and mechanisms of many RNA-regulatory proteins [6].

Here we describe a detailed protocol to perform tethering assay in early zebrafish embryos [7]. This protocol utilizes λN peptide and its target boxB stem-loop [5, 8] and focuses on the analysis of the effects of proteins of interest on translational output. To this end, reporter and effector mRNAs need to be synthesized in vitro and injected into fertilized eggs. Protein synthesis from the reporter mRNA in the presence of various effector constructs can be monitored by measuring luciferase activities. The assay can be performed with ten gastrula stage embryos, thus enabling quick and parallel evaluation of the post-transcriptional activity of proteins of interest in the context of vertebrate development.

2 Materials

2.1 Plasmid Construction and mRNA Preparation

1. λN effector plasmid: pCS2 + HA-λN-MCS-globin 3′ untranslated region (UTR).
2. Renilla luciferase (Rluc) reporter plasmid: pCS2 + Rluc-boxB×5-SV40 3′ UTR.
3. Firefly luciferase (Fluc) internal control plasmid: pCS2 + Fluc-globin 3′ UTR.

   Details of these plasmids are provided in Fig. 2 and [7].

4. DNA purification columns (e.g., NucleoSpin Gel and PCR Clean-up, MACHEREY-NAGEL).
5. Capped mRNA synthesized with mMessage mMachine SP6 Transcription kit (ThermoFisher).

2.2 mRNA Injection

1. Stereomicroscope (e.g., Olympus SZX7).
2. Microinjector (e.g., Narishige IM300).
3. Manipulator (e.g., Narishige M-152).
4. Injection needle: Pull Glass tubes (Narishige G-1) with Puller (Narishige PC-100). Cut the tip of the needle by a razor blade under a microscope.
5. 0.01 mm microscope stage micrometer.
6. Mineral oil.

2.3 Luciferase Assay

1. Phosphate buffered saline (PBS): 137 mM NaCl, 2.7 mM KCl, 10 mM Na_2HPO_4, 1.76 mM KH_2PO_4, pH 7.4.
2. 1× Passive lysis buffer with protease inhibitor (Lysis buffer): Dilute 5× passive lysis buffer (Promega) and 25× protease inhibitor with distilled water. Prepare at time of use and keep at 4 °C.
3. 25× Protease inhibitor: Dissolve one tablet of complete EDTA-free Protease Inhibitor Cocktail (Merck) in 2 mL of distilled water to prepare 25× stock solution. Store at −30 °C.
4. Pellet pestles for 1.5-mL microtube.
5. Dual-Luciferase Reporter Assay System (Promega).

   Luciferase assay reagent II (LARII): Dissolve luciferase assay substrate in luciferase assay buffer II. Divide into aliquots and store at −80 °C.

   Stop & Glo reagent: Prepare at time of use by diluting 50× Stop & Glo substrate with Stop & Glo buffer and keep at 4 °C.
6. Luminometer.

3 Methods

Always handle RNA at 4 °C under RNase-free condition.

3.1 Plasmid Construction and mRNA Preparation

1. Clone open reading frame (ORF) of interest into the effector plasmid (pCS2 + HA-λN-MCS-globin 3′ UTR) by using restriction enzymes or other cloning strategies. Make sure that the ORF is fused to HA-λN in-frame (*see* Fig. 2 and **Note 1**).
2. Digest 5–10 μg of plasmids encoding reporters and effector. Use NotI for linearization in a 100 μL reaction and incubate at 37 °C for overnight (*see* **Note 2**).

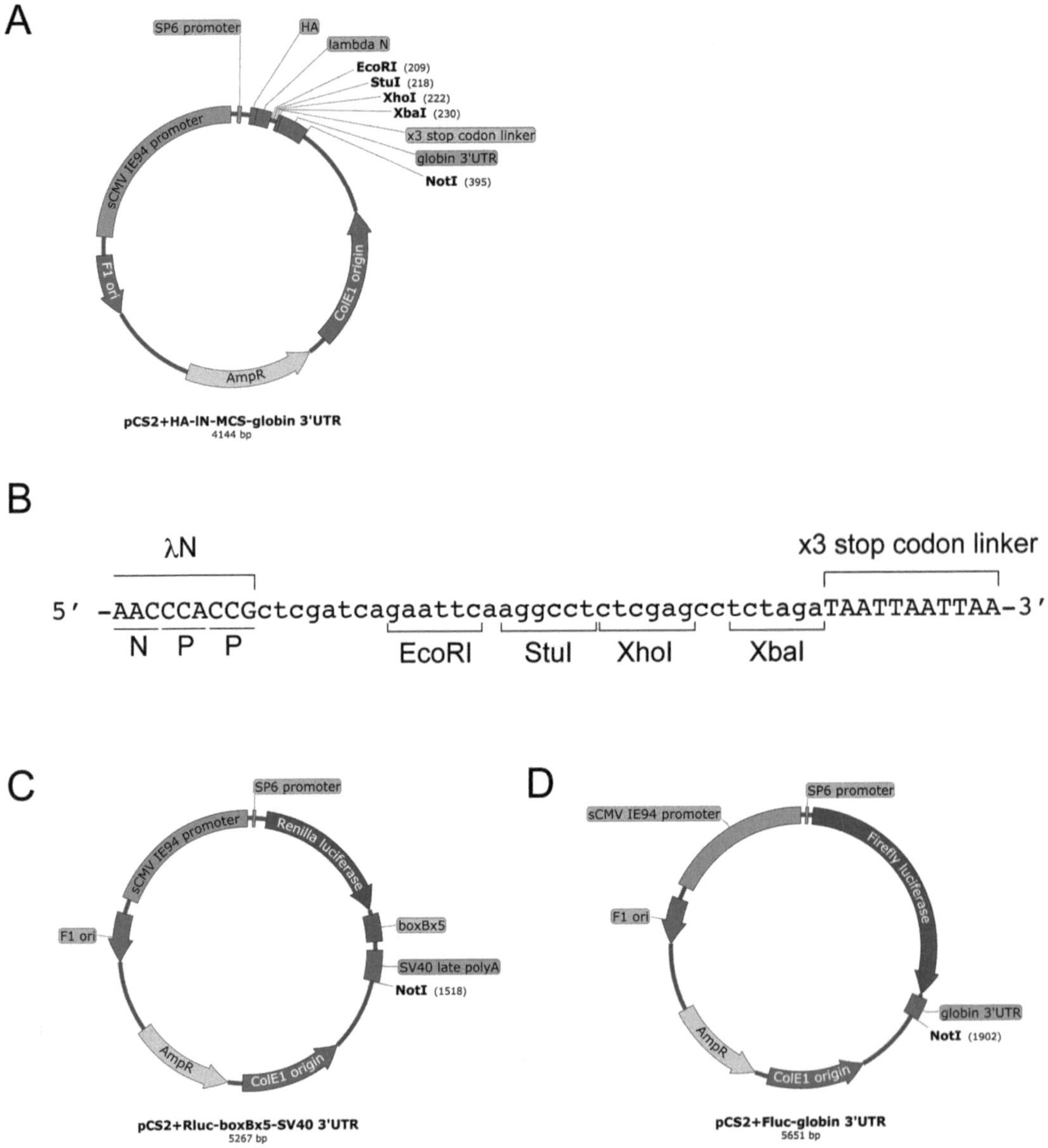

Fig. 2 Maps of plasmids used in this protocol. (**a**) Effector plasmid pCS2 + HA-λN-MCS-globin 3′ UTR. This plasmid is used to express effector protein fused to HA-tag and λN peptide at the N-terminus. (**b**) A detailed sequence of the multi-cloning site in pCS2 + HA-λN-MCS-globin 3′ UTR. EcoRI, StuI, XhoI, and XbaI sites downstream of λN are available for cloning ORFs. Stop codon linker is inserted downstream of multi-cloning site (MCS) so that partial ORF fragments lacking a stop codon can be cloned. (**c**) Reporter plasmid pCS2 + Rluc-boxB × 5-SV40 3′ UTR. Five copies of the boxB sequence are inserted downstream of the ORF of Renilla luciferase. (**d**) Internal control plasmid pCS2 + Fluc-globin 3′ UTR. For all plasmids, use NotI for linearization and SP6 polymerase for in vitro transcription

3. Purify linearized plasmid DNA with a microcolumn-based method. Elute DNA in 20 μL RNase-free water and measure the concentration. The concentration will be 100–500 ng/μL.

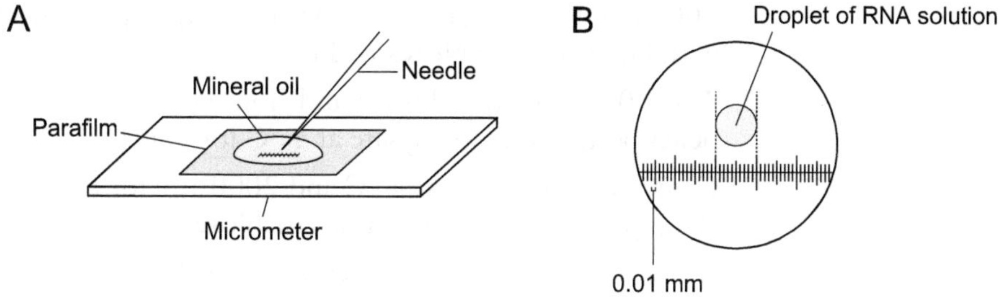

Fig. 3 Calibration before mRNA injection. (**a**) How to setup microscope micrometer for measuring injection volume. Tightly fit parafilm to micrometer and put mineral oil on the scale. Place the tip of an injection needle in the oil close to the scale. (**b**) A schematic representation of a field of view under the stereomicroscope. Injected mRNA solution forms a droplet in the oil. Make a droplet close to the scale and measure the diameter of the droplet immediately after injection

4. Using 500 ng to 1 μg of the purified DNA as templates, prepare capped mRNAs by in vitro transcription (*see* Subheading 2) (*see* **Note 3** and **4**).

3.2 mRNA Injection

1. Prepare a mixture of mRNA solution for injection at the following concentration with RNase-free water (*see* **Note 5**). Reporter Rluc-boxB×5 mRNA: 50 ng/μL.

   Control Fluc mRNA: 50 ng/μL.

   Effector mRNA: 50–200 ng/μL
2. Fill the injection needle with mRNA solution.
3. Place a piece of parafilm on a micrometer and put mineral oil on it so that the oil covers a scale (Fig. 3a).
4. Inject mRNA solution into the oil and measure the diameter of the droplet of mRNA solution under the microscope. Calibrate the injection setting so that the diameter of the droplet becomes 100 μm (Fig. 3b). The volume of the droplet is approximately 500 pL with this diameter.
5. Collect zebrafish embryos by natural breeding. ~20 embryos are required per injection.
6. Inject 1000 pL mRNA solution into the cytoplasm of 1-cell stage embryos. Place the tip of the needle at the center of the cytoplasm and add pressure twice to inject 1000 pL (*see* **Note 6**).
7. Repeat **steps 2–6** to compare multiple effector constructs (*see* **Note** 7).
8. Culture embryos at 28.5 °C in the system water.

3.3 Luciferase Assay

1. At 6 h post-fertilization (hpf), collect ten embryos in a 1.5-mL tube. Remove system water and rinse the embryos twice with

500 μL of PBS. Remove PBS carefully and snap-freeze embryos with liquid N_2 (*see* **Notes 8–11**).

2. Add 100 μL of lysis buffer and homogenize embryos with a pellet pestle. Keep the lysate at 4 °C for 5 min.
3. Centrifuge the lysate at 4 °C and 2000 × *g* for 5 min and take the supernatant to a new tube (*see* **Note 12**).
4. In case of single tube measurement, take 20 μL of the lysate to a new transparent tube. Add 100 μL of LAR II, mix by vortex, and measure the luminescence of Fluc by a luminometer (*see* **Note 13**).
5. Add 100 μL of Stop & Glo reagent, mix by vortex, and measure the luminescence of Rluc by a luminometer. Repeat **steps 4** and **5** to analyze multiple samples (*see* **Note 14**).
6. Normalize the activity of Rluc by the activity of Fluc. Normalize the data with the result of a control effector construct if necessary (*see* **Notes 15** and **16**).
7. Perform western blotting with the rest of the lysate to confirm effector protein expression. Use anti-HA-tag antibody for the detection of the effector protein (*see* **Note 17**).

4 Notes

1. Control constructs are required to interpret the results correctly. An empty construct encoding only HA-λN is the easiest option, but this is not sufficient in most cases. Tethering any protein can potentially affect translation and mRNA stability. Mock control with no HA-λN construct and unrelated proteins that are assumed to have no post-transcriptional activity, such as GFP or β-galactosidase, should be included as controls at the beginning of the analysis. Subsequently, analyze deletion or point-mutation versions and compare with the wildtype protein to confirm if a specific domain of the protein mediates the effect.
2. If the NotI site is present in the effector ORF, Asp718 is an alternative option.
3. Synthesized mRNA should be purified with columns (e.g., RNeasy Mini, QIAGEN). Do not use phenol/chloroform, which perturbs embryogenesis.
4. In vitro transcribed mRNA lacks a poly(A) tail, and the poly (A) tail is added in the cytoplasm after injection [9, 10]. Tethered proteins might affect polyadenylation and affect translation as a consequence. To test this possibility, reporter mRNA can be polyadenylated before injection by using poly (A) polymerase (e.g., Poly(A) Tailing Kit, ThermoFisher).

5. Each protein has differential expression level and stability. It is therefore important to normalize the expression of the effector proteins [11]. For the initial attempt, it is recommended to titrate the amount of the effector mRNA. When comparing effectors with different ORF length (e.g., deletion constructs), include the equimolar amount of effector mRNAs.
6. While mRNA injection is often carried out by injecting into the yolk, injection into the cytoplasm usually produces more stable results in our hands. It is also crucial to carry out injection during the one-cell stage, as translation activity seems to change over time after fertilization.
7. When comparing multiple effector constructs, it is important to obtain new eggs for every construct to carry out injection during the one-cell stage. Differences in the timing of injection might affect the results. For instance, injecting one construct at the 1-cell stage and another at the 4-cell stage is not recommended.
8. Use only fertilized eggs and do not mix with unfertilized eggs. Although unfertilized eggs retain translation activity, the results can be variable between fertilized and unfertilized eggs.
9. It is possible to perform the assay with fewer embryos (3–5 embryos). However, it is better to use ten embryos or more to average the variability of injection and embryos.
10. For the initial trial, it is better to perform time course experiments such as at 4, 6, and 8 hpf.
11. Embryos can be stored at −80 °C for at least 1 month.
12. Lysates can be stored at −80 °C, but do not freeze and thaw multiple times.
13. For the initial trial, titrate the amount of the lysate and confirm that the luminescent signal does not reach saturation.
14. Alternatively, use a multi-well plate reader with automatic injection system.
15. Since the injection is not as accurate as pipetting, comparison of the activity of Rluc without normalization is challenging and not recommended.
16. Make sure that the difference in the Fluc activity is minimum. The results tend to become less reproducible if there are several-fold differences in the Fluc activity. In case firefly luciferase activity is not constant, carefully titrate the ratio of reporter and effector mRNAs, calibrate injection amount, and check the reproducibility of injection.
17. If there are differences in effector protein expression, try to normalize the expression level by adjusting the amount of effector mRNA.

Acknowledgments

This work was supported by KAKENHI JP18H02370 to Y.M.

References

1. Teixeira FK, Lehmann R (2019) Translational control during developmental transitions. Cold Spring Harb Perspect Biol 11:a032987
2. Mishima Y (2012) Widespread roles of microRNAs during zebrafish development and beyond. Develop Growth Differ 54:55–65
3. Coller JM, Gray NK, Wickens MP (1998) mRNA stabilization by poly(A) binding protein is independent of poly(A) and requires translation. Genes Dev 12:3226–3235
4. Carroll JS, Munchel SE, Weis K (2011) The DExD/H box ATPase Dhh1 functions in translational repression, mRNA decay, and processing body dynamics. J Cell Biol 194:527–537
5. De Gregorio E (1999) Translation driven by an eIF4G core domain in vivo. EMBO J 18:4865–4874
6. Bos TJ, Nussbacher JK, Aigner S, Yeo GW (2016) Tethered function assays as tools to elucidate the molecular roles of rna-binding proteins. Adv Exp Med Biol 907:61–88
7. Mishima Y, Fukao A, Kishimoto T, Sakamoto H, Fujiwara T, Inoue K (2012) Translational inhibition by deadenylation-independent mechanisms is central to microRNA-mediated silencing in zebrafish. Proc Natl Acad Sci U S A 109:1104–1109
8. Gehring NH, Neu-Yilik G, Schell T, Hentze MW, Kulozik AE (2003) Y14 and hUpf3b Form an NMD-activating complex. Mol Cell 11:939–949
9. Mishima Y, Giraldez AJ, Takeda Y, Fujiwara T, Sakamoto H, Schier AF, Inoue K (2006) Differential regulation of germline mRNAs in soma and germ cells by zebrafish miR-430. Curr Biol 16:2135–2142
10. Mishima Y, Tomari Y (2016) Codon usage and 3′ UTR length determine maternal mRNA stability in zebrafish. Mol Cell 61:874–885
11. Fukaya T, Tomari Y (2011) PABP is not essential for microRNA-mediated translational repression and deadenylation in vitro. EMBO J 30:4998–5009

Chapter 28

Massively Parallel Analysis of Regulatory RNA Sequences

Michal Rabani

Abstract

The stability of RNA transcripts is regulated by signals within their sequences, but the identity of those signals still remain elusive in many biological systems. Recently introduced massively parallel tools for the analysis of regulatory RNA sequences provide the ability to detect functional cis-regulatory sequences of post-transcriptional RNA regulation at a much larger scale and resolution than before. Their application formulates the underlying sequence-based rules and predicts the impact of genetic variations. Here, we describe the application of UTR-Seq, as a strategy to uncover cis-regulatory signals of RNA stability during early zebrafish embryogenesis. The method combines massively parallel reporter assays (MPRA) with computational regression models. It surveys the effect of tens of thousands of regulatory sequences on RNA stability and analyzes the results via regression models to identify sequence signals that impact RNA stability and to predict the in vivo effect of sequence variations.

Key words RNA stability, Massively parallel reporter assays, Post-transcriptional regulation, Zebrafish embryogenesis, *Cis*-regulation, UTR-Seq

1 Introduction

The stability of RNA molecules inside the cell is determined by regulatory interactions between cis-acting elements present on the RNA and trans-regulators that bind them, such as RNA-binding proteins (RBPs) [1–3] and micro-RNAs (miRNAs) [4, 5]. For example, during early development, 3' untranslated regions (3'UTR) sequences serve as degradation-inducing binding sites for the fly SMAUG protein [2] and for the zebrafish microRNA miR-430 [4]. However, even though sequence elements that regulate RNA stability have been identified [1, 2, 4, 6–11], a systematic and predictive understanding of the regulatory information within RNAs remains elusive.

Recently, massively parallel reporter assays (MPRAs) were introduced as a compelling tool for answering these questions at a much larger scale and resolution than before. They present significant advantages over other existing strategies, including computational [12, 13], in vivo [9, 14], and in vitro [15] methods, which

Roland Dosch (ed.), *Germline Development in the Zebrafish: Methods and Protocols*, Methods in Molecular Biology, vol. 2218, https://doi.org/10.1007/978-1-0716-0970-5_28, © Springer Science+Business Media, LLC, part of Springer Nature 2021

are limited by the difficulty in obtaining genome-scale data, degeneracy of sequence elements, and multiple or combinatorial interactions. MPRAs systematically introduce tens of thousands of artificial sequences into the biological system and read the regulatory output via high-throughput sequencing. Their application successfully identified functional cis-signals for mRNA transcription [16–18], splicing [19], and stability [20–23] in different biological systems. In a few examples, analysis of the resulting comprehensive MPRA datasets that was performed with dedicated computational tools, formulated sequence-based rules for transcriptional enhancers [18], alternative splicing frequencies [19], and mRNA stabilities [24]. Those predictive modeling approaches also successfully predicted the effect of genetic variations in the regulatory sequences.

In this chapter, we describe UTR-Seq (Fig. 1), a recent MPRA-based methodology for decoding cis-regulatory signals of RNA stability in the zebrafish embryo in vivo. Early embryos [25] are a powerful system to study post-transcriptional RNA regulation in the absence of de novo transcription. At the onset of development, fertilized embryos are transcriptionally silent and exclusively rely on post-transcriptional programs encoded by maternally provided mRNAs and proteins. In zebrafish, maternal programs drive rapid cleavage divisions, patterning, and cell movements [26], until newly synthesized zygotic mRNAs gradually replace destabilized maternal transcripts starting at 3 h post fertilization (hpf), and establish distinct zygotically encoded developmental programs [25]. UTR-Seq implements a dynamic MPRA in zebrafish embryos and provides dedicated computational models for its analysis. Its application identifies sequence signals that regulate RNA stability, deduces its sequence-based rules, and predicts the effect of sequence variations on RNA stability.

2 Materials

2.1 Reagents

1. DNA oligonucleotides to be used as primers (Table 1).
2. A vector containing the full-length universal adaptor sequence (as a cloning site) and encoding for an RNA transcript that is driven by an SP6 promoter.
3. High-fidelity DNA polymerase, with suitable reaction buffer and dNTP mix.
4. EcoRV restriction enzyme and suitable reaction buffer.
5. A DNA cleanup silica column that allows elution at a volume of 10 μL.
6. Gibson assembly master mix.
7. Electrocompetent cells with suitable recovery media.

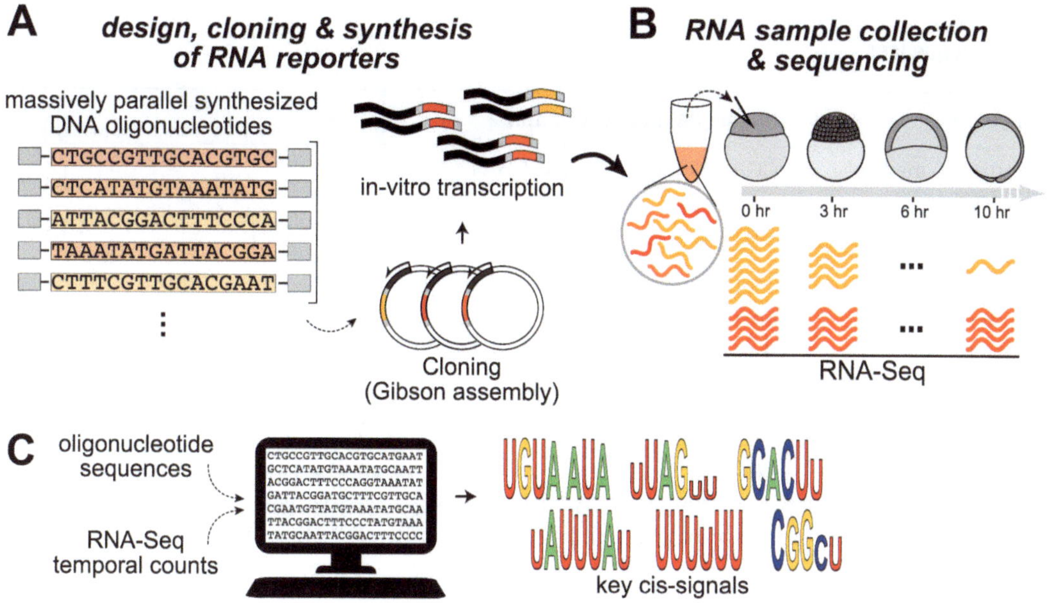

Fig. 1 A massively parallel reporter assay survey of RNA stability. (**a**) In-vitro-synthesized DNA fragments of RNA regulatory sequences, with two flanking terminal adaptors (gray) are cloned into a vector and in vitro transcribed to generate RNA reporters with different regulatory sequences. (**b**) RNA reporters are microinjected into one-cell stage zebrafish embryos, and RNA samples are collected at multiple developmental stages. Temporal samples are sequenced and the number of times that each reporter was sequenced is measured. Stable RNAs maintain a similar level in all samples (dark orange), while levels of unstable RNAs decrease over time (light orange). (**c**) The set of reporter sequences and their counts in RNA-Seq data is used as input for the UTR-Seq computational analysis tool, which defines short sequence elements (motifs) that positively or negatively affect the stability of RNA reporters

8. Petri dishes (ampicillin).
9. Mini or midi-prep plasmid purification kit.
10. Sp6 polymerase in vitro transcription kit.
11. Zebrafish embryo culture medium: 5.03 mM NaCl, 0.17 mM KCl, 0.33 mM $CaCl_2$, 0.33 mM $MgSO_4$, 0.1% methylene blue.
12. Pronase solution: 2 mg/mL Pronase in zebrafish embryo culture medium.
13. Liquid nitrogen.
14. Total RNA isolation kit.
15. Reverse-transcriptase enzyme with suitable reaction buffer and dNTP mix.
16. Solid phase reversible immobilization (SPRI) magnetic beads for DNA purification.

2.2 Equipment

1. Thermal cycler.
2. Electroporator (suitable for bacterial transformation).

Table 1
Primer sequences

ID	Sequence and description
Upstream universal adaptor	GGAGATCTGAGTTCAAGGAT
Downstream universal adaptor	ATCTAGAACTATAGTGAGTC
Universal adaptor sequence	**GGAGATCTGAGTTCAAGGAT**ATCTAGAACTATAGTGAGTC The 20 nt upstream adaptor sequence (bold) and 20 nt downstream adaptor sequence
Reversetranscription (RT) primer	CTACACGACGCTCTTCCGATCTNNNNNNNN GACTCACTATAGTTCTAGAT Illumina SBS-3 sequence (22 nt), a random UMI (8 nt, underline) and the reverse-complement of the downstream adaptor sequence (20 nt)
Upstream PCR primer	CAAGCAGAAGACGGCATACGAGATNNNNNNN GTGACTGGAGTTCAGACGTGTGCTCTTCCGATCT **GGAGATCTGAGTTCAAGGAT** Illumina P7 sequence (24 nt), sequencing index (7 nt, underline), Illumina SBS-12 sequence (34 nt) and the upstream universal adaptor (20 nt, bold)
Downstream PCR primer	**AATGATACGGCGACCACCGAGATCT**ACACTCTTTCCCTACACGACGCTCTTCC Illumine P5 sequence (25 nt, bold) and Illumina SBS-3 sequence (22 nt). Underlined sequence is designed to match the RT primer

3. Bacterial plating instrument and growth incubator (37 °C).
4. Qubit fluorometer (Invitrogen).
5. TapeStation (Agilent).
6. Zebrafish microinjection equipment:
   (a) Picopump.
   (b) Micromanipulator.
   (c) Binocular.
   (d) Glass needles.
7. Illumina high-throughput sequencing platform.

2.3 Software

1. Adaptor trimming software, such as Cutadapt [27] or AlienTrimmer [28].
2. High-throughput sequencing alignment tool, such as Bowtie2 [29].
3. MATLAB (Mathworks).
4. The UTR-Seq software package for MATLAB, available for download from https://github.com/rabanilab/UTR-Seq.

3 Methods

3.1 Design and Synthesis of the Oligonucleotide Library

1. Design a set of short RNA sequences that cover all the sequence variations that you would like to test in parallel (*see* **Note 1**).
2. Add two 20-nt universal adaptors—an upstream adaptor and a downstream adaptor—to each of the sequences in your set (Table 1). These will be used for primer extension and cloning of the oligonucleotide pool.
3. Synthesize your oligonucleotide library. Libraries are typically synthesized by massive parallel oligonucleotide synthesis at pmole scale.

3.2 Reporter Library Cloning and Transcription

1. Select a vector that encodes for an RNA transcript that is driven by an SP6 promoter. The transcript sequence must contain the full-length universal adaptor sequence (Table 1) as a cloning site, positioned at the precise location where sequences from the oligonucleotide library should appear. A restriction site at the end of the transcript sequence is also recommended.
2. To prepare the backbone fragment, cut the vector between the two adaptor sequences via enzymatic restriction with EcoRV. Clean the enzymatic restriction product on a DNA cleanup column and elute at a volume of 10 μL.
3. To prepare the insert fragments, amplify the oligonucleotide library using 22 cycles of PCR reaction, with primers that match the universal upstream and downstream adaptor sequences and a high-fidelity DNA polymerase in a 50-μL reaction volume. Clean the PCR product on a DNA cleanup column and elute at a volume of 10 μL.
4. Clone the oligonucleotide library into the vector via a two fragments Gibson assembly reaction. Add a 5:1 molar ratio of the insert and vector fragments to the Gibson assembly master mix. After the reaction is complete, clean the product on a DNA cleanup column, and elute at a volume of 10 μL.
5. Perform bacterial transformation by electroporation according to the manufacturer's recommendations. Transform 1 μL of the clean reaction product per 20 μL of electrocompetent cells, reaching a total of at least 5 μL of the clean reaction product, to ensure at least 10× coverage of the oligonucleotide library.
6. Prepare a 300× dilution plate in order to estimate the transformation efficiency. Dilute 4 μL of electroporated cells with 1200 μL of recovery media, and plate 200 μL on a 10-cm pre-warmed petri dish (ampicillin).

7. Plate the entire volume of electroporated cells on 10-cm pre--warmed petri dishes (ampicillin). Plate 200 μL of cells per petri dish, and incubate overnight. We recommend avoiding liquid cultures because those will generate larger biases in representations of different colonies.
8. Harvest the colonies. Use a sterile cell spreader to scrape colonies off cells, wash with PBS, and collect bacteria into tubes.
9. Extract plasmids with a standard mini- or midi-prep plasmid extraction kit.
10. Linearize plasmids at the end of the RNA transcript by enzymatic restriction with the appropriate enzyme, and use the Sp6 in vitro transcription kit to generate a library of RNA reporters from the linearized plasmids.
11. Quantify the RNA levels of the resulting library of RNA reporters by Qubit fluorometric quantification. Validate their length by TapeStation RNA analysis screen tape (Agilent).

3.3 Control RNA Cloning and Transcription

1. Design a set of five sequences of the same length as the sequences in your oligonucleotide library. Sequences should not resemble any of the sequences in the oligonucleotide library.
2. Repeat the cloning and in vitro transcription steps (Subheading 3.2) for the five additional sequences independently of the entire oligonucleotide library.
3. Mix the five control RNAs at 2-fold increments (highest 50 fg, lowest 3.125 fg).
4. The resulting control RNA mix will contain five transcripts with identical structure as RNA reporters in the library, which will be added to samples and will be used for temporal normalization.

3.4 Fish Microinjection and Sample Collection

1. Start by defining the number of embryos to inject and the developmental stages in which to collect RNA samples (*see* **Note 2**).
2. Follow standard procedures for growth and maintenance of embryos and adult zebrafish [30]. Collect the appropriate number of one-cell stage zebrafish embryos. Remove embryos from their chorion by incubating in Pronase solution for 5 min and washing thoroughly with culture medium (*see* **Note 3**).
3. Inject 10–100 pg of RNA reporters into each one-cell stage zebrafish embryo.
4. Grow embryos at 28 °C in culture medium and stage according to standard procedure, until reaching the appropriate developmental stage.
5. Collect embryos into a 1.5-mL tube and freeze in liquid nitrogen. Store at −80 °C until ready for RNA extraction and downstream processing of samples.

3.5 RNA-Seq Library Construction and Sequencing

1. Isolate total RNA from embryo samples using a total RNA isolation kit. Add 120 fg of control RNA (Subheading 3.3) into each RNA sample during the initial lysis step.
2. Carry out reverse transcription from total RNA samples using the reverse transcription primer (Table 1) at 50 °C for 30–50 min.
3. Amplify 1/10 of the final volume of the reverse transcription reaction using 18 cycles of PCR, with upstream and downstream PCR primers (Table 1) and a high-fidelity DNA polymerase, in a 50-μL reaction volume.
4. Clean amplified DNA with 1.0× volume of SPRI magnetic beads.
5. Quantify the resulting RNA-Seq library by Qubit fluorometric quantification and validate their length by TapeStation D1000hs screen tape. Expected length is 130 nt longer than the original oligonucleotide sequences in your library.
6. Sequence the resulting library on an Illumina high-throughput sequencing platform (*see* **Note 4**).

3.6 RNA-Seq Data Processing

Use the following steps to quantify RNA-Seq data collected for each sample.

1. *Trimming*. The universal adaptor sequences are retained in sequencing and should be trimmed before alignment. Use one of several available tools, such as Cutadapt [27] or AlienTrimmer [28]. If unique molecular identifiers (UMIs) are introduced during reverse transcription, the first N bases of the read (before adapter trimming) represent the UMI and should also be trimmed using these tools. UMI sequences should remain associated with the read for downstream calculations.
2. *Filtering*. Filter reads that do not contain the universal adaptor sequences (with up to ten mismatches) or that are shorter or longer than the expected insert length by more than 20%.
3. *Alignment*. Align retained reads to a reference set of all synthetic oligonucleotide sequences in your library, including the five control sequences. Use a standard alignment tools such as Bowtie2 [29] with "end-to-end" and "very sensitive" parameters. Require an edit score of $\leq$20 and quality score of $\geq$10 for all retained alignments.
4. *Counting*. Count the number of reads or UMIs (if available) that were mapped to each oligonucleotide sequence in the library. If UMI counts are used, adjust them to represent the expected number of RNA molecules in the sample [31].

3.7 Extracting Sequence Elements That Regulate RNA Stability

Use the UTR-Seq software package to analyze RNA-Seq counts from your samples and to define sequence elements that affect the stability of your RNA reporters.

1. *Prepare the input files.* Prepare two tab-delimited input files with a single line for each of the RNA reporters: (a) a sequence file: an id and oligonucleotide sequence (Fig. 2a), (b) an RNA-Seq count file: an id and read or UMI counts per temporal sample (Fig. 2b). The ids of the file control RNAs should be in the format SPIKE_<number>, and their read counts should appear in the RNA-Seq count file. Do not include the five control sequences in the sequence file. Place the files in a directory of your choice, which will be called the running directory.
2. *Prepare the running environment.* Download the UTR-Seq software package for MATLAB from https://github.com/rabanilab/UTR-Seq, and install according to instructions on the website.
3. *Run the software package using MATLAB.* In MATLAB, browse to the running directory and execute the analysis of your files by running the following command:

```
utrseq_fit_all (3, '<input sequence file name>', '<input
RNA-Seq counts file name>', [<sample #1 time> <sample #2
time> ...], '<preferred output name>');
```

   Provide times in hours and a full path for all files.

   This package will run in three steps (Fig. 2c). Step 1 will estimate the degradation rate for each reporter from its RNA-Seq counts. Step 2 will find an optimal "weight" for each short sequence (3–7 nt long), which defines its contribution to the degradation rate of a reporter. Finally, step 3 will find short sequences with maximal contribution to degradation rates ("peaks") and group them into motifs based on their sequence similarity. Some of the steps can be very slow, especially as the number of reporter sequences in your library and their length increase.
4. *View the results.* The output of each of the three steps of the analysis is available in a separate sub-directory with a different suffix within your running directory: step 1 directory has the suffix "`_fit_degradation`", step 2 directory has the suffix "`_kmer_regression`", and step 3 directory has the suffix "`_fit_peaks`". The step 3 directory will include a list of peak positions within your reporter sequences and a JPG logo file for each of the peak motifs that was predicted.

A ***Input sequence file*** *(reporter id, sequence)*

```
SO_M_T1006   TAGATAGAGATCATCTTTACAGTTCCTCGGGAAAATGTGCTTGTGACCATTTTGTATGGCTTGTCTGGAAACACATTTTGTGCCATTAAAGCTCTCACAAAACTTTCTAT
SO_M_T1013   TGATTCTAGTATATAATATTTTTGTCACGCACCTGCTGACTTAGGACTACATTTTGCTCTGTGAACTATCCCTCTCTGCAGCAAAGAGTTGCCCCTTCTTTCTACTCTAT
SO_M_T1022   AGCCGGTGGCATAAATATTCAGAATAACAAACTGATTCTTTGTAGATGTGGACCACCAAGGGGTTGTTGAAACATTTTAAATTGCTGCTACATACTGTCTTCTTAACTAG
SO_M_T1027   TTGCCTGAATAATTTCTCAGGATATTTTTGCACTGCTGTTTGATGGACATCTGTGATGGTTGCCTGTACACTTAATTGTTTTTAGTTGTCTAAATAAAATGCCTCTAGTT
SO_M_T1039   CCCATGCAATTAAATATCTTTTTCACATATGTTCATTGACTCAATCTTCTGTTGGATCACGGGAAATCAAGAGTTTAGTTTTTAAAGATGGAACTCCACCCTTTGCTGTG
SO_M_T1047   GTGTTCATTGTACAGTTTGTTTACACACGTGCAATTTGACTTTGTGAATACATATGTGTATATAAGAGGTTAATCTTGGTTGGACGTTTGTTTTCTCCTTAGTTGCTTGT
SO_M_T1048   CTATCACTGGCCAGAGTTTGTACCCTGTGTTTTACCTTCTTGGATATTTTATGATTTTACTTGAATGTGCCTGGTTCTATGATATTTGTAGTGCCGGGGCTCGCACTATG
SO_M_T1050   ACAACAAAATTAAAAAAAAAATAGATGTGCATTCAGTTTACAGAAGCTAAACCCTTCCCCACATGTGGTAGACTTATGTATTTTGATATACCATTGTTTGAACTTCTCCT
SO_M_T1051   TACAACTCTCCAGCCACTCTCTTCTACGCTGCCTCAACATGTATAAATGTTATCTGATATTGCTCTTAGATGTTTTAGCATTAAACTGTTGAACTTTTTTTTGTACAAGC
SO_M_T1052   CTTAACTGAATACTGTAGAAATGCACTTTGCTCAGTAATTCTGTAGGTTTATTTGTTGCATACTTTGACTTAGTAATGTTGTACTTATAATTCTGTACAGAAATGAGTCT
SO_M_T1059   ATGAATTTATTAAATCTAGTTGTCACTTTATTTTGGACGAACAAATTACAAATGAACTTTCTTGCACTTGCTCAGAGCTTTGTATTTTTGTTACGTACTGTTTCTTCCCA
SO_M_T1072   CGTAATGTTTTCCGTTCTTCCACAGAAATATGCAAAACAAACCAAATGTATAGAGCTTTGGGTGTAGGATATTCATTTGTATTTATACTTAAAAACACACAATGTTCAAA
SO_M_T1078   CTCTCTTTATTTTTTAACTCTATTACATTTTATTTTGTATGCCAGACTTCTAAAATAAATGTTTTGGAATTCTAGCAAAGTGTTTTAATCCACGGTTGTGCCTTATCAAG
```

B ***Input RNA-Seq counts file***
(reporter id, counts in sample #1, ..., counts in sample #11)

id	AR01	AR02	AR03	AR04	AR05	AR06	AR07	AR08	AR09	AR10	AR11	
SO_M_T1006	144	122	69	74	76	40	52	46	59	31	9	
SO_M_T1013	120	97	86	85	72	40	63	48	80	21	4	
SO_M_T1022	124	90	66	65	70	58	64	69	86	31	3	
SO_M_T1027	124	70	86	88	104	51	53	78	62	19	5	
SO_M_T1039	188	115	71	61	72	40	43	45	34	14	5	
SO_M_T1047	366	323	227	236	269	133	131	100	76	24	13	
SO_M_T1048	316	235	182	185	151	70	85	93	103	40	3	
SO_M_T1050	484	395	312	309	347	170	173	220	245	123	27	
SO_M_T1051	134	102	86	85	112	34	56	39	49	19	2	
SO_M_T1052	443	353	337	342	435	161	79	59	51	29	11	
SO_M_T1059	551	389	352	353	416	149	108	121	98	35	20	
SO_M_T1072	231	188	177	192	196	159	129	126	106	43	12	
SO_M_T1078	87	54	65	52	75	19	19	95	128	49	4	

C

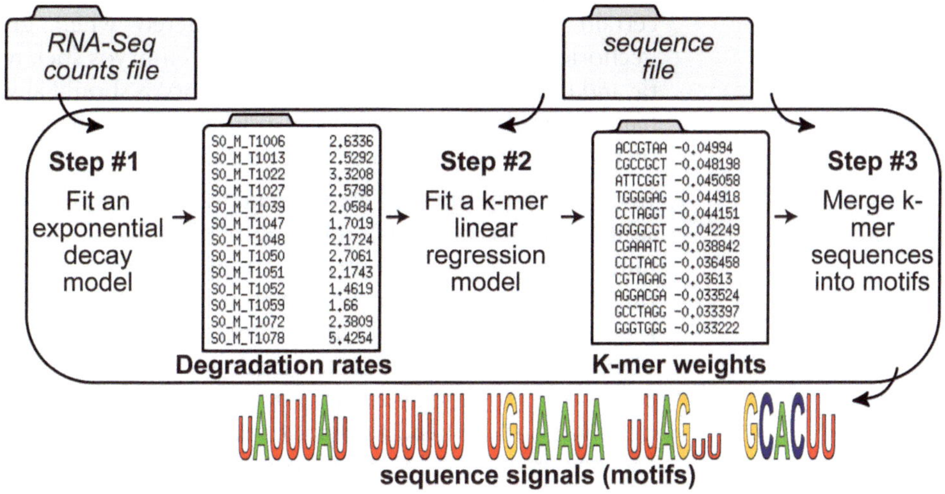

Fig. 2 The UTR-Seq computational analysis tool. (**a**) An input sequence file that contains a unique oligonucleotide sequence for each reporter id (rows). (**b**) An input RNA-Seq count file that contains the number of times that each unique oligonucleotide sequence (rows) was detected in each temporal sample (columns). (**c**) Both files (top) are used as input to the UTR-Seq computational analysis tool. This tool runs in three steps (left to right): step 1 estimates the degradation rate of each reporter from its RNA-Seq counts; step 2 defines the contribution ("weight") of each short (3–7 nt long) sequence within a reporter sequence to the overall degradation rate of the reporter; step 3 finds a set of short sequences with maximal contribution to degradation rates ("peaks") and group them into motifs based on their sequence similarity. These motifs are the final output of the analysis (bottom)

4 Notes

1. In designing your library of oligonucleotide sequences, it is important to ensure that sequences are different enough to allow accurate identification from short high-throughput sequencing reads. Select synthesis methods and scale that fit the number of sequences in your library and their overall length (including two 20-nt adaptors).
2. A minimum of ten injected embryos per sample is recommended. Using more embryos is recommended when possible and can help to minimize variability between timepoints that arises from differences in manipulation, developmental staging, or genetic background. A minimum of two RNA samples from two developmental stages per experiment is required, in order to determine the degradation rate of each RNA reporter. Sampling more stages is recommended when possible and can improve robustness of degradation rate quantification and provide a higher temporal resolution to identify changes in RNA stability that happen during development.
3. When raised in embryo culture medium at 28 °C, zebrafish embryos develop normally outside of their chorions. However, certain precautions should be considered when growing dechorionated embryos. Dechorionated embryos stick to plastic and dissolve; thus, dechorionated embryos should always be handled in glass dishes or plastic dishes coated with a thin layer of 1% agarose in embryo culture medium.
4. In choosing the appropriate sequencing platform, consider the scale (number of reads) and length of reads. The expected number of reads should provide enough depth to measure all the oligonucleotide sequences in your library. Reads should be long enough to measure a portion of sequence from each oligonucleotide that allow its unique identification within the entire oligonucleotide library. Determine whether single-end or paired-end sequencing is preferred. Note that a 20-nt adaptor sequence is also included from each end of the read. In the current design, the UMI (if used) is read before the downstream adaptor sequence.

References

1. Wharton RP, Struhl G (1991) RNA regulatory elements mediate control of Drosophila body pattern by the posterior morphogen nanos. Cell 67(5):955–967
2. Tadros W et al (2007) SMAUG is a major regulator of maternal mRNA destabilization in Drosophila and its translation is activated by the PAN GU kinase. Dev Cell 12 (1):143–155
3. Pique M et al (2008) A combinatorial code for CPE-mediated translational control. Cell 132 (3):434–448
4. Giraldez AJ et al (2006) Zebrafish MiR-430 promotes deadenylation and clearance of maternal mRNAs. Science 312(5770):75–79

5. Guo H et al (2010) Mammalian microRNAs predominantly act to decrease target mRNA levels. Nature 466(7308):835–840
6. Barckmann B, Simonelig M (2013) Control of maternal mRNA stability in germ cells and early embryos. Biochim Biophys Acta 1829 (6-7):714–724
7. Voeltz GK, Steitz JA (1998) AUUUA sequences direct mRNA deadenylation uncoupled from decay during Xenopus early development. Mol Cell Biol 18 (12):7537–7545
8. Kedde M et al (2007) RNA-binding protein Dnd1 inhibits microRNA access to target mRNA. Cell 131(7):1273–1286
9. Geisberg JV et al (2014) Global analysis of mRNA isoform half-lives reveals stabilizing and destabilizing elements in yeast. Cell 156 (4):812–824
10. Mishima Y, Tomari Y (2016) Codon usage and 3′ UTR length determine maternal mRNA stability in zebrafish. Mol Cell 61(6):874–885
11. Bazzini AA et al (2016) Codon identity regulates mRNA stability and translation efficiency during the maternal-to-zygotic transition. EMBO J 35(19):2087–2103
12. Rabani M, Kertesz M, Segal E (2008) Computational prediction of RNA structural motifs involved in posttranscriptional regulatory processes. Proc Natl Acad Sci U S A 105 (39):14885–14890
13. Ray D et al (2013) A compendium of RNA-binding motifs for decoding gene regulation. Nature 499(7457):172–177
14. Hafner M et al (2010) Transcriptome-wide identification of RNA-binding protein and microRNA target sites by PAR-CLIP. Cell 141(1):129–141
15. Tuerk C, Gold L (1990) Systematic evolution of ligands by exponential enrichment: RNA ligands to bacteriophage T4 DNA polymerase. Science 249(4968):505–510
16. Arnold CD et al (2013) Genome-wide quantitative enhancer activity maps identified by STARR-seq. Science 339(6123):1074–1077
17. Kheradpour P et al (2013) Systematic dissection of regulatory motifs in 2000 predicted human enhancers using a massively parallel reporter assay. Genome Res 23(5):800–811
18. Grossman SR et al (2017) Systematic dissection of genomic features determining transcription factor binding and enhancer function. Proc Natl Acad Sci U S A 114(7): E1291–E1300
19. Rosenberg AB et al (2015) Learning the sequence determinants of alternative splicing from millions of random sequences. Cell 163 (3):698–711
20. Zhao W et al (2014) Massively parallel functional annotation of 3′ untranslated regions. Nat Biotechnol 32(4):387–391
21. Oikonomou P, Goodarzi H, Tavazoie S (2014) Systematic identification of regulatory elements in conserved 3′ UTRs of human transcripts. Cell Rep 7(1):281–292
22. Shalem O et al (2015) Systematic dissection of the sequence determinants of gene 3′ end mediated expression control. PLoS Genet 11 (4):e1005147
23. Yartseva V et al (2017) RESA identifies mRNA-regulatory sequences at high resolution. Nat Methods 14(2):201–207
24. Rabani MPL, Chew GL, Schier AF (2017) Massively parallel reporter assay of 3'UTR sequences identifies in vivo rules for mRNA degradation. Mol Cell 68(6):1083–1094.ep
25. Jukam D, Shariati SAM, Skotheim JM (2017) Zygotic genome activation in vertebrates. Dev Cell 42(4):316–332
26. Kane DA, Kimmel CB (1993) The zebrafish midblastula transition. Development 119 (2):447–456
27. Marcel M (2011) Cutadapt removes adapter sequences from high-throughput sequencing reads. EMBnet journal 17(1):10–12
28. Criscuolo A, Brisse S (2014) AlienTrimmer removes adapter oligonucleotides with high sensitivity in short-insert paired-end reads. Front Genet 13(5):130
29. Langmead B, Salzberg SL (2012) Fast gapped-read alignment with Bowtie 2. Nat Methods 9 (4):357–359
30. Alestrom P, Dangelo L, Midtlyng PJ, Schorderet DF, Schulte-Merker S, Sohm F et al (2019) Zebrafish: housing and husbandry recommendations. Lab Anim 2019:23677219869037
31. Islam S et al (2014) Quantitative single-cell RNA-seq with unique molecular identifiers. Nat Methods 11(2):163–166

Chapter 29

Exploring Translational Control of Maternal mRNAs in Zebrafish

Cecilia Lanny Winata, Maciej Łapiński, Hisyam Ismail, Sinnakaruppan Mathavan, and Prabha Sampath

Abstract

The study of translational regulation requires reliable measurement of both mRNA levels and protein synthesis. Cytoplasmic polyadenylation is a prevalent mode of translational regulation during oogenesis and early embryogenesis. Here the length of the poly(A) tail of an mRNA is coupled to its translatability. We describe a protocol to identify translationally regulated genes and measure their translation rate in the early zebrafish embryo using genome-wide polysome profiling. This protocol relies on the isolation of mRNA by means of an rRNA depletion strategy, which avoids capture bias due to short poly(A) tail that can occur when using conventional oligo(dT)-based methods. We also present a simple PCR-based method to measure the poly(A) tail length of selected mRNAs.

Keywords Maternal RNA, Translational regulation, Polysome profiling, Cytoplasmic polyadenylation, Zebrafish

1 Introduction

In the oocyte and early embryo, maternally deposited mRNAs direct biological processes up to the point of zygotic genome activation. In the absence of zygotic genome transcription, these maternal mRNAs undergo translational regulation by means of cytoplasmic polyadenylation in order to regulate their function at the appropriate time frame of development [1–5]. Previous genomics studies in the early zebrafish embryo identified a large cohort of maternal mRNAs that undergo cytoplasmic polyadenylation during the course of embryogenesis, leading to midblastula transition (MBT) [6]. Cytoplasmic polyadenylation occurs almost exclusively prior to MBT and was shown to correlate with the recruitment of such mRNAs into polysomes [5].

In order to study the early embryonic transcriptome and the mechanism of its regulation, it is necessary to employ a reliable method that can measure not only mRNA levels but also protein

Roland Dosch (ed.), *Germline Development in the Zebrafish: Methods and Protocols*, Methods in Molecular Biology, vol. 2218, https://doi.org/10.1007/978-1-0716-0970-5_29, © Springer Science+Business Media, LLC, part of Springer Nature 2021

levels. Proteomics method is very challenging in the zebrafish egg and early embryo due to the lack of sensitive techniques and the presence of high-molecular-weight proteins in the yolk. Thus, there is currently poor knowledge on the actual translation activity of maternal transcripts. Polysome profiling [7, 8] offers the possibility to provide crucial information on proteins that are synthesized from maternal transcripts. In addition to this, the prevalence of cytoplasmic polyadenylation in the early embryonic stages renders it essential to consider the various degrees of polyadenylation state in different transcript species. However, conventional methods of mRNA isolation still rely on poly(A)-dependent enrichment, which causes inaccurate measurements of transcript levels.

To examine the correlation between cytoplasmic polyadenylation and polysome recruitment of specific mRNA species, it is useful to assess the elongation of the poly(A) tail. Various methods have been described to assay the length of the poly(A) tail of an mRNA. These methods range from assessing poly(A) tail length of a few specific genes to measurements on the scale of the complete transcriptome. The classical methods to determine the length of the poly(A) tail employ RNAse H and a transcript-specific oligonucleotide to digest the mRNA of interest, leaving the 3′ terminus with a length that ensures efficient gel resolution and detection by northern blotting (typically 100–300 nucleotides) [9, 10]. This approach, however, is time-consuming and reserved only for highly abundant mRNA particles [11, 12].

The advent of PCR-based methods allowed for the much-improved sensitivity of the analysis. Two of the techniques rely on anchored oligo(dT) primers to specifically amplify the 3′ termini of the chosen mRNA molecules. The **r**apid **a**mplification of **c**DNA **e**nds **p**oly(**A**) **t**est (RACE-PAT) involves performing the reverse transcription reaction with an oligo(dT) primer linked to a G/C rich anchor sequence [4]. Hybridization of the oligo(dT) sequence can occur anywhere along the poly(A) tail. Subsequently, the 3′ terminus of interest is amplified with sequence-specific forward primer and the oligo(dT) anchor as the reverse primer. The amplification results can be visualized on ethidium bromide-stained agarose gels. A more sensitive modification of this technique is the **l**igation-**m**ediated **p**oly(**A**) **t**est (LM-PAT) where the poly(A) tail is first saturated with phosphorylated oligo(dT) particles in the presence of T4 DNA ligase. The anchored oligo(dT) is then targeted to the 3′ end fragment of the tail [13]. This allows for a higher resolution assessment of the homopolymer length. A similar technique is the anchored RT-PCR using T4 RNA ligase [14]. Here, the poly(A) mRNA is ligated to the RNA oligonucleotide that contains a 3′-amino blocking group, such that the ligation reaction occurs only between the 3′ end of the mRNA and the 5′ end of the oligonucleotide. The ligated oligonucleotide then serves as a priming site for reverse transcription and PCR reaction with the additional target-specific sequence. To add a priming site anchor right after the poly(A) 3′ homopolymer track, one could also utilize Klenow fragment of *Escherichia coli* DNA polymerase I, as it is performed in the **e**xtension **p**oly(**A**) **T**est method (ePAT) [15]. Another PCR-based method is circularization RT-PCR (cRT-PCR) [16]. This technique utilizes RNA ligase to circularize the decapped mRNA before reverse transcription. The subsequent PCR amplification is performed over the joint, with primers specific to both the 5′ and 3′ end of the mRNA, which enables the analysis of each of those fragments in the same molecule. Yet another transcript-specific technique employing the PCR amplification and a method, the modification of which is presented in this review is the **hi**gh-**re**solution **p**oly(**A**) **t**est (Hire-PAT) [17]. In brief, total RNA is isolated, followed by guanosine (G) and inosine (I) addition to the 3′ end by poly(A) RNA polymerase. The newly added G/I stretch serves as an anchor site for reverse transcription. Subsequently, PCR is performed using a transcript-specific oligo and a fluorescent-labeled (6-FAM) oligo complementary to the anchor sequence. The PCR products can be resolved *via* capillary electrophoresis or standard agarose gel.

This protocol describes the isolation of polysome fractions from early (up to gastrulation stage) zebrafish embryos and preparation of RNA-seq library by rRNA depletion strategy for studying the transcriptome and their translational regulation in the early embryonic development. It also includes a protocol to assay poly (A) tail length using a simple PCR-based method.

2 Materials

2.1 Zebrafish Maintenance and Embryonic Staging

1. Zebrafish embryos grown to desired developmental stage.
2. 100-mm culture dish.
3. Egg water: 60 μg/ml Instant Ocean® sea salt in diH_2O.
4. Dissecting-type microscope (Leica M125C or equivalent).
5. Incubator at 28 °C.

2.2 Cell Lysate Isolation and Polysome Fractionation

1. Cycloheximide.
2. Lysis buffer: 10 mM Tris–HCl, 5 mM $MgCl_2$, 3 mM KCl, 1% Triton X-100, 1% sodium deoxycholate, 4 mM DTT, 400 μg/ml CHX, RNAse inhibitor, ice cold.
3. Sucrose gradient solutions: 5% or 45% (w/v) sucrose; 20 mM HEPES, pH 7.5; 100 mM KCl; 5 mM $MgCl_2$; 0.5 mg/ml heparin; 100 μg/ml cycloheximide, filtered.
4. 4-ml thin-wall polypropylene tube (Beckman-Coulter).
5. Nutator.

6. Gradient Master (Biocomp).
7. Gradient fractionator (Brandel).
8. UV detector (ISCO UA-6 or equivalent).
9. Ultracentrifuge with SW60Ti rotor (Beckman).

2.3 RNA Isolation and Quality Assessment

1. GeneChip™ Eukaryotic Poly-A RNA Spike-in mix or equivalent.
2. Qiagen RNeasy Mini Kit.
3. Tabletop microcentrifuge with cooling system.
4. UV spectrophotometer.
5. NanoDrop 8000 spectrophotometer.
6. Agilent 2100 Bioanalyzer.
7. RNA 6000 Nano reagents.
8. Zymo Clean and Concentrator Kit.

2.4 Ribo-Depleted Sequencing Library Preparation

1. TruSeq Stranded Total RNA Library Prep Gold Kit.

2.5 Poly(A) Tail Length Assay

1. Yeast poly(A) polymerase.
2. Guanosine 5′-triphosphate sodium salt.
3. Inosine 5′-triphosphate trisodium salt.
4. Prepare a GTP/ITP stock (5 mM each) in RNA-seq free water. Aliquot into small volumes and store at −20 °C, avoid multiple freeze-thaw cycles.
5. C10T2 DNA primer (5′-CCC CCC CCC CTT-3′), prepare 10 μM stock solution, and aliquot into small volumes and store at −20 °C.
6. M13RC10T2 primer (5′-CA GGA AAC AGC TAT GAC CCC CCC CCC CTT-3′), prepare 10 μM stock solution, and aliquot into small volumes and store at −20 °C.
7. Superscript IV First-strand Synthesis system.
8. RiboLock RNAse inhibitor.
9. dNTP mix (10 mM each).
10. DreamTaq DNA Polymerase.
11. Agarose powder (Sigma, Cat. No. A9539).
12. Tris–acetate EDTA (TAE) buffer: 40 mM Tris, 20 mM acetic acid, 1 mM EDTA.
13. PCR thermocycler.
14. Gel electrophoresis apparatus.
15. Gel documentation system.

3 Methods

Cellular lysates are prepared from zebrafish embryos at the desired developmental stages. To stop the process of translation, the embryos are treated with cycloheximide. Isolated lysates are then loaded onto sucrose gradients before being subjected to ultracentrifugation. Procedures for harvesting and maintaining zebrafish embryos are well established [18, 19]. Embryos are staged according to standard criteria [20]. Figure 1 summarizes the workflow of polysome isolation.

3.1 Isolation of Cell Lysates from Zebrafish Embryos

1. Collect zebrafish eggs in a 100-mm culture dish containing egg water and keep at 28 °C in an incubator until desired developmental stage is reached.
2. Observe embryos periodically under a dissecting-type microscope to determine developmental stage.
3. Set tabletop microcentrifuge to 4 °C.
4. Harvest 300 embryos from each desired developmental stage and place them in a 2-ml microcentrifuge tube.
5. Replace embryo medium with 200 μg/ml cycloheximide in 100 ml egg water and incubate for 10 min at room temperature with gentle tilting of the tube on a nutator.
6. Add 500 μl of 2× lysis buffer and mechanically dissociate embryos using a 1-ml pipette tip. The mechanical dissociation should be performed in two stages, by using a 1-ml tip, which is cut ~1 mm from the tip, followed by using uncut 1-ml tips (*see* **Note 1**).
7. Centrifuge lysates for 3 min at full speed in a microcentrifuge at 4 °C.
8. Transfer the supernatant into a new microcentrifuge tube.
9. Dilute 5 μl supernatant and 495 μl nuclease-free water and measure the optical density (OD) at 260 nm using a UV spectrophotometer.
10. Obtain a volume of lysate equivalent to 21–22 OD units and volume-adjust to 300 μl with water. Set aside the rest of the lysate (~200 μl) and keep at −80 °C for unfractionated total RNA (*see* **Note 2**).

3.2 Preparation of Sucrose Gradients

1. Position 4-ml ultracentrifuge tubes on a tube rack. Draw a line halfway along the tube.
2. Make sucrose gradient with 5% and 45% sucrose:
   (a) Layer 5% sucrose solution until slightly above the halfway mark.

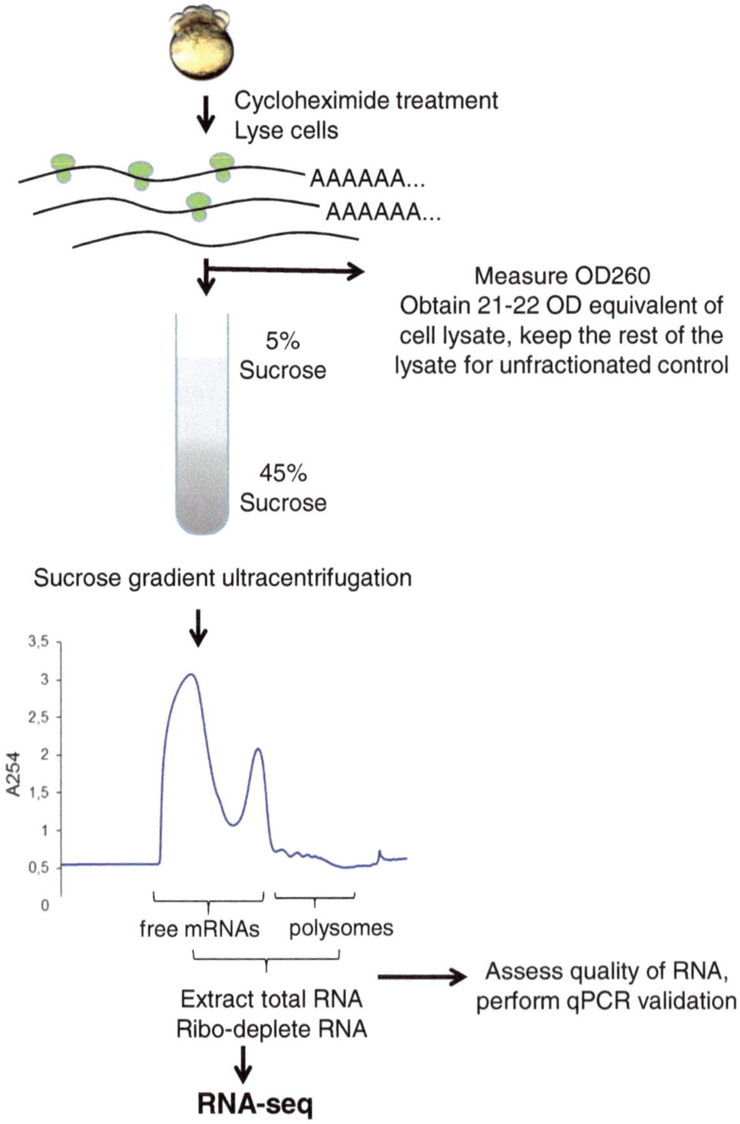

Fig. 1 Workflow of polysome fractionation based on sucrose density gradient centrifugation. Zebrafish embryos were treated with cycloheximide to fix ribosomes on mRNAs and lysed. Cell lysates were subjected to sucrose gradient ultracentrifugation and loaded onto a gradient fractionator to isolate polysome fractions. Fractions that contained polysome-unbound and polysome-bound mRNAs can be distinguished based on their A254 profile

(b) Fill in 45% sucrose solution starting from the bottom of tube, carefully to avoid disturbing the interface.

(c) Cap tube, load onto gradient master machine, and spin for 1 min. Keep tubes at 4 °C until samples are ready.

3. Load samples on top of the sucrose gradient.

4. Ultracentrifuge samples at 165,000 × *g* for 1.5 h at 4 °C in a SW40-Ti swinging bucket rotor.
5. Prepare microcentrifuge tubes for every sample containing 50 μl of ammonium acetate and 25 μl of 1% SDS.
6. Run sample in a gradient fraction collector and obtain 8× 500 μl fragments using prepared tubes. Immediately transfer the samples to ice. Monitor polysome profiles at 254 nm on the UV detector to determine the fractions containing mRNA bound by different numbers of ribosomes.
7. Transfer to 15-ml Falcon tubes containing 2.5 ml of 100% EtOH and add spike-in mix according to manufacturer's instructions.
8. Store at −80 °C or proceed with RNA isolation.

3.3 Total RNA Isolation from Polysome Fractions

1. Precipitate each individual or pooled fractions with 2.5 ml of 100% ethanol for RNA isolation.
2. Isolate RNAs using Qiagen RNA extraction kit incorporating the on-column DNAseI digestion with two additional rounds of column-washing step.
3. Perform RNA quality control on a Nanodrop 8000 spectrophotometer (*see* **Note 3**).
4. RNA integrity is assayed using Agilent 2100 Bioanalyzer or equivalent systems according to the manufacturer's protocol (*see* **Note 4**).
5. Perform quantitative reverse transcription polymerase chain reaction (qRT-PCR) to validate expression dynamics of representative genes in polysome fractions.

3.4 Preparation of RNA-Seq Library and Sequencing

Due to the prevalence of cytoplasmic polyadenylation process occurring in early embryonic stages, particularly prior to MBT, it is important to employ mRNA isolation methods that are not based on poly(A) capture, as this will cause bias toward the capture of mRNAs with longer poly(A) tail [6]. We thus recommend using the ribosomal RNA depletion strategy.

1. Perform rRNA depletion on 1–5 μg of isolated total RNA with the Ribo-Zero Magnetic Gold Kit (Human, Mouse, Rat) according to the manufacturer's protocol.
2. Quantify rRNA-depleted mRNA samples using the Qubit system and assay the depletion of rRNAs with the RNA 6000 Nano chip on the Agilent 2100 Bioanalyzer or equivalent systems. If the rRNA depletion is successful, the rRNA peaks should disappear as seen in the bioanalyzer electropherogram profile (*see* **Note 5**).

3. Perform RNA-seq on Illumina next generation sequencing platform according to manufacturer's protocol. For RNA-seq, we found that 100 million paired-end reads is sufficient to represent the zebrafish early embryonic transcriptome and for reliable calculation of translation rate [5].

3.5 Assessment of Translation Rate and Translational Regulation

1. Perform processing of sequencing reads and basic quality check according to Illumina guidelines.
2. Map sequencing reads to the latest zebrafish Refseq genome assembly and generate raw read counts using standard algorithms, e.g., STAR [21] and Salmon [22].
3. Perform normalization of read counts according to the library size by calculating normalization factor in R using DeSeq2 method [23].
4. Polysome association is expressed as a function of polysome-bound over whole fraction expression values. Calculate transcription (ΔTS) and translation (ΔTL) rates as the fraction of total and polysome-bound expression values over time. Translational regulation can be defined as a non-linear relationship between ΔTS and ΔTL.

3.6 Poly(A) Tail Length Assay

Here we describe a PCR-based method for poly(A) tail measurement that works best in our hands and is less time- and resource-consuming than Northern Blot-based assays. This assay can be performed using the commercially available Poly(A) Tail Length Assay Kit (USB Corporation). However, it is also possible to purchase components individually and set up the assay on your own (for example, *see* [24]). In principle, guanosine and inosine (G/I) residues are added to the end of the poly(A) tail by yeast poly (A) polymerase and subsequently converted into cDNA using the newly added G/I tails as priming sites. Measurement of poly(A) tail was performed by PCR using a gene-specific forward primer and a universal reverse primer followed by running the PCR products in 2% agarose gel.

3.6.1 Isolate Total RNA from Embryos

We typically use the Trizol method followed by column purification, which gives a consistently good quality of RNA (in terms of purity and integrity). However, other alternative methods are available such as using the Qiagen RNeasy Mini Kit. Here we describe the RA isolation method based on Trizol protocol.

1. Harvest embryos at desired stages as described in Subheading 3.1 **steps 1–4** (*see* **Notes 6** and 7).
2. Homogenize in 500 μl Trizol in a microcentrifuge tube using plastic pestle (Sigma), then add another 500 μl Trizol; mix well and incubate for 5 min.

3. Add 200 μl chloroform, shake vigorously for 15 s, and incubate for 3 min.
4. Centrifuge at 12,000 × *g* for 15 min at 4 °C.
5. Carefully take the clear aqueous layer (~400 μl) and transfer to a new tube (*see* **Note 8**).
6. Add 500 μl isopropanol and incubate for 30 min.
7. Centrifuge at 12,000 × *g* for 10 min at 4 °C.
8. Wash pellet with 80% ethanol and centrifuge for 5 min at 4 °C.
9. Remove all supernatant, and air-dry pellets.
10. Redissolve RNA in 50 μl nuclease-free water.
11. Assess quality using nanodrop and ensure the standard qualities are met (*see* **Note 9**).
12. RNA cleanup using column incorporating the on-column DNAseI digestion (*see* **Note 10**).
13. Measure concentration using nanodrop and ensure the standard qualities are met (*see* **Note 11**).

3.6.2 Perform Poly(A) Tail Length Assay

1. For each transcript, prepare two parallel reactions for experiment (tail-specific PCR) and control (gene-specific PCR) in 0.2-μl PCR tubes.
2. Place 1–2 μg of total RNA in the PCR tubes and add RNAse-free water to make a total volume of 15.5 μl.
3. Heat at 65 °C for 5 min and immediately transfer to ice to eliminate RNA secondary structures.
4. Add 5 μl of 5× poly(A) polymerase reaction buffer, 2.5 μl of GTP-ITP mix, 1 μl of RNAse inhibitor, and 1 μl of yeast poly (A) polymerase (*see* **Note 12**).
5. Incubate this mixture for 20 min at 37 °C and terminate the reaction by heating at 65 °C for 10 min.
6. Purify the G/I-tailed RNA using column (Qiagen or Zymo) according to manufacturer's protocol. Elute the final product in 10 μl RNAse-free water.
7. Synthesize cDNA using Superscript IV First-strand Synthesis system according to the manufacturer's protocol:
   - Mix 9 μl of the GI-tailed RNA with 1 μl of C10T2 primer (10 μM) in a sterile 0.2-ml PCR tube.
   - Heat this mixture at 65 °C for 5 min, then 42 °C, and hold for 5 min.
   - Prepare the reverse transcription mixture in a separate tube: 4 μl of 5× SSIV buffer, 1 μl of DTT, 1 μl of 10 mM dNTP mix, 1 μl RNAse inhibitor, and 1 μl of SSIV reverse transcriptase enzyme.

- Transfer the reverse transcription mixture into the denatured RNA-primer mix.
- Incubate for 30 min at 50 °C and heat at 80 °C for 10 min to inactivate the reaction.
- Dilute the mixture by adding 20 μl sterile water.

8. Set up PCR reactions (tail-specific and gene-specific in parallel) in sterile 0.2-μl PCR tubes.

	Tail-specific, μl	Gene-specific, μl
10× DreamTaq buffer	5	5
10 mM dNTPs	1	1
10 μM Gene-specific primer (forward)	1	1
10 μM Gene-specific primer (reverse)	–	1
10 μM M13RC10T2 primer	1	–
cDNA	2	2
DreamTaq polymerase enzyme	0.4	0.4
Water	39.6	39.6

9. Perform PCR on a thermocycler with the following conditions:

95 °C for 2 min	Repeat 30 cycles
95 °C for 30 s	
55 °C for 30 s	
72 °C for 1 min	
72 °C for 10 min	
4 °C hold	

10. Perform gel electrophoresis on a 2% agarose gel in TAE buffer.
11. Obtain gel image on a gel documentation system and assess poly(A) tail length by comparison to the control (gene-specific PCR). Refer to Fig. 2b for a representative image (*see* **Note 13**).

4 Notes

1. Take care not to introduce too much foaming of the samples.
2. Volume = OD260 × 21/5 (e.g., if OD260 = 0.403, 21 OD units is 262 μl).

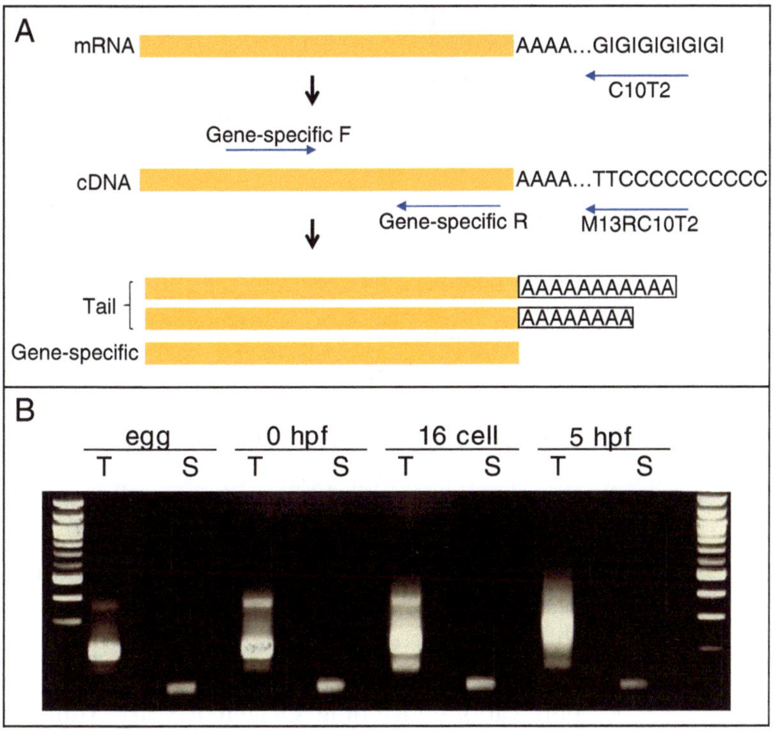

Fig. 2 Poly(A) tail length assay. (**a**) Schematics of the poly(A) tail assay. Total RNA was G/I-tailed and reverse-transcribed using the C10T2 as a primer. cDNAs were subsequently PCR-amplified with the gene-specific forward primer (gene-specific *F*) and either the M13RC10T2 reverse primer (for tail reaction) or gene-specific reverse primer (gene-specific reaction). (**b**) Measurement of *dnaja2* poly(A) tail length in the egg, one-cell, 16-cell, and 5.3-hpf stages in the tail-specific reaction (T) and a gene-specific control reaction with *dnaja2*-specific F and R primers (S). The increasing poly(A) tail appears as a smear that increases gradually from the one-cell stage to the 5.3-hpf stage, indicating the presence of an extended poly(A) tail over time

3. The *A*260/280 ratio should be above 2.0 and *A*260/230 ratio should be at least 1.8.
4. Ideally, RIN number should be at least 8.0.
5. The Ribo-Zero Magnetic Gold Kit has been incorporated into the Illumina TruSeq Stranded Total RNA library Prep Gold Kit (Illumina, USA), the end result of which is the RNA-seq library.
6. One wild-type embryo at pre-MBT stage typically yields 100 ng total RNA. For routine RNA preparations for isolating cDNA or RNA-seq library preparation, usually 100 embryos is sufficient. For normalization purposes, especially when differential expression is the main interest, it is ideal to use the same number of embryos in all samples and then use a spike-in control.

7. Regarding spike-ins: In polysome profiling experiment, we used the GeneChip™ Eukaryotic Poly(A) RNA Control Kit (Thermo Fisher Scientific). We have also used kanamycin (Promega; $c = 0.5$ μg/μl) for qPCR experiments:
   - Dissolve 1 μl of Promega stock in 99 μl MQ (100 μl of 5 ng/μl stock).
   - 1 μl of 5 ng/μl stock dissolve in 99 μl MQ and prepare 5 × 20 μl aliquots
   - Working concentration is 0.25 ng/ml (5 μl into 1 ml Trizol or 0.5 μl into 75 μl RLT).
   - Primers for kanamycin (120 bp fragment):
     - F: CGAGTGATTTTGATGACGAGCGT
     - R: CGACTGAATCCGGTGAGAATGG.
8. It is absolutely critical to avoid contamination from the lower phase as this will spoil the isolated RNA purity.
9. *A*260/280 of at least 1.8, and *A*260/230 at least 1.5; if this value is low, re-do the protocol starting from the chloroform step.
10. We have had good results using various types of RNA column purifications including Qiagen RNeasy Mini Kit and Zymo RNA Clean and Concentrator Kit.
11. After the column purification, A260/280 should be at least 1.8. We typically obtain around 200–300 μg/μl of RNA from the earlier stages (up to 30% epiboly) and slightly more from the later stages (up to gastrulation).
12. It is important to use the yeast poly(A) polymerase due to its ability to incorporate non-A residues [25]. The use of *E. coli* poly(A) polymerase is not recommended.
13. The size of PCR product generated using universal reverse primer is compared to that generated using gene-specific reverse primer designed at the 3′ end of the mRNA sequence. A smear indicates the presence of poly(A) tail, while a distinct band indicates the lack of poly(A) tail.

Acknowledgments

M.Ł. is supported by the DIAMENTOWY scholarship from the Polish Ministry of Science and Higher Education. The project no. POIR.04.04.00-00-1AF0/16-00/ is carried out within the First TEAM programme of the Foundation for Polish Science co-financed by the European Union under the European Regional Development Fund. This work was supported by funding from A*STAR and BMRC, Singapore.

References

1. Evsikov AV, de Vries WN, Peaston AE, Radford EE, Fancher KS, Chen FH, Blake JA, Bult CJ, Latham KE, Solter D, Knowles BB (2004) Systems biology of the 2-cell mouse embryo. Cytogenet Genome Res 105(2–4):240–250. https://doi.org/10.1159/000078195
2. Oh B, Hwang S, McLaughlin J, Solter D, Knowles BB (2000) Timely translation during the mouse oocyte-to-embryo transition. Development 127(17):3795–3803
3. Potireddy S, Vassena R, Patel BG, Latham KE (2006) Analysis of polysomal mRNA populations of mouse oocytes and zygotes: dynamic changes in maternal mRNA utilization and function. Dev Biol 298(1):155–166. https://doi.org/10.1016/j.ydbio.2006.06.024
4. Salles FJ, Darrow AL, O'Connell ML, Strickland S (1992) Isolation of novel murine maternal mRNAs regulated by cytoplasmic polyadenylation. Genes Dev 6(7):1202–1212. https://doi.org/10.1101/gad.6.7.1202
5. Winata CL, Lapinski M, Pryszcz L, Vaz C, Bin Ismail MH, Nama S, Hajan HS, Lee SGP, Korzh V, Sampath P, Tanavde V, Mathavan S (2018) Cytoplasmic polyadenylation-mediated translational control of maternal mRNAs directs maternal-to-zygotic transition. Development 145(1). https://doi.org/10.1242/dev.159566
6. Aanes H, Winata CL, Lin CH, Chen JP, Srinivasan KG, Lee SG, Lim AY, Hajan HS, Collas P, Bourque G, Gong Z, Korzh V, Alestrom P, Mathavan S (2011) Zebrafish mRNA sequencing deciphers novelties in transcriptome dynamics during maternal to zygotic transition. Genome Res 21(8):1328–1338. https://doi.org/10.1101/gr.116012.110
7. Sampath P, Lee QY, Tanavde V (2011) Identifying translationally regulated genes during stem cell differentiation. Curr Protoc Stem Cell Biol Chapter 1:Unit1B 8. https://doi.org/10.1002/9780470151808.sc01b08s18
8. Sampath P, Pritchard DK, Pabon L, Reinecke H, Schwartz SM, Morris DR, Murry CE (2008) A hierarchical network controls protein translation during murine embryonic stem cell self-renewal and differentiation. Cell Stem Cell 2(5):448–460. https://doi.org/10.1016/j.stem.2008.03.013
9. Mercer JF, Wake SA (1985) An analysis of the rate of metallothionein mRNA poly(A)-shortening using RNA blot hybridization. Nucleic Acids Res 13(22):7929–7943. https://doi.org/10.1093/nar/13.22.7929
10. Sheets MD, Fox CA, Hunt T, Vande Woude G, Wickens M (1994) The 3′-untranslated regions of c-mos and cyclin mRNAs stimulate translation by regulating cytoplasmic polyadenylation. Genes Dev 8(8):926–938. https://doi.org/10.1101/gad.8.8.926
11. Murray EL, Schoenberg DR (2008) Assays for determining poly(A) tail length and the polarity of mRNA decay in mammalian cells. Methods Enzymol 448:483–504. https://doi.org/10.1016/S0076-6879(08)02624-4
12. Nilsen TW (2015) Measuring the length of poly(A) tails. Cold Spring Harb Protoc 2015 (4):413–418. https://doi.org/10.1101/pdb.prot081034
13. Salles FJ, Strickland S (1999) Analysis of poly (A) tail lengths by PCR: the PAT assay. Methods Mol Biol 118:441–448. https://doi.org/10.1385/1-59259-676-2:441
14. Rassa JC, Wilson GM, Brewer GA, Parks GD (2000) Spacing constraints on reinitiation of paramyxovirus transcription: the gene end U tract acts as a spacer to separate gene end from gene start sites. Virology 274 (2):438–449. https://doi.org/10.1006/viro.2000.0494
15. Janicke A, Vancuylenberg J, Boag PR, Traven A, Beilharz TH (2012) ePAT: a simple method to tag adenylated RNA to measure poly(A)-tail length and other 3′ RACE applications. RNA 18(6):1289–1295. https://doi.org/10.1261/rna.031898.111
16. Couttet P, Fromont-Racine M, Steel D, Pictet R, Grange T (1997) Messenger RNA deadenylylation precedes decapping in mammalian cells. Proc Natl Acad Sci U S A 94 (11):5628–5633. https://doi.org/10.1073/pnas.94.11.5628
17. Bazzini AA, Lee MT, Giraldez AJ (2012) Ribosome profiling shows that miR-430 reduces translation before causing mRNA decay in zebrafish. Science 336(6078):233–237. https://doi.org/10.1126/science.1215704
18. Westerfield M (2000) The zebrafish book. A guide for the laboratory use of zebrafish

(*Danio rerio*). University of Oregon Press, Eugene, OR

19. Alestrom P, D'Angelo L, Midtlyng PJ, Schorderet DF, Schulte-Merker S, Sohm F, Warner S (2019) Zebrafish: housing and husbandry recommendations. Lab Anim 2019:23677219869037. https://doi.org/10.1177/0023677219869037
20. Kimmel CB, Ballard WW, Kimmel SR, Ullmann B, Schilling TF (1995) Stages of embryonic development of the zebrafish. Dev Dyn 203(3):253–310. https://doi.org/10.1002/aja.1002030302
21. Dobin A, Davis CA, Schlesinger F, Drenkow J, Zaleski C, Jha S, Batut P, Chaisson M, Gingeras TR (2013) STAR: ultrafast universal RNA-seq aligner. Bioinformatics 29(1):15–21. https://doi.org/10.1093/bioinformatics/bts635
22. Patro R, Duggal G, Love MI, Irizarry RA, Kingsford C (2017) Salmon provides fast and bias-aware quantification of transcript expression. Nat Methods 14(4):417–419. https://doi.org/10.1038/nmeth.4197
23. Anders S, Huber W (2010) Differential expression analysis for sequence count data. Genome Biol 11(10):R106. https://doi.org/10.1186/gb-2010-11-10-r106
24. Patil DP, Bakthavachalu B, Schoenberg DR (2014) Poly(A) polymerase-based poly (A) length assay. Methods Mol Biol 1125:13–23. https://doi.org/10.1007/978-1-62703-971-0_2
25. Martin G, Keller W (1998) Tailing and 3′-end labeling of RNA with yeast poly(A) polymerase and various nucleotides. RNA 4(2):226–230

Index

Roland Dosch (ed.), *Germline Development in the Zebrafish: Methods and Protocols*, Methods in Molecular Biology, vol. 2218, https://doi.org/10.1007/978-1-0716-0970-5, © Springer Science+Business Media, LLC, part of Springer Nature 2020

MIX
Papier aus verantwortungsvollen Quellen
Paper from responsible sources
FSC® C105338

If you have any concerns about our products,
you can contact us on
ProductSafety@springernature.com

In case Publisher is established outside the EU,
the EU authorized representative is:
Springer Nature Customer Service Center GmbH
Europaplatz 3, 69115 Heidelberg, Germany

Printed by Libri Plureos GmbH
in Hamburg, Germany